ELEMENTS OF

MECHANICAL ENGINEERING

With Gratitude to God for the Inspiration

and

With Gratefulness to Mama for his Guidance

This Book is Dedicated

To My Dear Students, Past, Present and Future

Preface

This book is written to help the freshers to Engineering course in getting some insight into the field of Mechanical Engineering.

Very good books in this area are available, but the present attempt is aimed at giving the knowledge with the help of easy-to-draw sketches and schematics. Drawing is said to be the language of engineers, but to be useful, it must be as simple as possible. The book aims precisely at this task of simplification not sacrificing the important details.

The highlights of the book are:

In part I, energy sources have been discussed and the student is made to understand the world scenario today vis-a-vis energy availability from different sources. The dwindling resource of petroleum as a source of energy must catch the immediate attention of the would-be engineers taking up Mechanical Engineering as their specialization. It is their duty to exploit alternatives like Ocean Thermal Energy, Wind and Wave Power, Geothermal Energy etc., not forgetting about the prospects of fusion energy, which is round the corner.

Part I also discusses about Boilers, the main component of Thermal and Nuclear Plants. The gradual development of the boilers has been traced starting from Simple Vertical Boiler to present day High Pressure, High Temperature Modern Boilers, examining only those types, that have been designed and developed with distinct advantages.

In part II, production alternatives have been discussed with an eye on productivity, quality and surface finish characteristics. The basic machine tools have been described. Other topics are Refrigeration and Air-conditioning and a brief write-up on CAM & CIM. Topics on Engineering Materials emphasies Composites and the latest in manufacturing trend *viz*., Robotics.

In Glossary Question bear and some useful definitions and data are included for thorough understanding of text.

Suggestions are most welcome to improve this book and also pointers to mistakes, which might have escaped the attention of the author.

I record with gratitude the encouragement and guidance given to me by my Gurus, Dr. S.H. Ibrahim, Principal, Md. Sathak Engineering College, Kilakarai and Prof. A. Krishnan, Principal, Sri Venkateswara Engineering College, Madras. I acknowledge with many thanks the support given to me by Dr. V. Shanmuganathan, former Director of Mookambigai College of Engineering and Dr. M. Ponnavaikko, Head of the Department of Computer Science and Engineering, MCE, but for whom this venture would not have been a success.

T.S. Rajan

Contents

PART II: PRODUCTION PROCESSES

Syllabus

Subject Code : 14EME 14/14EME 24
Hours/Week : 04
Total Hours : 50

IA Marks : 25
Exam. Hours : 03
Exam. Marks : 100

Course Objectives: Students belonging to all branches of Engineering are made to learn certain fundamental topics related to mechanical engineering so that they will have a minimum understanding of mechanical systems, equipment and process.

Module – 1

Energy Resources: Non-renewable and renewable energy resources, **Petroleum based** solid, liquid and gaseous fuels, Calorific values of fuels, Combustion and combustion products of fuels, **Solar Power**: Solar Radiation, Solar constant (definition only), Solar Thermal energy harvesting, ex: liquid flat plate collectors, solar ponds (principle of operation only), Solar photovoltaic principle. **Wind Power**: Principle of operation of a typical windmill. **Hydro Power**: Principles of electric power generation from hydropowerplants, **Nuclear Power**: Principles of Nuclear power plants, **Biofuels**: introduction to biofuels, examples of various biofuels used in engineering applications, Comparison of biofuels with petroleum fuels in terms of calorific value and emission. **Steam Formation and Properties:** Classification of boilers, Lancashire boiler, Babcock and Wilcox boiler, boiler mountings and accessories (No sketches for mountings and accessories), wet steam, saturated and superheated steam, specific volume, enthalpy and internal energy.

(No numerical problems in this module) **10 Hours**

Module – 2

Turbines and IC Engines and Pumps

Steam turbines: Classification, Principle of operation of Impulse and reaction turbines, Delaval's turbine, Parson's turbine. *(No compounding of turbines).*

Gas turbines: Classification, Working principles and Operations of Open cycle and closed cycle gas turbines.

Water turbines: Classification, Principles and operations of Pelton wheel, Francis turbine and Kaplan turbine.

Internal Combustion Engines: Classification, I.C. Engines parts, 2 Stroke and 4 stroke Petrol engines, 4 stroke diesel engines. P-V diagrams of Otto and Diesel cycles. Problems on indicated power, brake power, indicated thermal efficiency, brake thermal efficiency, mechanical efficiency, and specific fuel consumption, *[numericals on IC Engines].* **10 hours**

Module – 3

Machine Tools and Automation

Machine Tools Operations: Turning, facing, knurling, Thread cutting, Taper Turning by swivelling the compound rest, Drilling, Boring, Reaming, Tapping, Counter Sinking, Counter Boring, Plane milling, End milling, Slot milling. *(No sketches of Machine tools, sketches to be used only for explaining operations. Students to be shown the available machine tools in the Machine Shop of the college before explaining the operations).*

Robotics and Automation:

Robotics: Introduction, classification based on robots configuration; Polar, cylindrical, Cartesian Coordinate and spherical. Application, Advantages, and disadvantages

Automation: Definition, types–Fixed, Programmable and Flexible automation, NC/CNC machines: Basic elements with simple block diagrams, advantages and disadvantages. **10 hours**

Module – 4

Engineering Materials and Joining Processes

Engineering Materials: Types and applications of Ferrous and Nonferrous metals and alloys,

Composites: Introduction: Definition, Classification and applications (Aircraft and Automobiles)

Soldering, Brazing and Welding: Definitions, classification and method of soldering, Brazing and welding. Differences between soldering, Brazing and Welding. Description of Electric Arc Welding and Oxy-Acetylene Welding. **10 hours**

Module – 5

Refrigeration, Air-Conditioning

Refrigerants: Properties of refrigerants, list of commonly used refrigerants. Refrigeration – Definitions – Refrigerating effect, Ton of Refrigeration, Ice making capacity, COP, Relative COP, Unit of Refrigeration. Principle and working of vapor compression refrigeration and vapour absorption refrigeration: Principles and applications of air conditioners, Room air conditioner. **10 hours**

Course outcomes:

Students shall demonstrate the Knowledge associated with,

1. Various Energy sources, Boilers, Prime movers such as turbines and IC engines, refrigeration and air-conditioning systems
2. Metal removal process using Lathe, Drilling, Milling Robotics and Automation.
3. Fair understanding of application and usage of various engineering materials.

Scheme of examination

- Two full questions (with a maximum of four sub questions) of twenty marks each to be set from each module. Each question should cover all the contents of the respective module.
- Students have to answer five full questions choosing one full question from each module

Part-I

Power Plants

INTRODUCTION

Right from the dawn of civilization, man has been continually searching for work simplification by employing energy available by whatever means to substitute or reduce muscle power or human toiling. In the process, he has mastered harnessing of various sources of energy and today we have power plants generating thousands of Mega Watts (MW) of electric power. This has meant a great leap forward in Industrial front as well as in many other activities.

We shall see in the chapter to follow what sources of energy are available, how much of each is available and how do these meet the present and future needs of energy.

Among the many manifestations of energy, the most favoured is the electrical energy. Because of its versatility, electrical energy or simply electricity finds many applications, for example, it can be used for heating a room or it can also be used to air-condition a room with cooling comfort. It is used to run the wheels of many types of industries. It provides energy for transportation, communication, lighting and many other activities. It is also the energy behind computers crunching data and making decisions, the robots and artificial intelligence. This is the reason why we see all over the world huge power plants being set up to generate large quantity of electrical energy.

Also, electricity can be transmitted over long distances without much difficulty. Transmission is almost instantaneous.

Again, it can be stored and used much better than thermal or mechanical energy.

Chapter- 1

Energy and its Sources

Objective

In this chapter we shall recall the definition of energy, types of energy and the units in which energy is used in calculations. We shall examine the traditional energy sources, their abundance and economics. We shall finally see the prospects of these traditional sources in the foreseeable future and examine the alternatives —the unconventional energy sources for possible solution to the evergrowing demands of energy.

1.1 DEFINITION OF ENERGY

Energy, as is commonly understood, is the capacity to do work. Work implies movement of a body by the application of a force. Work may be perceptible vividly, as for example, in raising of a weight, walking, moving of a vehicle etc. Work may also manifest in the most imperceptible way as in the case of reaction of a photographic plate by an X-ray or gamma ray. Energy, can therefore be termed as that which can bring about a change and is the cause of all dynamic manifestations.

1.2 TYPES OF ENERGY

For the purpose of convenience, we may classify all types of energy into six categories.

(*i*) Mechanical Energy
(*ii*) Electrical Energy
(*iii*) Electromagnetic Energy
(*iv*) Chemical Energy
(*v*) Nuclear Energy
(*vi*) Thermal Energy

Mechanical Energy is found in the motion of an object. For example, a moving vehicle or a rotating fly wheel has the mechanical energy.

Electrical Energy is the result of flow of electrons in a conductor or collection of charge in a condenser. It can be seen or felt in its transformed form like light, heat, etc.

Electromagnetic energy is the energy coming out of electromagnetic radiation. The gamma radiation, X-rays, light rays are all examples of this form of energy. This energy is linked to the frequency of radiation 'f' in the formula $E = hf$, where E is the energy (J) and 'h' is the Planck's constant ($= 6.626 \times 10^{-34}$ Js).

Chemical energy is released out of the reaction taking place between elements/molecules to form a more stable compound. An example is the combustion of fuel, where the fuel combines with the oxygen of the atmospheric air and stable compounds like CO_2, H_2O are formed.

Nuclear energy is the energy released by either fission or fusion process. Here again the resulting products assume a more stable condition.

Thermal energy is the result of molecular agitation causing rise in temperature. It is the basic energy form, in the sense all other forms of energy can be completely converted into thermal energy. The other way, *i.e.*, the complete conversion of thermal into other forms of energy is not possible and is governed by the Second law of thermodynamics.

1.3 UNITS OF ENERGY

Depending on convenience, we use different units of energy, but they are all mutually convertible.

Joule (J) is the most favoured unit to represent work or Mechanical Energy. With the introduction of S.I. units, (Standard International) the heat or thermal energy is also represented in Joules instead of the calories and British Thermal Units.

The Joule is defined as the work done (or the energy consumed) when a force of 1 Newton, acting on a body of mass 1kg, moves the body through a distance of 1 m. Therefore,

$$1 \text{ Joule} = 1 \text{ N.m.}$$

The smaller unit of work Erg (dyne.cm) is equated to Joule in

$$1 \text{ Joule} = 10^7 \text{ ergs.}$$

A very small unit of energy eV or MeV (Million electron volt) is normally used to express energy released by electromagnetic radiation. It is also used in expressing the energy released by atomic reactions.

$$1 \text{ J} = 6.242 \times 10^{12} \text{ MeV.}$$

Electrical power (rate of energy consumed or generated) is expressed in Watts (1W = 1 J/s). Therefore, electrical energy is expressed in kWh = 3.6×10^6J.

1.4 CLASSIFICATION OF ENERGY SOURCES

Energy sources may be classified as traditional and non-traditional sources. Traditional sources are (*i*) Coal, (*ii*) Oil and gas, (*iii*) Water falls and (*iv*) Nuclear fuels.

These are called traditional in the sense we have been using these sources for quite some time and are recognised to be commercially viable sources used for large scale production of electrical energy.

As against these, we have non-traditional sources like (*i*) Solar energy, (*ii*) Wind energy, (*iii*) Tidal energy and Wave energy, (*iv*) Ocean Thermal energy and (*v*) Geothermal energy. These sources are not commercially viable as on date. However, some of these sources, like sun and wind have been put to use almost over the past 2000 years. For example, wind has been used to power the sails of ships and rotate wind mills to pump water. Sun's energy has been used to dry farm produce in the village side. But these are not found good enough to produce large scale electrical power because of economic considerations. Energy sources may be further classified as renewable and non renewable sources. Renewable sources are the Sun, Wind, Waterfalls, Tidal, Wave and Ocean Thermal sources, for the reason they are perpetual and non-depletable. Against this group, the traditional ones like coal, oil and gas and also nuclear sources are nonrenewable, because these reserves are getting depleted once and for all, as we use them.

Energy types may also be classified as (*i*) Transitional and (*ii*) Stored energy.

Example of Transitional Energy: Under mechanical energy we talk of kinetic energy, which is transitional, *i.e.*, it is the energy possessed by a body by virtue of its motion. Under electrical energy, the flow of current could be considered as transitional energy.

Example of Stored Energy: Stored mechanical energy is referred to as potential energy. A body placed at a height and acted upon by gravitational field, a fluid confined in a container under pressure, a spring or torsion bar under strain — are all examples of stored or potential energy. Similarly, a charged condenser is an example of stored electrical energy.

1.5 TRADITIONAL SOURCES OF ENERGY (CONVENTIONAL)

As of now, the following are considered the traditional sources of energy.

(*i*) Coal
(*ii*) Oil and natural gas
(*iii*) Waterfalls
(*iv*) Nuclear energy

Wood, animal and agricultural wastes are also traditional sources but are not of commercial value.

Coal, oil and natural gas are called fossil fuels, because they are formed out of fossils of vegetation and animal matter. The fossils of vegetation burried deep under the bowels of the earth for millions of years have turned into coal under intense pressure and temperature. Similarly the animal matter has turned into oil and gas.

(i) Coal

Under the category of coal, we have peat, lignite, bituminous coal and anthracite. Peat is the lowest grade of coal in the sense it is in the first stages of transformation from the organic matter. Lignite is the next and so on—all in the increasing order of magnitude of time-temperature-pressure influence. Anthracite is of the highest grade and would have undergone the maximum transformation.

The total coal reserve in the world is estimated to be 5×10^{12} tonne.

It is interesting to note that although human civilization has been several thousand years old, the clamour for energy started about 250 years ago, perhaps coinciding with the beginning of industrial revolution. In this short period we have consumed about 5% of available coal. This figure may not appear quite alarming, but looking at Fig. 1.1 we see that the consumption rate, although very steady and small up to about 1910, has started going up thereafter almost exponentially. Models of growth rate predict that the apparently large balance of reserves existing as on date is going to be consumed

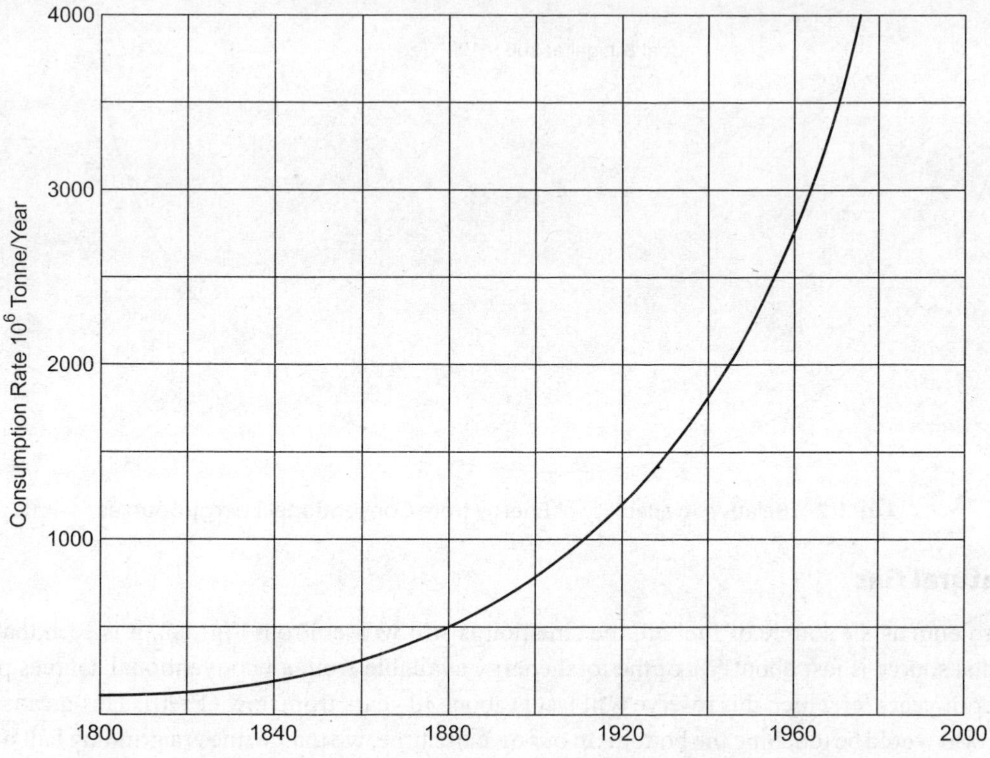

Fig. 1.1 Model of Consumption Pattern of Coal Depicting the Alarming Trend

in the next 150 years — a fact which drives chill in to the minds of the society in general and of scientific community in particular. The aggravating factor is that once the reserve tapers off, the cost escalates and the availability thus gets reduced.

Realising the situation, we should take every step to conserve the available fuel reserve. The first step in this direction is to use the fuel efficiently. This requires the knowledge of the characteristics of fuel, which helps design of efficient combustion equipment.

Coal is analysed using either the method of Proximate analysis or Ultimate analysis.

Proximate analysis gives the amount of Fixed Carbon (F.C.) Volatile Matter (V.M.), Moisture and Ash content, whereas Ultimate analysis gives the % composition of each element like carbon, hydrogen, sulphur etc.

In the order of formation of coal, from peat to anthracite, fixed carbon increases from peat to anthracite and the volatile matter decreases. Anthracite, the highest grade of coal contains maximum F.C. and minimum V.M.

In addition to knowing the constituents of coal, we should know the heating value (calorific value), grindability index, caking property, ash softening temperature — all those, that describe the coal completely.

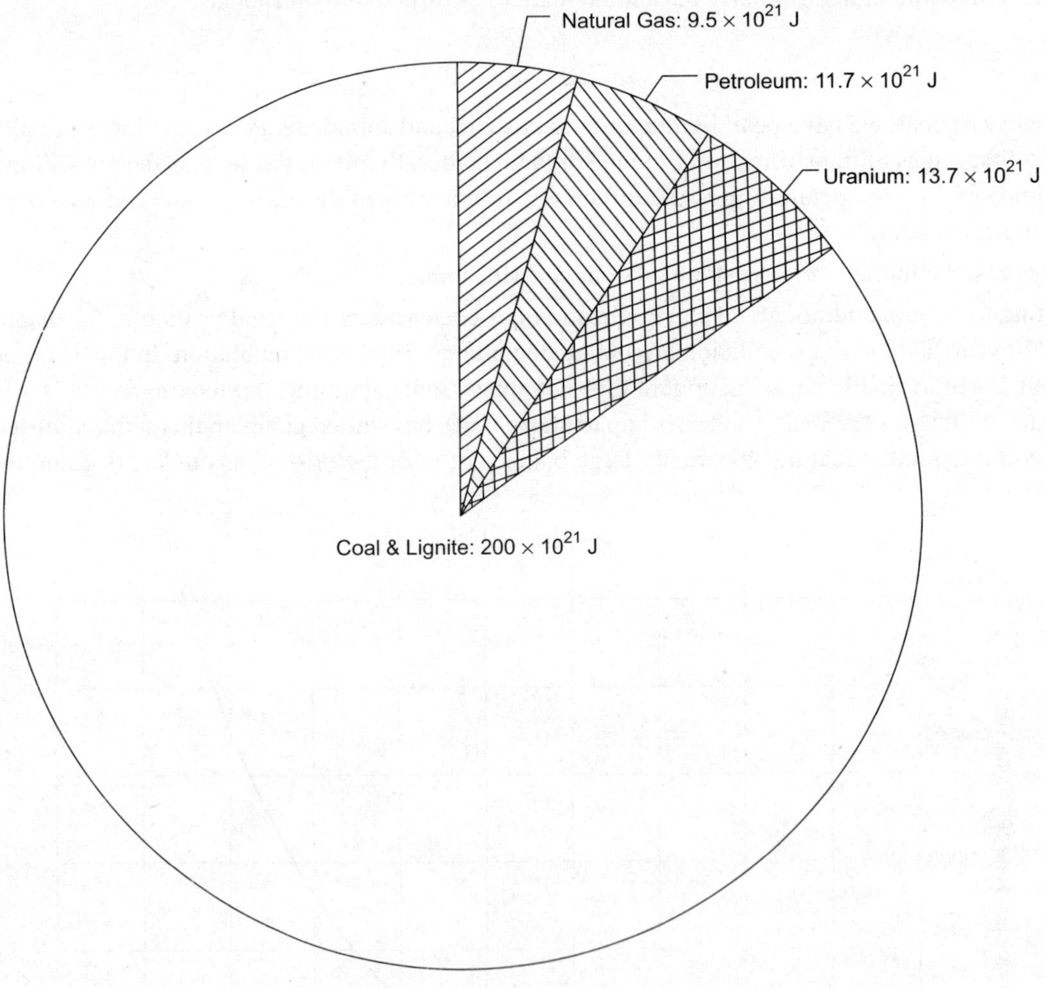

Fig. 1.2 Relative Availability of Energy from Conventional Energy Sources

(ii) Oil and Natural Gas

Considering Petroleum as the source of fuel oil, the situation is still worse. From Fig. 1.2, it is seen that the availability of energy from this source is just about 5% of the total energy available from all conventional sources put together. The estimated number of years for which this reserve will last is about 30 years from now (1990). This means that by 2020 the oil wells of the world would be touching the bottom. In our own life time, we may witness a grinding halt of all automobiles using the traditional I.C. engines, shut down of many industries and a sort of death knell everywhere.

The estimated global reserve of petroleum is 0.4×10^{12} kL. Since the petroleum is also used as a raw material in many chemical and pharmaceutical industries, its use as fuel should be curtailed and an all round effort made to conserve this meagre resource.

Petroleum or crude oil as it is called, is a dark brown liquid. This crude petroleum is fractionated in fractionating columns in the refineries and several fractions like petrol, diesel, aviation fuel, lubricating oils, wax, asphalt etc. are obtained. Each has its own characteristics and is used for specific applications. Many of these are used as fuels to burn in I.C. engines and furnaces.

The characteristics of importance, to use any fuel oil are: (*i*) flash point, (*ii*) fire point, (*iii*) pour point, (*iv*) specific gravity and heating value.

Advantages of oil as fuel: It is easier to handle, store and transport. It burns easily and the combustion equipment is very much simplified. It leaves very little ash.

However, the sulphur and vanadium content of some of the oils cause corrosion problems.

Natural gas: Natural gas issues along with crude oil and its global reserve is estimated at 250×10^{12} m^3. It has methane as the primary combustible constituent.

There are certain other fuel gases, which are artificially made, but they are all made using fossil fuels.

Following are the important ones.

(*a*) *Producer gas*: Made by burning coal or coke in a restricted quantity of air. The gas contains mainly carbon- monoxide as the combustible gas.

(*b*) *Liquified petroleum gas* (LPG) (also called refinery gas) is composed of lighter fractions of petroleum — mostly propane and butane. It is stored and transported under-pressure of 4 – 20 atm. depending on atmospheric temperature.

(*c*) *Water gas*: Produced by passing alternately steam and air over incandescent coke. The steam reacts with the hot coke to produce hydrogen and carbon monoxide.

(*d*) *Synthetic Natural Gas* (SNG): It is produced from coal and in one process, pressurised hydrogen is made to react with the coal at 900°C.

(*e*) *Blast furnace gas*: This is the by-product of the steel industry and contains mostly carbon monoxide.

Sewage gas: It is obtained by digesting in a digestor the animal and vegetable wastes. It consists of methane essentially. Methane is produced by bacterial action on the bio-mass (animal and vegetable wastes) in the absence of air. The process is called anaerobic digestion. The gas produced in this way is called bio-gas. In India, gobar gas plants are used for this purpose. They use mostly the dung of animals to produce the gas. The gas is used for cooking, lighting and also running of I.C. engines.

Advantages of gaseous fuels: Combustion is almost instantaneous and burns effortlessly and continuously. Practically no ash is left behind. Transportation is easily done through pipe lines.

The only disadvantage, perhaps, is storage of large quantity of gas requires lot of space.

(iii) Water Falls

Among all the sources being exploited commercially, for producing electrical power, waterfalls or water storage at high elevations, provides a source totally renewable and non-polluting. The world's water potential is 3×10^6 MW and in India it is about 40,000 MW.

(iv) Nuclear Energy

The energy scenario has been brightened up by the discovery of nuclear energy. The state-of-art of nuclear technology—the fission process — has added already a substantial quantity of energy the world over. The estimated global reserve of U-235, the only fissile material, is around 300,000 tonne. The possibility of Thorium 232 and U-238 (fertile materials) getting converted into fissile materials in the breeder reactors, further augments the prospects of this energy source.

In India the U-235 reserve is estimated at 33,000 tonne and the availability of thorium is fairly large. Thorium is found in the monazite beach sand of Kerala.

1.6 THE UNCONVENTIONAL SOURCES OF ENERGY AND THEIR UTILIZATION

With the prospects of conventional energy sources fading out in the near future, scientists all over the world are looking into alternative energy sources. The alternatives are: (*i*) Solar energy, (*ii*) Wind energy, (*iii*) Tidal and Wave energy, (*iv*) Geo-thermal energy and (*v*) Ocean thermal energy.

1.6.1 Solar Energy

Practically every conceivable source of energy could be linked with energy from Sun, called solar energy. Sun (Ref. Fig. 1.3), which is several times larger than earth, is a burning mass. The abundant energy release in the Sun is attributed to the continuous fusion reaction taking place in the Sun.

Sun sends out energy in the form of electromagnetic radiation at the rate of 3.7×10^{20} MW. The energy intercepted by the earth is about 1.85×10^{11} MW. This is several times more than all the energy produced and consumed in the world today.

Inspite of the vast quantity of Sun's energy being received by the earth, the commercial exploitation of this energy is rendered difficult for the following reasons.

(*i*) The energy is very dilute and spread out. The average power obtained per m^2 area is about 1 kW.

(*ii*) The intensity of radiation is weather-dependent. On a cloudy day the intensity is very low. Dust, fog and smog also affect the intensity of radiation.

(*iii*) The day and night cycle interrupts the continuous flow of energy.

(*iv*) Added to this, the seasonal variations affect the quantum of radiation received in a given place.

Nevertheless, necessity to seek alternative sources of energy has turned the attention of research workers to see how best this energy could be exploited.

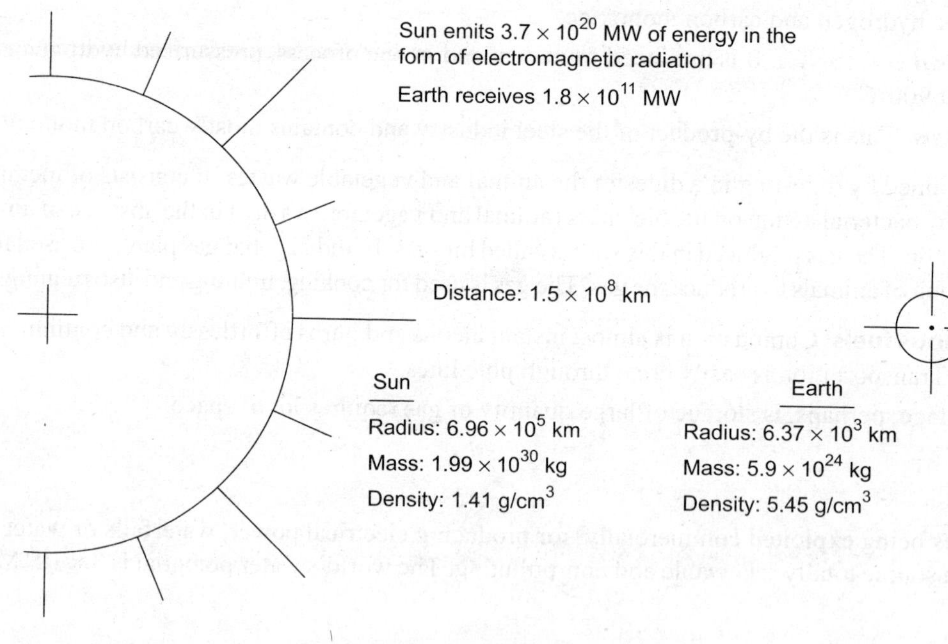

Fig. 1.3 Sun and Earth–Some Facts

The Advantages of Solar Energy

(*i*) It is a renewable source of energy and is inexhaustible.

(*ii*) It is totally free from pollution problem.

(*iii*) It can be converted into electrical energy directly using photo-voltaic cells.

Although the success of commercialization of solar energy is still far off, its use for certain applications has become a reality. For example, (*i*) The satellites necessarily make use of photo-voltaic panels to continuously generate electricity to power the instruments and equipment on board. The advantages are: It has high power to weight ratio and there is no payload of energy to be carried.

(*ii*) Solar water heater application to provide hot water in houses, hostels and hotels for cooking and cleaning has proved economical.

(*iii*) Solar driers have been very useful for drying agricultural produce.

(*iv*) In remote areas, where power from conventional power plants can't reach, solar energy can power the street lights, pump water etc.

(*v*) Solar energy is also used in desalination of water by distillation.

Solar power systems: In an effort to make use of solar energy following three systems are currently being tried out.

(*i*) Low temperature system using flat plate collectors.

(*ii*) Medium temperature system using concentrated collectors.

(*iii*) High temperature solar power tower system.

Low temperature solar system: Figure 1.4 shows a low temperature solar system in which a fluid like butane, boiling at low temperature, is used as working medium. The fluid is heated in a heat exchanger using water heated by solar energy and its vapours are used to drive a turbine to provide mechanical power. The turbine may be connected to a generator to produce electricity or connected to a water pump. Here the solar energy collecting device is a flat plate collector, which raises the temperature of water to about 80°C.

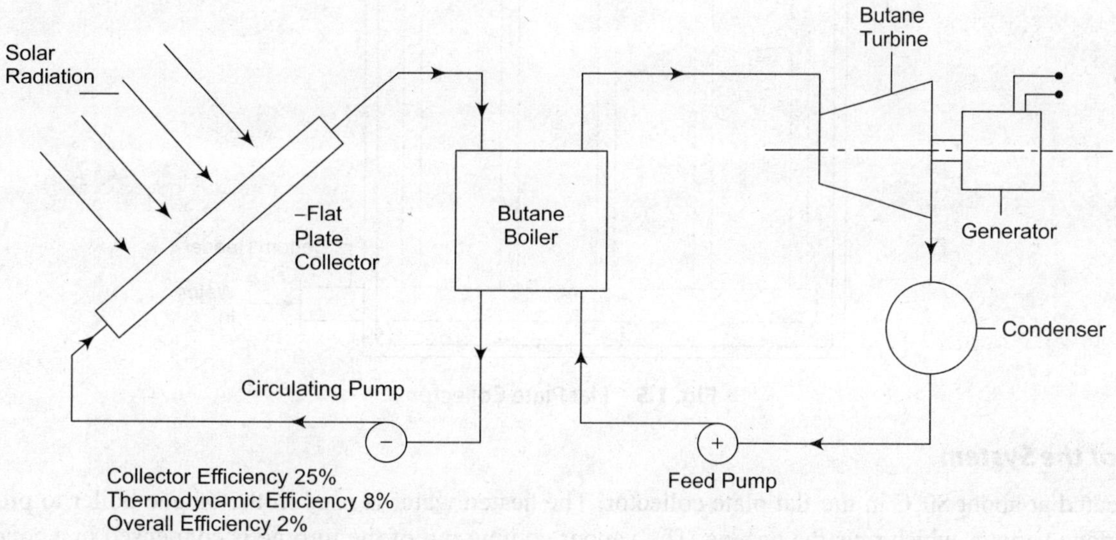

Fig. 1.4 Low Temperature System for Power Generation Using Butane as Working Fluid

Description of the Flat Plate Collector (Fig. 1.5)

This is a box like structure, in which a flat plate coated with black paint having good absorptivity is used to face the solar radiation. Steel, copper or aluminium tubings in required numbers and size connecting bottom and top headers are placed in close contact with the flat plate. The heat absorbed by the flat plate is transferred to the tubings carrying water. The heated water rises up the tubings, passes through the heat exchanger (Fig. 1.4) and gets recirculated with the help of the circulating pump. Thermal insulation is provided at the bottom of the tubings and also at the sides of flat plate to minimise the heat loss. The insulation material is generally fibre glass or mineral wool. In order to trap the heat as it is received by the flat plate, a glass cover is provided. The glass, capable of letting the light radiation of smaller wave lengths, cannot pass the heat radiation, which are of longer wave length. The heat is thus trapped and the temperature inside the box increases. This effect is called the Green House effect. Instead of one glass plate, two glass plates could be used to make the system more efficient.

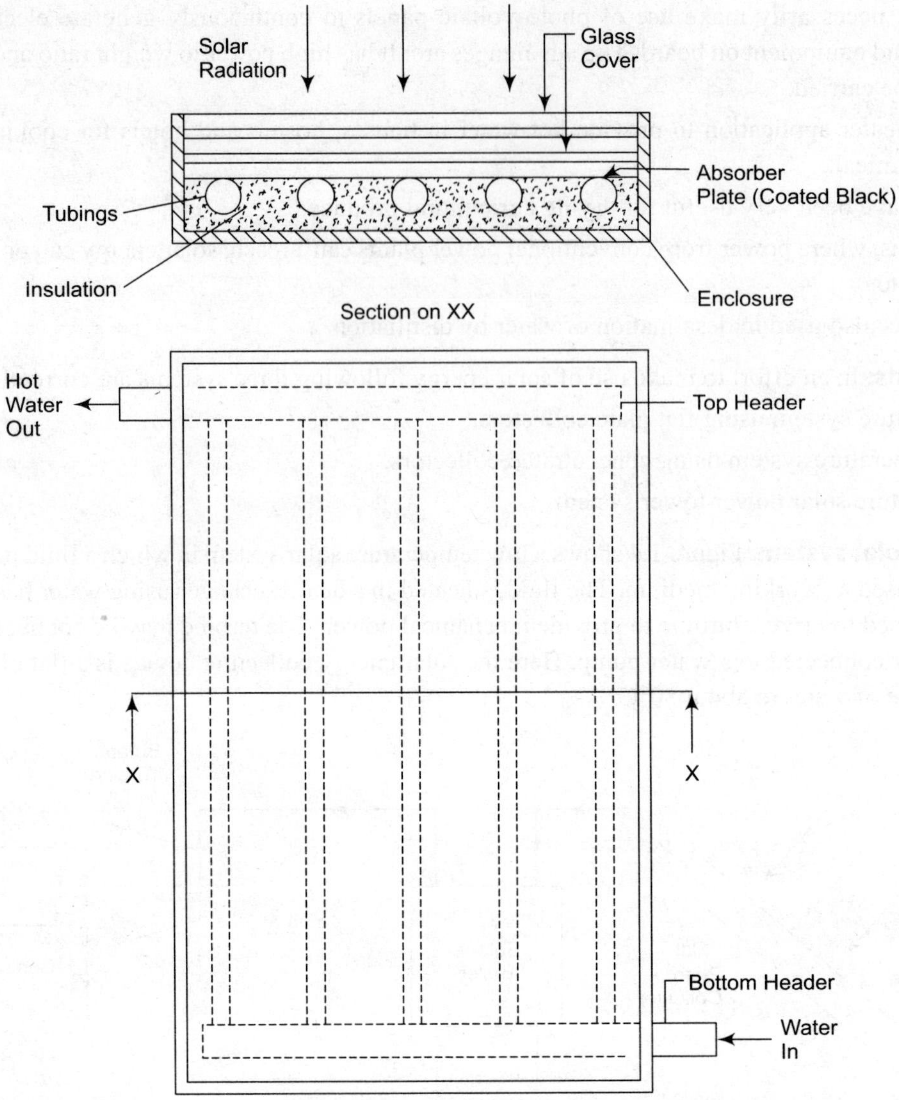

Fig. 1.5 Flat Plate Collector

Working of the System

Water is heated at about 80°C in the flat plate collector. The heated water is used in the butane boiler to produce high pressure butane vapour, which runs the turbine. The vapour coming out of the turbine is condensed in a condenser and sent back for recirculation with the help of a feed pump.

Medium temperature system (Fig. 1.6): This operates at temperatures ranging from 260 to 650°C. It uses concentrating collector such as parabolic cylindrical unit to give the high temperature. The Sun's radiation is focussed along a line about which the absorber tube is located. Because of the concentration of the energy the temperature attained in this system is much higher than that in the flat plate collector. In this case, conventional steam turbine could be used for power generation.

Solar power tower: Figure 1.7 shows the arrangement of collecting solar energy on a large scale. Here, a large number of tracking mirrors called 'heliostats' are arranged to focus solar energy on to a receiver kept on a tower of sufficient height. The water in the absorber-receiver is converted into steam, which runs a conventional turbine-generator set.

A pilot plant of 50 kW, built and being operated in Italy, has a tower of 500 m high and generates steam, the temperature of which may go up to 1700°C.

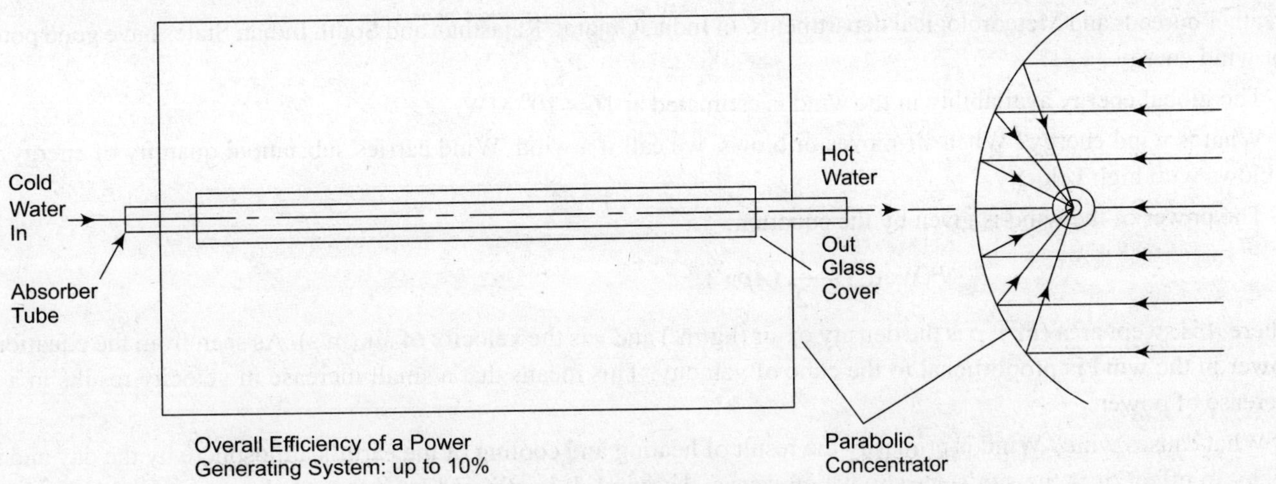

Fig. 1.6 Parabolic Focussing Collector for A Medium Temperature System

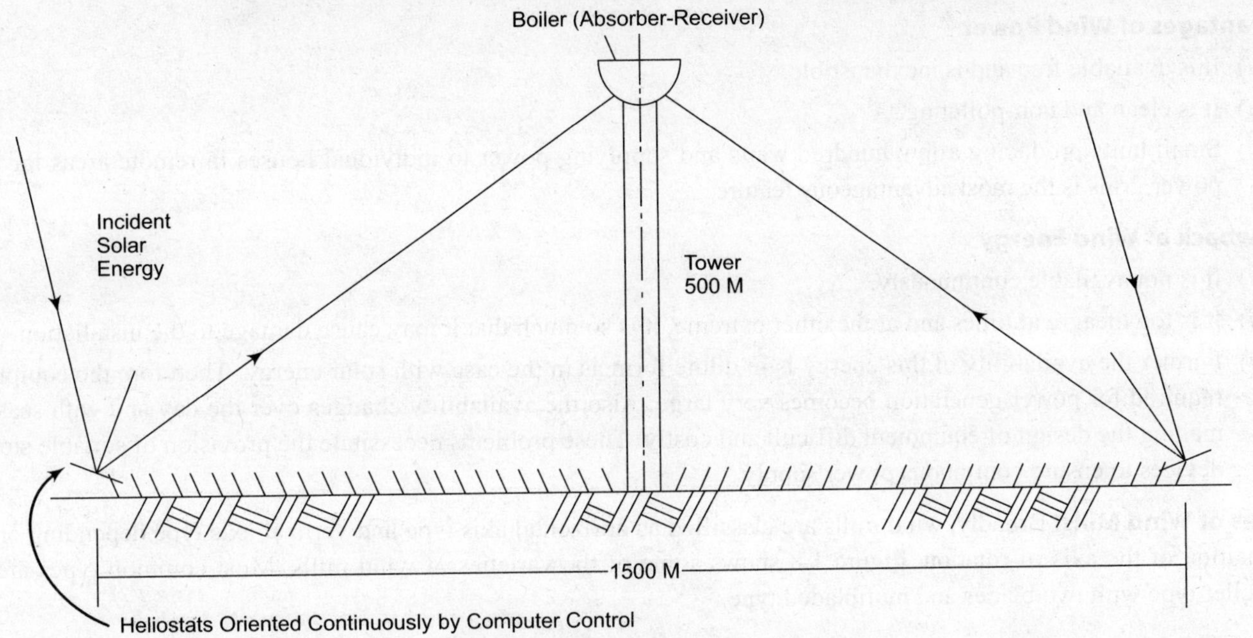

Fig. 1.7 High Temperature System

1.6.2 Wind Energy

The power of wind is glaringly seen in the devastation it causes in the form of cyclones, hurricanes and tornadoes. But even during normal times, the wind has sufficient potential and intelligently, our ancestors have used wind power to power the sails of the ships and rotate wind mills to lift water and grind corns. In the recent past small wind mills have been used to generate electricity. In Germany and United States notably, many small wind mills generating a few kW of energy, are used in individual houses.

It is only now, when the cost of energy has started escalating at a rapid rate, research workers all over the world are trying to harness the wind in a better and bigger way. Wind farms are being created to generate large quantity of electricity to augment the supply of energy. In India, National Aeronautics Ltd. have developed several prototypes. Near Thirunalveli coast, a wind farm (called Kaithar project) has been developed and a number of wind mills of Danish type have been operating and supplying electrical energy to the Tamilnadu grid.

Wind mills can't be put up anywhere and everywhere. Places must be selected where there is an assured flow of wind for most part of the year with fairly good velocity (greater than 10 km/hr). Information concerning this is available in weather bureaus and Meteorological departments. In India, Gujarat, Rajasthan and South Indian States have good potential for wind energy.

The global energy availability in the wind is estimated at 16×10^6 MW.

What is wind energy? When air moves or blows, we call it a wind. Wind carries substantial quantity of energy when it blows with high velocity.

The power of the wind is given by the equation:

$$P(\text{Watts}) = \frac{1}{2}(A\rho v^3)$$

where A is swept area (m^2), ρ is the density of air (kg/m^3) and v is the velocity of air (m/s). As seen from the equation, the power in the wind is proportional to the cube of velocity. This means that a small increase in velocity results in a large increase of power.

What causes wind? Wind is primarily the result of heating and cooling of the earth's atmosphere by the day and night cycle. In effect, it is the sun's energy, which causes the wind. It is affected by seasonal changes brought about by the earth's rotation round the sun and also other disturbances both natural and manmade.

Advantages of Wind Power

(*i*) It is available free and is inexhaustible.

(*ii*) It is clean and non-polluting.

(*iii*) Small units producing a few hundred watts and supplying power to individual houses in remote areas far from power grids is the most advantageous feature.

Drawback of Wind Energy

(*i*) It is not available continuously.

(*ii*) It is too meagre at times and at the other extreme, it is so much that it may cause damage to the installation.

(*iii*) Further the availability of this energy is in dilute form as in the case with solar energy. Therefore the equipment required for power generation becomes very large. Also the availability changes over the day and with seasons, making the design of equipment difficult and costly. These problems necessitate the provision of suitable storage devices to ensure continuous power supply.

Types of Wind Mills: Broadly, wind mills are classified as horizontal axis type and vertical axis type depending on the orientation of the axis of rotation. Figure 1.8 shows some of the varieties of wind mills. Most common types are the propeller type with two blades and mutibladed type.

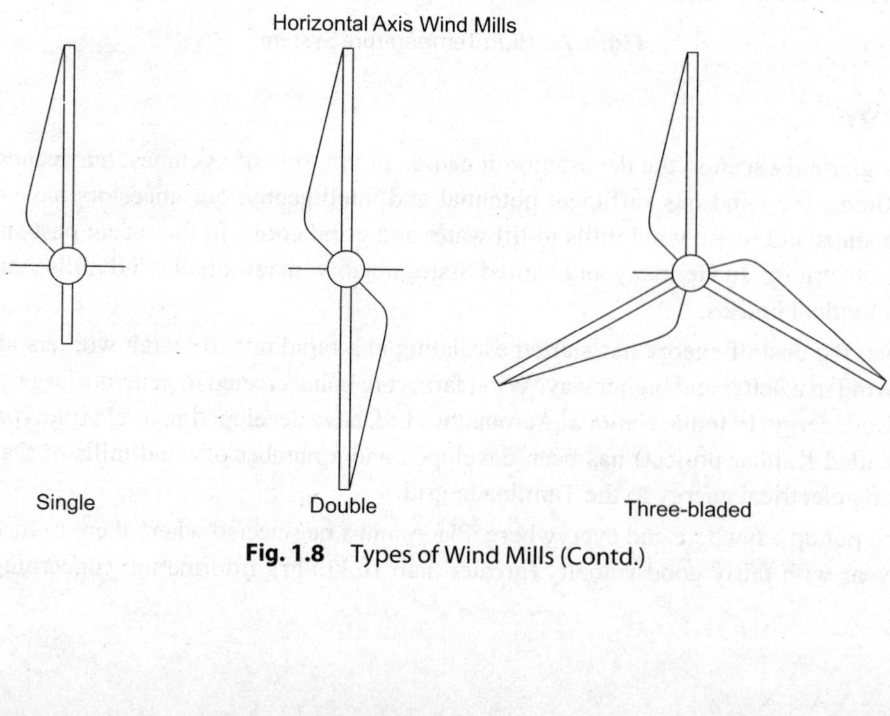

Fig. 1.8 Types of Wind Mills (Contd.)

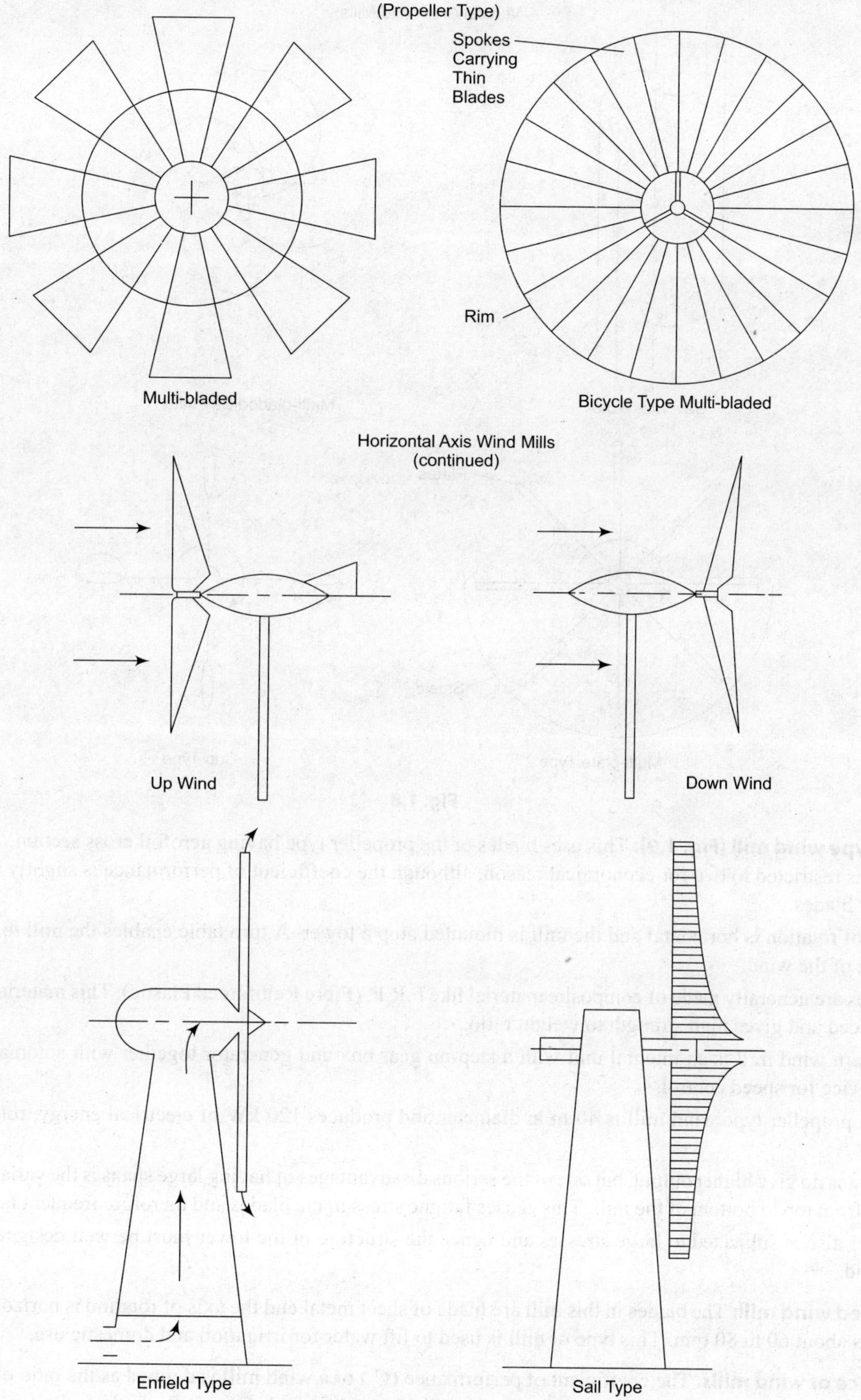
(Propeller Type)
Spokes Carrying Thin Blades
Rim
Multi-bladed
Bicycle Type Multi-bladed
Horizontal Axis Wind Mills (continued)
Up Wind
Down Wind
Enfield Type
Sail Type

Fig. 1.8 (Contd.)

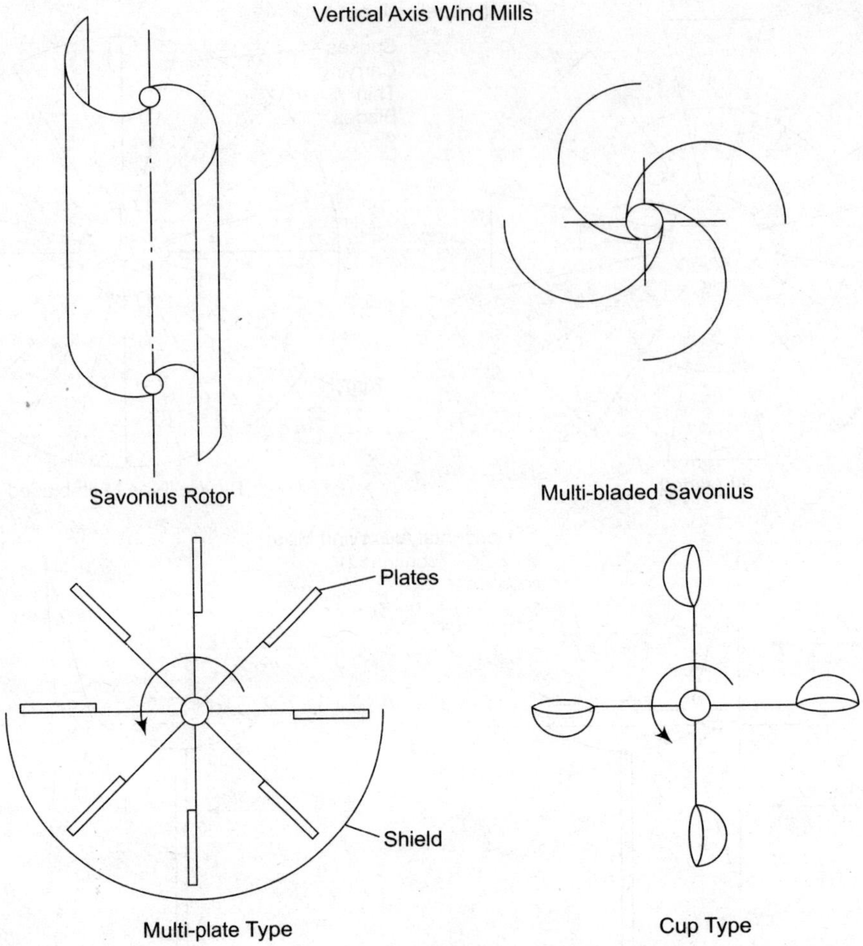

Fig. 1.8

Propeller type wind mill (Fig. 1.9): This uses blades of the propeller type having aerofoil cross section. The number of blades used is restricted to two for economical reason, although the coefficient of performance is slightly improved with three or four blades.

The axis of rotation is horizontal and the mill is mounted atop a tower. A turn table enables the mill to orient itself in the direction of the wind.

The blades are generally made of composite material like F.R.P. (Fibre Reinforced Plastic). This material is less costly, easily produced and gives high strength to weight ratio.

The modern wind mill is an integral unit with a step-up gear box and generator together with automatic blade pitch adjusting device for speed control.

A typical propeller type wind mill is 40 m in diameter and produces 120 kW of electrical energy, rotating at 300 to 400 rpm.

Larger spans do give higher output, but one of the serious disadvantages of having large spans is the variation in velocity of the wind from top to bottom of the mill. This causes fatigue stress in the blades and therefore frequent failure of blades.

The tower also is subjected to large stresses and hence the structure of the tower must be well designed to withstand the wind load.

Multi-bladed wind mill: The blades in this mill are made of sheet metal and the axis of rotation is horizontal. The speed is low and is about 60 to 80 rpm. This type of mill is used to lift water for irrigation and domestic use.

Performance of wind mills: The coefficient of performance (C_p) of a wind mill is defined as the ratio of the power (P) delivered by the rotor to the maximum power (P_{max}) available in the wind. $C_p = P/P_{max}$ where $P_{max} = (1/2)\ (\rho A v^3)$ as already explained.

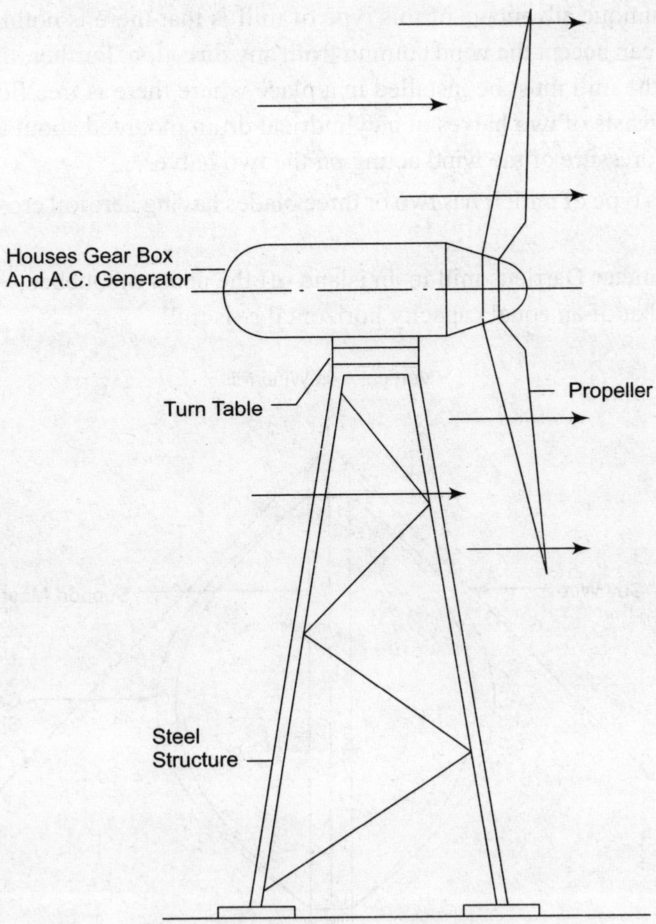

Fig. 1.9 Propeller Type Wind Mill

Figure 1.10 shows the variation in C_p with the tip speed to wind speed ratio. The graphs show that there is some optimum tip speed for each type of wind mill giving maximum C_p. The propeller type exhibits the maximum C_p (about 0.45) at the tip speed ratio of about 6.

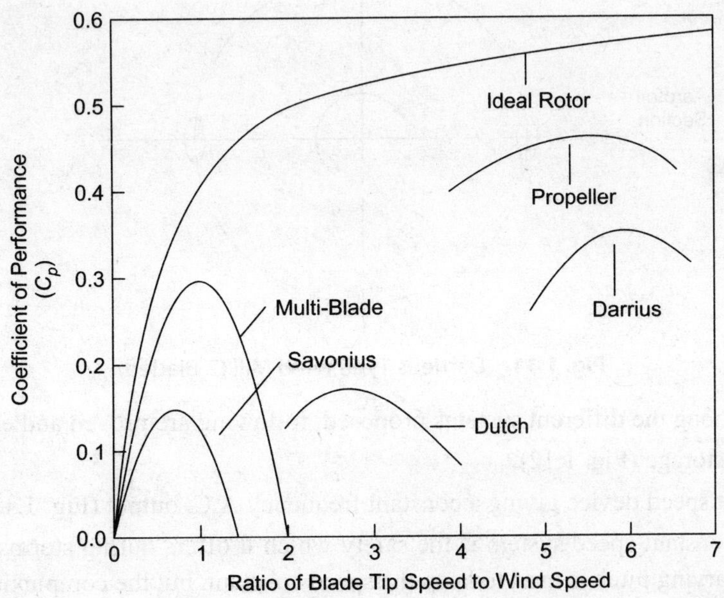

Fig. 1.10 Coefficient of Performance vs Tip Speed to Wind Speed Ratio

Vertical axis wind mill: The unique advantage of this type of mill is that there is nothing like orienting the mill in the direction of the wind. The mill can accept the wind coming from any direction. Further, there is no need for tall structures to support the mill. However, the mill must be installed in a place where there is free flow of air. In Fig. 1.8 is shown a Savonius wind mill rotor. It consists of two halves of a cylindrical drum mounted about a vertical axis as shown. Torque is produced by the differential pressure of the wind acting on the two halves.

Figure 1.11 shows a Darrieus type of mill. It has two or three blades having aerofoil cross-section and bent into catenery shape.

Canada has built a 25 m diameter Darrieus mill in an island off the coast of Quebec, which generates 200 kW power. The cost is claimed to be half that of an equal capacity horizontal axis mill.

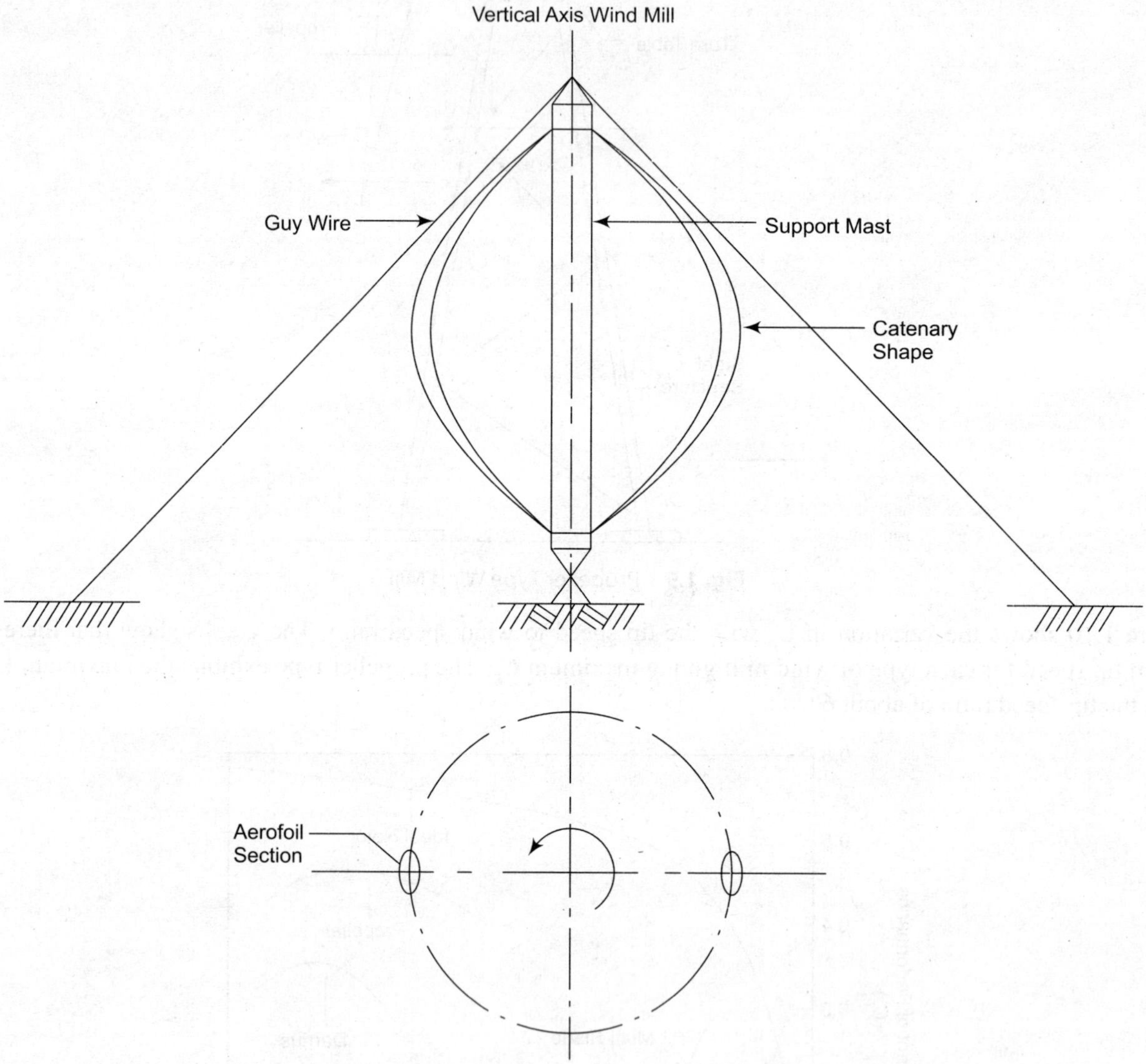

Fig. 1.11 Darrieus Type Wind Mill (2-Bladed)

Wind energy systems: Among the different systems proposed, following are proven and established ones.

1. System with battery storage (Fig. 1.12).
2. System with constant speed device giving a constant frequency A.C. output (Fig. 1.13).

A big advantage of the constant speed system is the safety which it offers during storms, when the wind speed may exceed 70 km/h. In this, a varying pitch is used to keep the speed constant, but the complexity of the mechanism adds to the cost.

Optimum size of wind mills: Experience shows that smaller units producing about 2 MW are more economical and safe. They are more reliable and have less maintenance problem.

Best location: Off shore, coastal areas and mountains, where there is unobstructed and free flow of air, are the best possible locations.

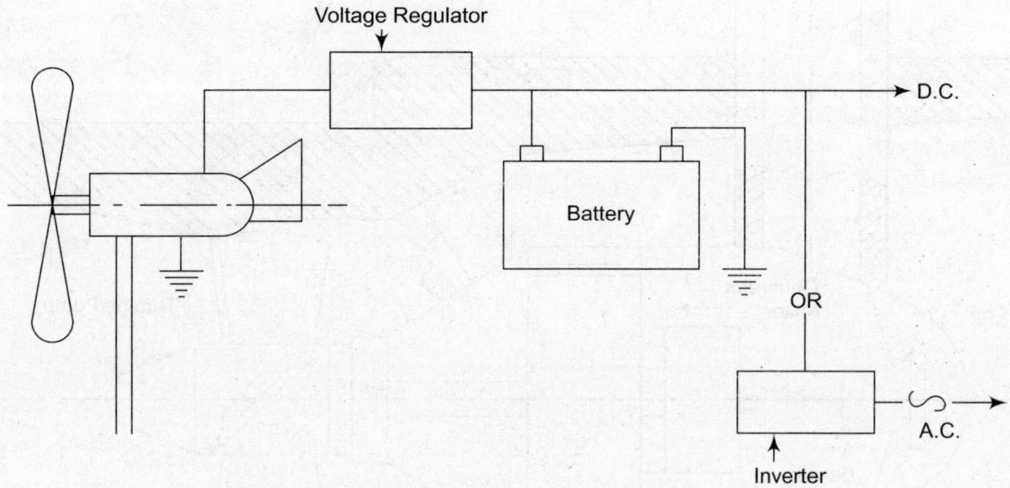

Fig. 1.12 Wind Energy System with Battery Storage

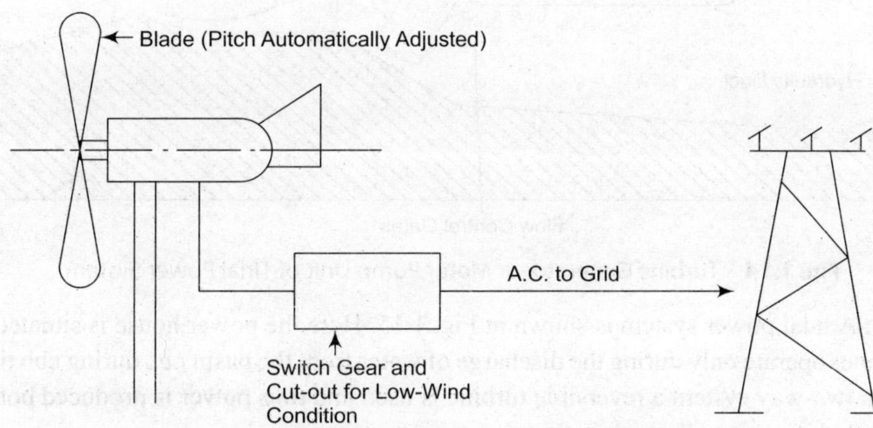

Fig. 1.13 Wind Energy System with Constant Speed Device

1.6.3 Tidal Energy

Tides are associated with moon's influence on the waters of the sea or ocean. Tide is periodic rise and fall in the level of the sea from the mean sea level.

The variation in the level between high tide (called flood tide) and the low tide (called ebb tide) may be as high as 10 m and thus provides sufficient head for exploiting tidal energy. The available head is more in an estuary (the outlet of river entering the sea) than in other places for the reason that water will be able to move up the narrow channel of the estuary. Power plants to tap tidal energy, are therefore, situated at the mouth of the estuaries.

The principle of obtaining tidal power is to store the water during the high tide in a basin formed by constructing a dam across the mouth of the estuary and release it through turbines during low tides.

A typical tidal power system has been constructed on the Rance estuary in France. This has 24 turbine units and provides 240 MW of electrical power. Schematic diagram of one of the turbine units is shown in Fig. 1.14. This plant is also used as a pumped-storage system. That is, when the demand for power is less, turbine-generator unit works in a reverse way, generator acting as a motor and turbine as a pump. This reversed system pumps water into the basin, later to be released during peak power demand.

The world's total tidal power potential is estimated to be 64,000 MW.

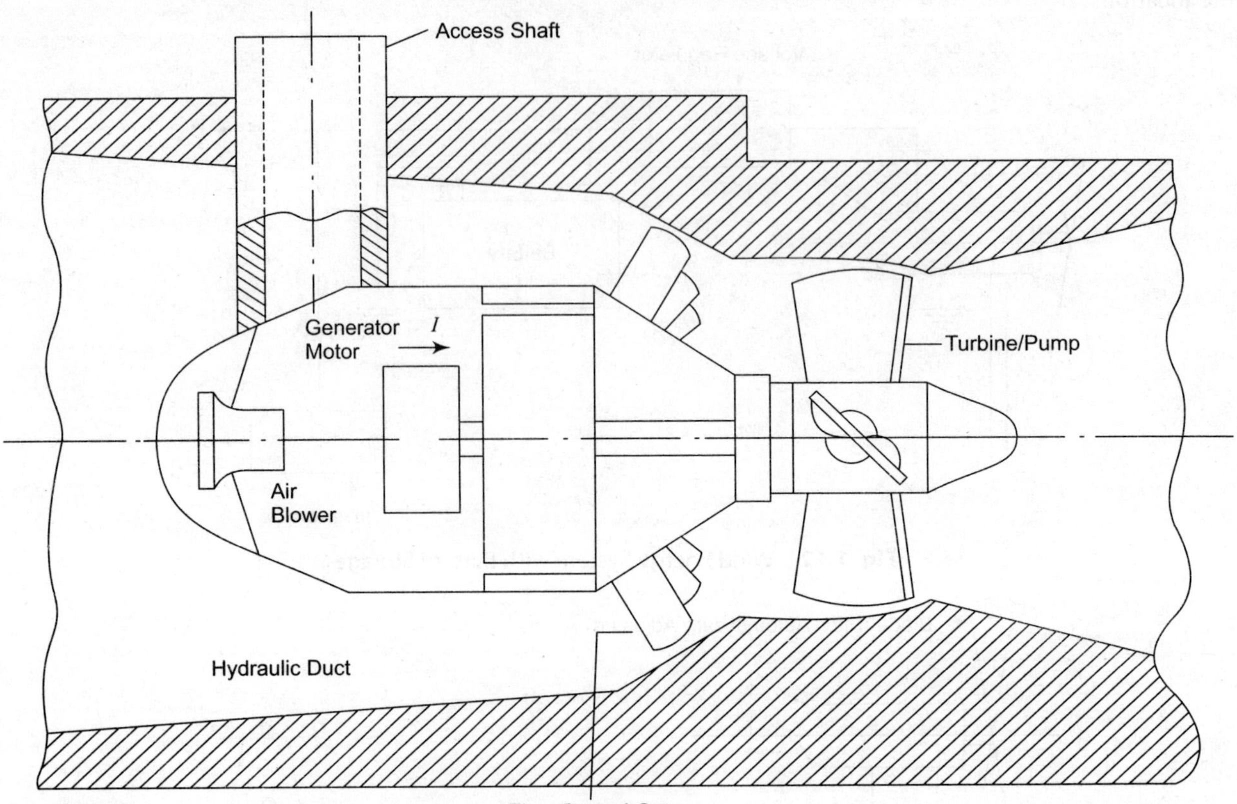

Fig. 1.14 Turbine Generator or Motor Pump Unit of Tidal Power System

Tidal Power Systems: A tidal power system is shown in Fig. 1.15. Here the power house is situated at the mouth of the estuary. In this the turbines operate only during the discharge of water from the basin *i.e.*, during ebb tide. This is a oneway system. In contrast, in a two-way system a reversible turbine is used and thus power is produced both during low tide as well as high tide. Capital cost is naturally high in this case.

Advantages

1. The energy is freely available.
2. It is assured all through the year and is independent of weather conditions.
3. The power produced may meet the small local needs of the coastal area.
4. It provides clean energy without any pollution.
5. It has the capability to meet the peak power demand efficiently in combination with thermal and hydel systems.

Disadvantages

1. The power available is limited and power production may be done on a small scale only.
2. The variation in the available head is a problem and it introduces complexity in the power equipment and hence the cost of plant goes up.
3. Not many locations are suitable for the exploitation of this energy.

Potential for tidal power plants in India: Gulf of Cambay, Gulf of Kutch in Gujarat and Sunderban areas in West Bengal are potential locations for establishing tidal power plants.

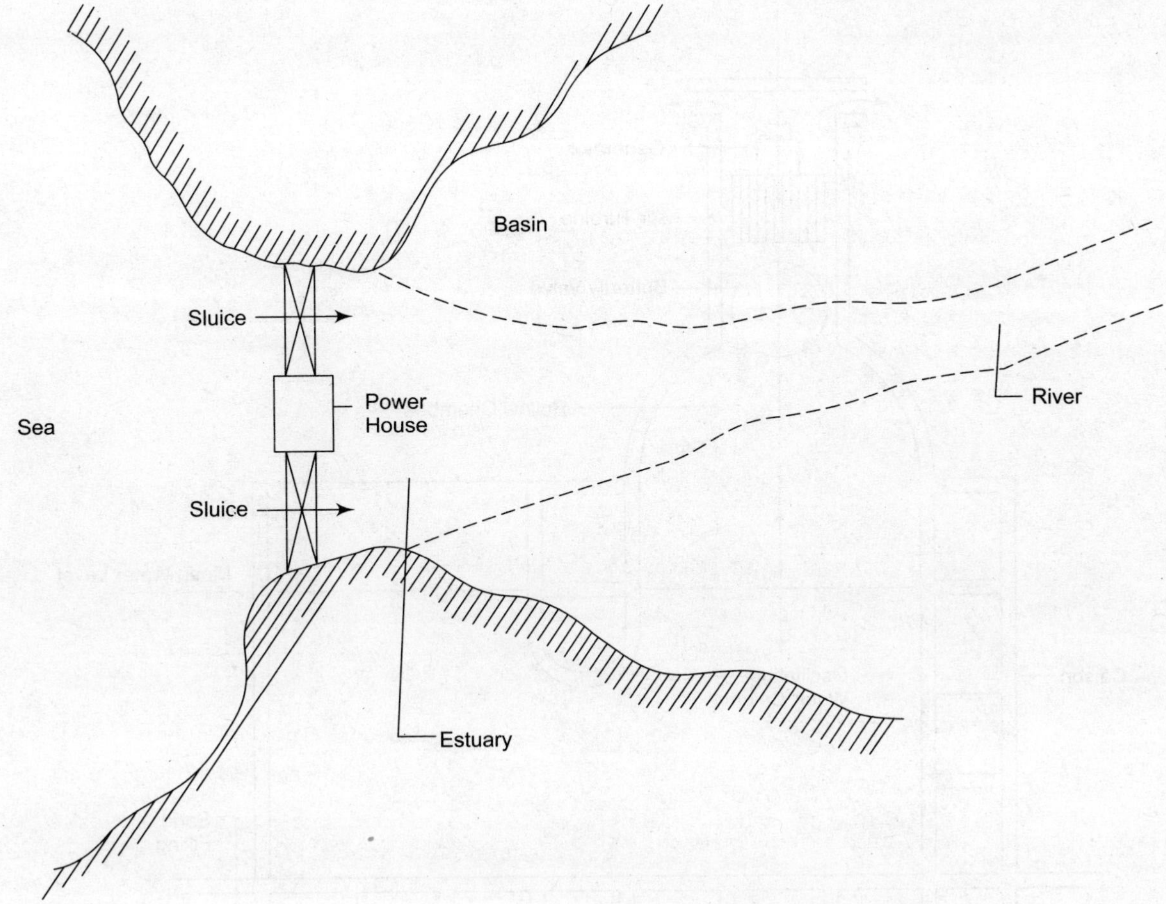

Fig. 1.15 Single Basin One-way System of Tidal Power Plant

1.6.4 Wave Energy

Considering the fact that the waves sweep across the oceans of the world, which cover almost 71% of the earth's surface, the energy available in the waves must be quite abundant.

The waves are caused by the interaction of the wind with the waters of the ocean. The wind gets about 1 to 5% of Sun's energy and part of this energy is transferred to the waves.

It is estimated that about 10 kW of energy is available for every one metre of wave front. In India the potential is estimated to be 40,000 MW.

Wave energy systems: Many wave energy extraction systems have been proposed and are being experimented upon. IIT, Madras is experimenting on a system using the principle of oscillating water column. This is considered best for the prevailing conditions in our country.

The basic principle of operation of this system consists of: (*a*) Causing the wave to enter a chamber and push up a column of air above it. (*b*) The air getting compressed thus, flows through a small opening where an air turbine-generator unit is installed. (*c*) When the wave recedes, it causes suction and the atmospheric air is drawn in to the chamber. Thus there is a regular oscillatory movement of the water column causing to and for motion of the air which in turn drives the turbine.

IIT system – a 150 kW wave energy power plant – is situated in Kerala. This is intended as a dual purpose system for reasons of economy. In the first place, it produces electrical energy and secondly it acts as a break-water, which protects the erosion of the sea shore or fisheries harbour.

The system (Fig. 1.16) consists of a caison which has a hollow chamber inside. The chamber provides space for the wave to rise up and fall down enabling an oscillating motion of water to be set up. The air above the water column, getting pushed up, drives the air turbine.

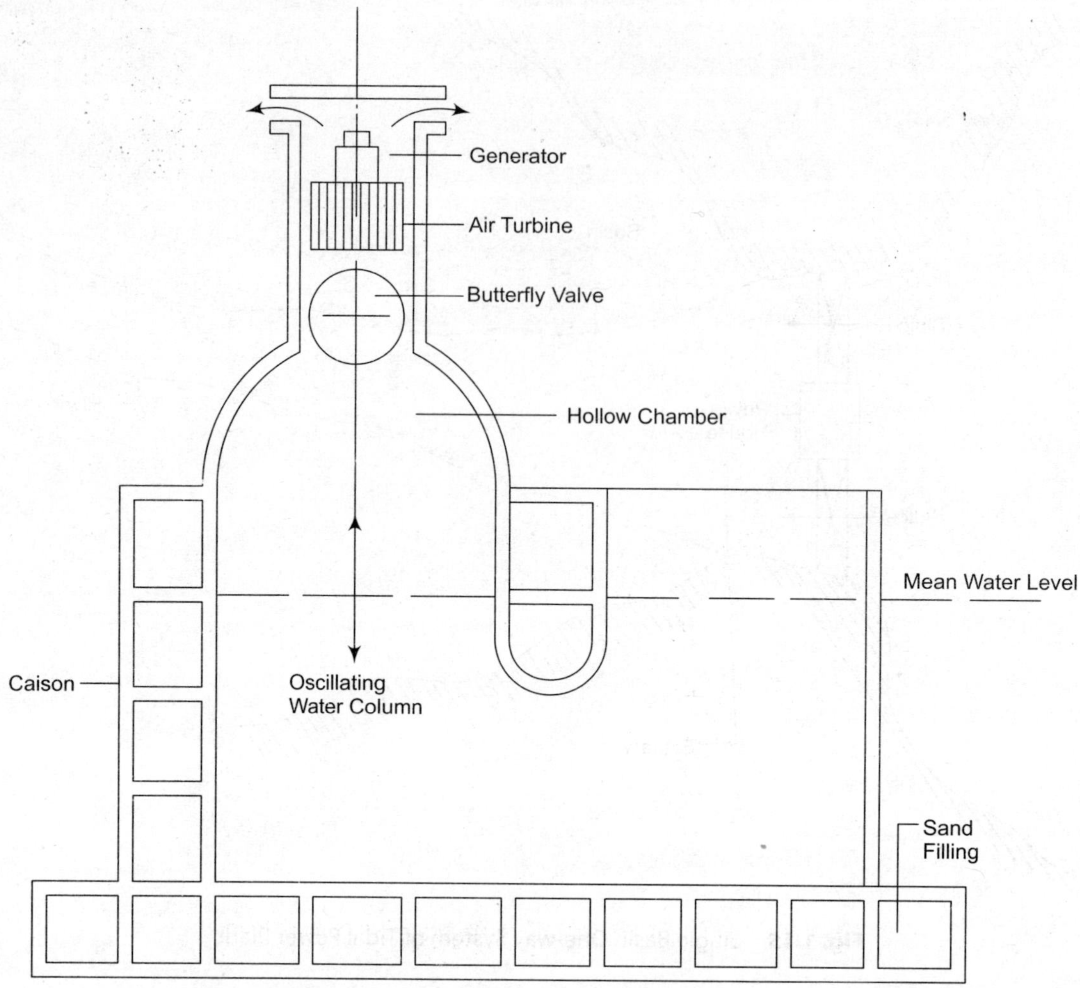

Fig. 1.16 Schematics of Wave Power Generating System (IIT, Madras)

A type of symmetrical aerofoil is used to enable unidirectional rotation of the air turbine, eventhough the flow of air takes place in both directions. An induction generator coupled to the air-turbine produces the electrical energy. The induction type generator overcomes the difficulty of rectification and inversion associated with variable speed alternators.

Advantages and Disadvantages

Energy is freely available and does not pose any pollution problems. The only drawback is that in any given location there is so much variation in the height and frequency of waves through out the year, that it badly affects the power production. The IIT system is estimated to generate about 15% only during December to March, when the wave potential is at the lowest.

1.6.5 Ocean Thermal Energy Conversion (OTEC)

The ocean serves as a big store house of solar energy. In the tropical regions of the earth, the surface of the water heated by the sun, is at an average temperature of around 25°C and at depths more than 100 m, it is about 5°C. Typical variation of temperature with ocean depths is shown in Fig. 1.17. The average difference in temperature of around 20°C may be made use of in running a thermodynamic cycle to yield mechanical power, which in turn could be used to generate electrical power.

Because of the very low temperatures involved, secondary working fluids such as ammonia, propane, R-12 etc., having low boiling point, are proposed to be used.

Scheme of a OTEC system is shown in Fig. 1.18.

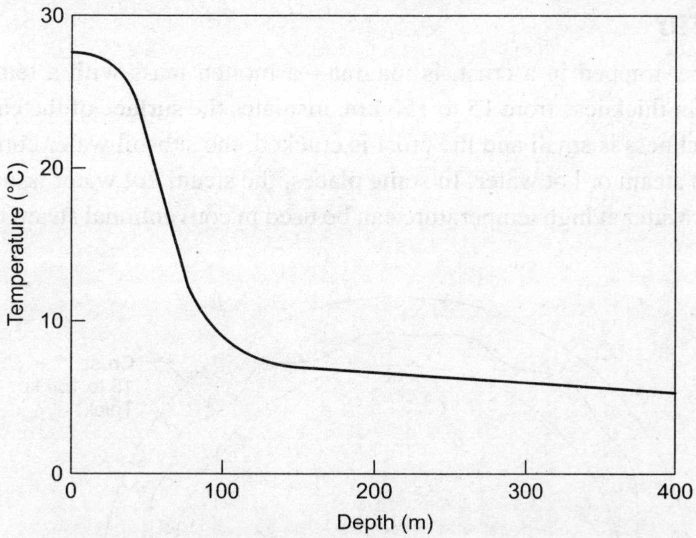

Fig. 1.17 Temperature Profile Across Depth of Ocean

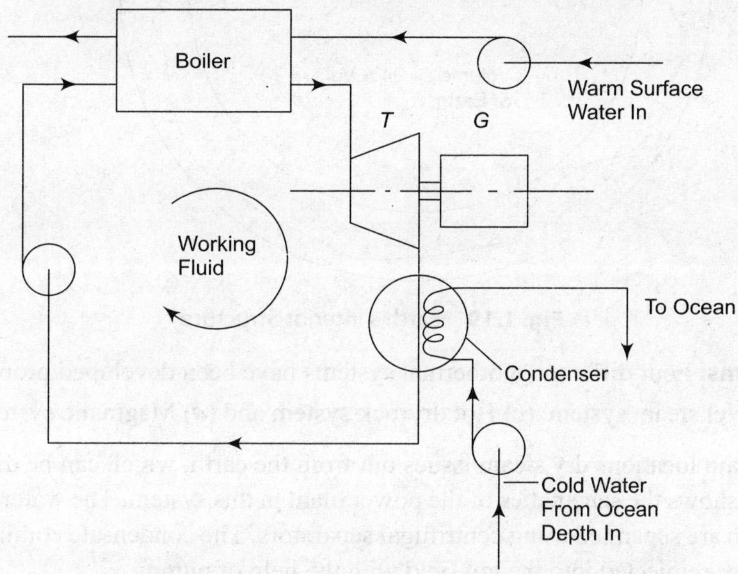

Fig. 1.18 Scheme of OTEC System

Advantages

1. Collection and storage of energy is done by nature and hence costs nothing.
2. This phenomenon of heating up of the ocean water occurs through out the year and so continuous power supply may be assured.

Drawback

1. The very small temperature difference necessarily means a low thermodynamic efficiency of the order of 2 to 3%.
2. The low levels of efficiencies in turn, call for handling large quantities of working fluid to obtain reasonable amount of power.
3. The heat exchangers — boilers and condensers —become necessarily big and capital cost goes up.
4. On-shore installations require long and big pipings, which add up to the cost. The proposal is therefore in favour of off-shore installations using floating platforms.
5. The materials suggested for heat exchangers are titanium or alloy of copper and nickel, which are resistant to corrosion. But then, the cost adds up.

1.6.6 Geothermal Energy

Underneath the earth's surface, trapped in a crust, is magma—a molten mass with a temperature of around 3000°C. (Fig. 1.19). The crust ranging in thickness from 15 to 150 km, insulates the surface of the earth and protects it. However, at certain locations, where thickness is small and the crust is cracked, the subsoil water comes in to contact with the hot matter and gets converted into steam or hot water. In some places, the steam/hot water issues under pressure. The steam coming under high pressure or water at high temperature can be used in conventional steam turbines for power generation or used for heating purposes.

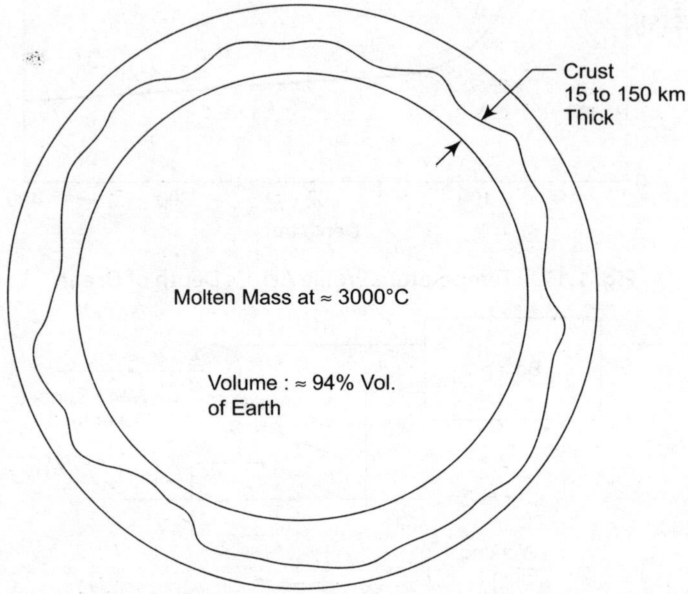

Fig. 1.19 Earth's Interior Structure

Geothermal power systems: Four different geothermal systems have been developed/proposed and they are: (*a*) Dry steam system, (*b*) Wet steam system, (*c*) Hot dryrock system and (*d*) Magmatic system.

Dry steam system: In certain locations dry steam issues out from the earth, which can be directly used in conventional steam turbines. Figure 1.20 shows the schematics of the power plant in this system. The water particles and solid particles coming along with the steam are separated using centrifugal separators. The condensate coming out of condenser is either sent for chemical recovery or reinjected into the hot field with the help of pump.

Wet steam system: In regions where hot water at high temperature and pressure issues out the hot water is flashed in a flash chamber to get the steam. The steam is then used as usual. Figure 1.21 shows the schematics of this type of system.

Wet steam binary system: Where the hot water issues out at lower temperature and pressure, it can't be flashed to steam. The heat is then used in a closed system as in Fig. 1.22. Here freon or isobutane is used as the working fluid and is kept in continuous circulation. The heat exchanger transports the heat from the hot water to the working fluid.

Hot dry rock system: Dry rock deep in the geothermal region holds lot of heat energy. 1 m^3 of rock with a temperature differential of 100°C may yield 10 kWh of energy if it could be effectively recovered.

For transmitting the heat to the surface, water is injected into the fractured rock and steam is brought up.

Magmatic-chamber system: The proposal is to extract heat from the volcanic chamber by lowering a heat exchanger in to the zone.

Advantages of geothermal energy

1. Energy is continuously available and is more dependable.
2. It has a good potential among the non-conventional energy sources.

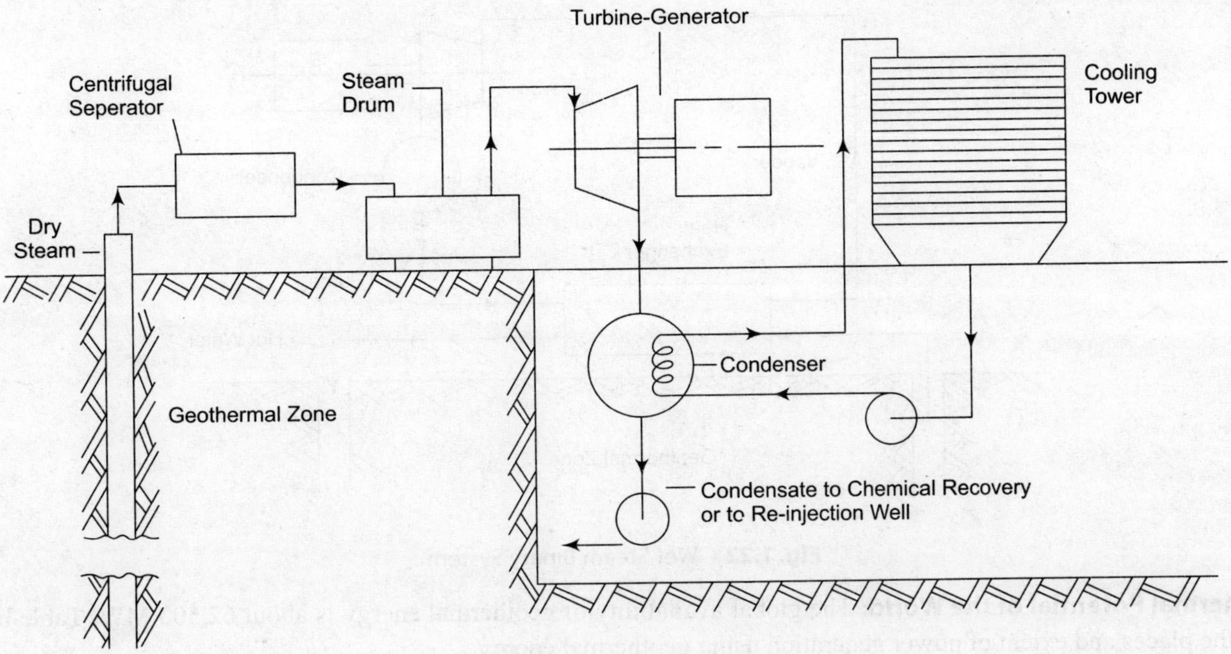

Fig. 1.20 Dry Steam System

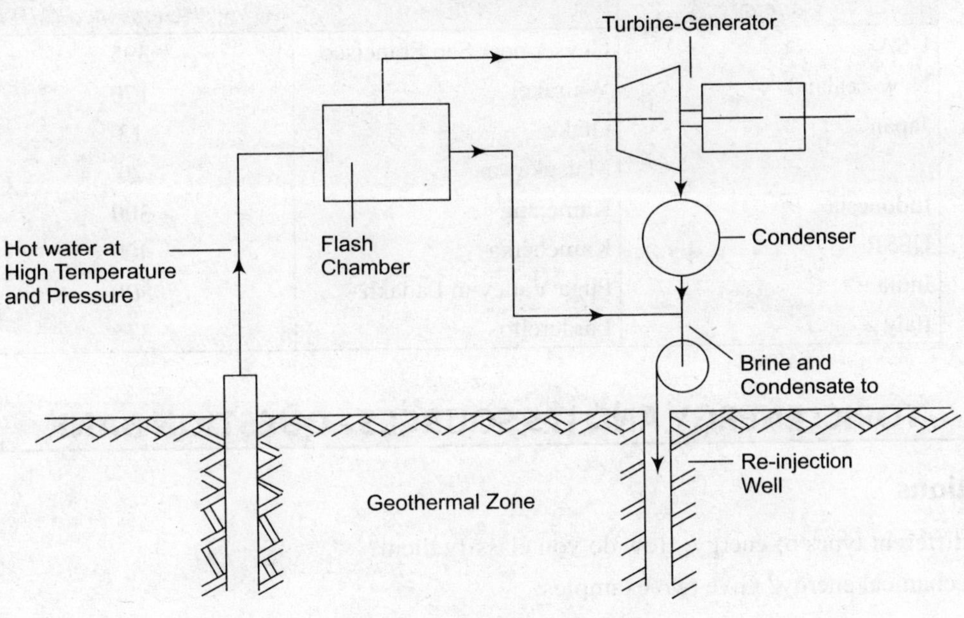

Fig. 1.21 Wet Steam System

3. Capital and generation cost is the lowest compared to nuclear and coal plants.
4. No solid pollutants and no radiation fall out.

Disadvantages

1. Problem of corrosion of components due to salt.
2. Thermal pollution due to effluent if not reinjected.
3. Gaseous effluent, especially hydrogen sulphide, is a nuisance.
4. Ground subsidence may occur.
5. Life span may be less compared with nuclear and coal plants.
6. The highly mineralised effluent may pollute ground water and hence requires reinjection into a well.

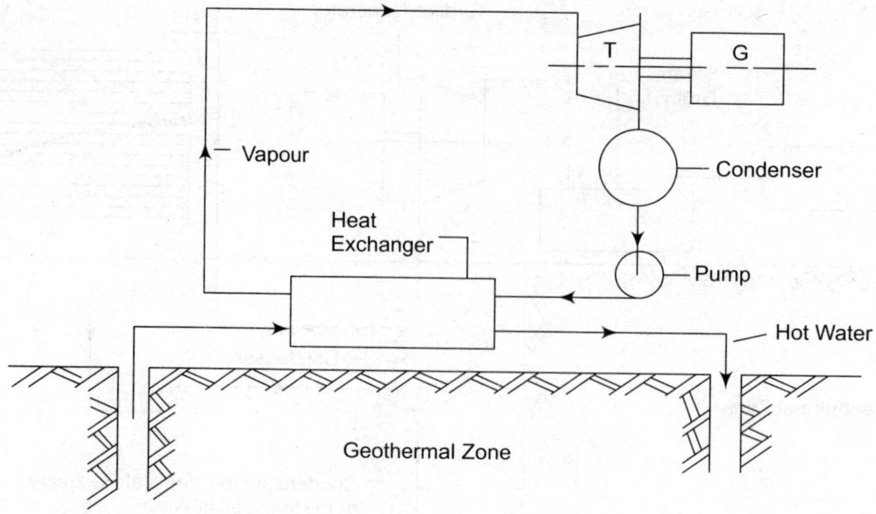

Fig. 1.22 Wet Steam Binary System

Geothermal Potential of the World: The global availability of geothermal energy is about 62,500 MW. Table below gives the places and extent of power generation using geothermal energy.

Country	*Place*	*Power Generated (MW)*
USA	Geyser near San Francisco	395
New Zealand	Wairakei	170
Japan	Otake	13
	Matsukawa	20
Indonesia	Kamejang	300
USSR	Kamchetka	400
India	Puga Valley in Ladakh	5(?)
Italy	Larderello	375

TOPIC: ENERGY AND ITS SOURCES - QUESTION BANK

2-Marks Questions

1. Name the different types of energy. How do you classify them?
2. What is mechanical energy? Give two examples.
3. What is electrical energy? It is a versatile energy. Why?
4. What is electromagnetic energy? Give two examples.
5. Define thermal energy. Why is it said to be the basic form of energy?
6. List the various non-traditional (unconventional) energy sources.
7. What are fossil fuels? Give examples.
8. What are the categories of coal? Name them in the decreasing order of Fixed Carbon content.
9. How is coal analysed? What are the properties of coal which interest a boiler design engineer?
10. What are the important characteristics of a fuel?
11. List the advantages of fuel oil over solid fuel such as coal.
12. Name the common fuel gases.
13. What is water gas? How is it produced?

14. What is producer gas? How is it produced?
15. What are the advantages of gaseous fuels?
16. What are the advantages of energy from waterfalls?
17. How is nuclear energy produced using fission process?
18. What is fusion process?
19. Why do you call thorium and U-238 as fertile materials?
20. What are fissile materials? Give examples.
21. What is the basis of solar energy?
22. List the ways in which solar energy is used.
23. Solar energy is yet to catch up with other forms of energy. Why?
24. What are the advantages of solar energy?
25. What are the compelling reasons to go in for solar energy?
26. Name the different solar power systems.
27. What is 'Green House Effect'?
28. What are 'Heliostats'?
29. What causes wind and what is wind energy?
30. What are the advantages of wind energy?
31. What are the limitations of wind energy?
32. Name some types of wind mills generally used.
33. Define 'Coefficient of Performance' of a wind mill.
34. What is the principle of obtaining tidal energy?
35. What are the advantages of tidal energy?
36. What is meant by pumped storage system?
37. How are waves caused? What are the advantages and limitations of wave energy?
38. What is the working principle of OTEC?
39. What are the advantages and limitations of OTEC?
40. What is geothermal energy? How is it used for power generation?

10-Marks Questions

1. What are conventional and non-conventional energy sources? Why is it important to turn our attention to non-conventional energy sources? Discuss.
2. Enumerate the various systems of solar energy collection and utilization. Explain the tower concept of solar energy collection and production of power. Make a neat sketch to explain.
3. Sketch and explain the working of a low temperature solar system for power generation.
4. Describe the working of a flat plate collector with a neat sketch.
5. (*a*) Sketch the arrangement of a propeller type wind mill and name its parts.

   (*b*) If the mill has to produce 1 kW of power and the mill is 7 m in diameter what should be the approximate velocity of wind? Take coefficient of performance = 0.4 (4.5 m/s)
6. Explain the working of Darrius rotor wind mill. Make a neat sketch.
7. Sketch and explain wind energy system with constant speed device.
8. Describe with sketches how tidal energy is harnessed for generating power.
9. Make a schematics of the wave power generating system as executed by IIT, Madras and explain its functioning.
10. Explain the working of OTEC with a neat sketch.
11. How is geothermal energy tapped for power generation (*i*) using Dry steam or (*ii*) Wet steam?

State True or False

1. Hydro-electric energy is pollution free.
2. Solar and wind sources are renewable sources.
3. Geothermal energy is non-polluting.
4. Coal reserve in India is much more than oil.
5. Solar cell has a conversion efficiency of about 15%.
6. Solar insolation (incident radiation) is about 10 kW/m^2.
7. Wind power for a given swept area varies directly as the square of the wind velocity.
8. Geothermal source produces clean steam.
9. OTEC converts solar energy directly into electricity.

Answers

1. T **2.** T **3.** F **4.** T **5.** T **6.** F **7.** F **8.** F **9.** F

Fill up the Blanks

1. For direct conversion of solar energy to electricity .... are used.
2. Conventional sources of energy are ...., ..., ..., ...
3. Renewable sources of energy are ...., ..., ..., ...
4. Solar insolation is about ....
5. The power available in the wind is proportional to the .... of wind speed.

Answers

1. photovoltaic cells
2. coal, oil, gas and waterfalls
3. solar, wind, tidal, wave, bio-mass
4. 500–900 W/m^2
5. cube

Chapter-2

Utilization of Conventional Energy: Power Plant

Generally, a power plant is one, which produces electric power on a large scale. Electric power, being very versatile, the energy available in different forms is converted to electrical energy in these power plants. Based on the type of energy converted, we have:

1. Steam Power Plant
2. Hydro-electric Power Plant
3. Nuclear Power Plant
4. Diesel Power Plant

2.1 STEAM POWER PLANT

Here we try to understand briefly how high pressure steam is produced to be utilized for mechanical power generation and ultimately, the most useful form of and much-sought form of energy - the electrical energy which finds application in almost all human activities like lighting, heating, cooling, transport etc.

Looking into Fig 2.1(*a*), depicting a Temperature - Enthalpy diagram, point *A* refers to condition of water at 0°C. When heated at constant pressure, temperature rises to point *B*, which is called saturation temperature.

Addition of further heat does not bring about a rise in temperature till point *C*, when all the water (liquid phase) would have been converted into gaseous state, and we call this steam as 'dry saturated steam'. From *B* to *C*, liquid and gaseous phases exist together and in this state the steam is called 'wet steam'. Addition of heat beyond *C* would result in increase of temperature and we call the steam as super heated steam. Thus we have wet steam, dry steam, and superheated steam.

Point *B* represents boiling point of water and the heat added between *B* and *C* is called latent heat, which makes the steam fully dry.

Some definitions connected with properties of steam:

(*i*) **Enthalpy of steam:** It is the sum of internal energy and product of pressure and volume.

(*ii*) **Specific volume:** It is the volume occupied by unit mass (1 kg).

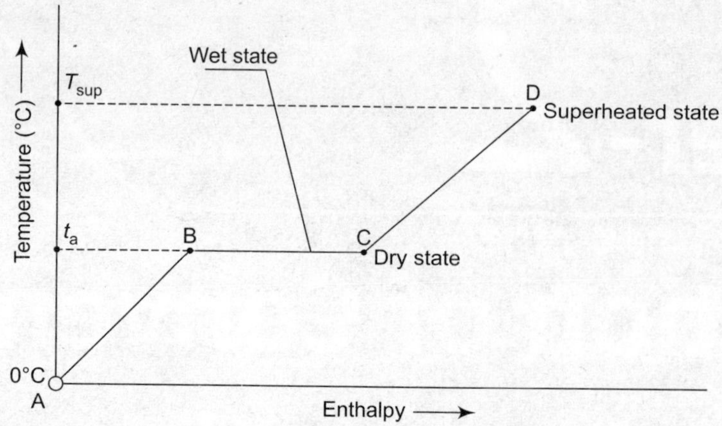

Fig. 2.1 *(a)* Temperature enthalpy diagram

(iii) **Internal energy:** It is the actual energy stored in steam, *i.e.*, $U = h - PV$.

(iv) **Saturation temperature:** It is the temperature at which water starts boiling. Steam table gives the saturation temperature at a given pressure called saturation pressure.

(v) **Sensible heat:** It is the heat required to raise the temperature of 1 kg of liquid (water) from an initial temperature (t) to final temperature (t) before water reaches saturation temperature.

(vi) **Super heat:** It is the heat added to saturated steam to raise the temperature to $t_{sup} = C\,(t_{sup} - t_s)$.

(vii) **Dryness fraction:** It is the amount of dry steam contained in a given quantity of wet steam.

(viii) **Latent heat:** It is the quantity of heat required to convert unit mass of a fluid at saturation temperature to vapour at saturated temperature..

Now referring to block diagram Fig. 2.1(*b*), we see how energy available in steam at high pressure is transformed into electrical energy.

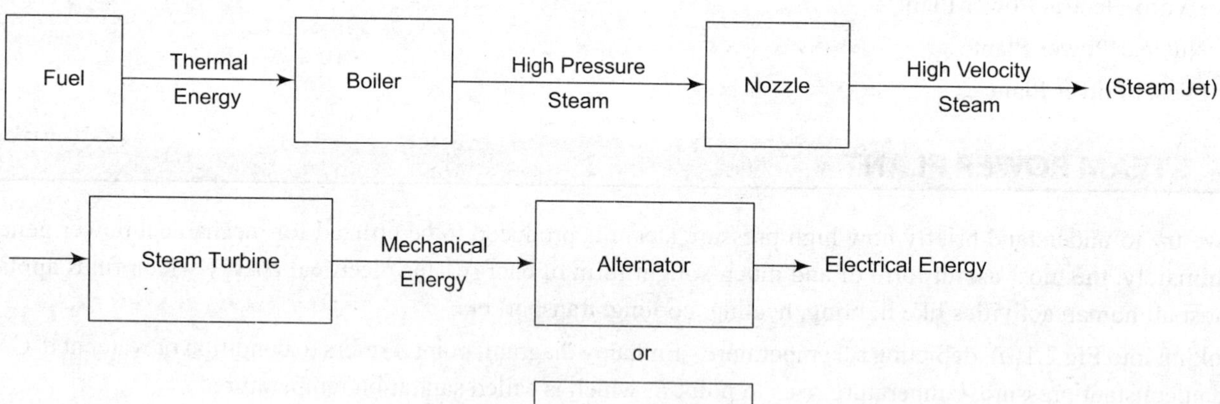

Fig. 2.1 *(b)* Energy Conversion in Steam Power Plant

Figure 2.2 is an elaborate schematics of a steam power plant using pulverised coal as fuel. Some of the important features of the power plant are:

1. Fuel system providing for coal storage and its preparation.
2. Heat recovery system consisting of economiser and air preheater.
3. Ash disposal system.
4. Boiler water circulation system and cooling water system.
5. Draft system consisting of induced draft fan and chimney.

Knowing the importance of uninterrupted supply of power, adequate and continuous supply of fuel must be ensured. In this, provision for storage is an important factor. Normally coal is stored to last for a month and the guiding factors in storage are (*a*) cost, (*b*) risk, (*c*) space and (*d*) deterioration in storage.

Heat recovery systems like economiser, air preheater etc., improve the thermal efficiency of the boiler.

Need for continuous and automatic disposal of ash needs no mention. Condensing the exhaust steam from the turbine and recirculating the condensate reduces the demand for water of boiler quality. Cooling water system provides continuous supply of water for condensing the exhaust steam.

The induced draft (I.D.) fan ensures controlled flow of combustion air and also aids disposing of the smoke and exhaust gases from the boiler through the chimney. In addition to these basic features, it has now become mandatory to install pollution control devices, like electro-static precipitators. They prevent dust and ash getting into the atmosphere making the plant eco-friendly.

2.1.1 Boilers

In the steam power plant, the boiler (also called steam generator) is the major component. Let us try to understand how the boiler has been developed over the years. It is important to remember, that as engineers, we strive, to aim at maximum efficiency with the view to economise on the resources used. At the same time it should not be forgotten that safety is of vital importance in dealing with a pressure vessel like boiler and therefore safety aspect must be given the tope priority. Equally important is maintenance and therefore adequate provision must be made to facilitate easy maintenance.

Types and Classification of Boilers

Following are the best known types of boilers:

1. Simple vertical boiler
2. Cochran boiler
3. Lancashire boiler (1844)
4. Cornish boiler
5. Locomotive boiler
6. Marine boiler
7. Babcock and Wilcox boiler (1867)
8. Stirling boiler (1918)
9. Modern high pressure boiler
10. Once — through boiler etc.

The classification is made as follows:

1. Smoke tube boilers and water tube boilers.

   In a smoke tube boiler, the hot flue gases pass through the smoke tubes which are surrounded by water being heated. In the case of a water tube boiler, water passes through tubes, over which the hot gases flow transferring the heat to water in the tubes.

   Smoke tube boilers are essentially of smaller capacity and serve the needs of small industries. Water tube boilers can be built to provide not only larger quantity of steam but also steam at very high temperature and pressure.
2. Boilers are classified based on the arrangement of the boiler shell. Accordingly, we have horizontal or vertical boiler depending on whether the axis of the shell is horizontal or vertical.
3. Based on capacity/pressure, we have low, medium or high capacity/pressure boilers.
4. Based on application, we have locomotive boiler (used on locomotives) marine boiler (used on ships), process steam boiler, power boiler, etc. When a boiler is used for both power generation and process steam purposes, the system is called cogeneration.

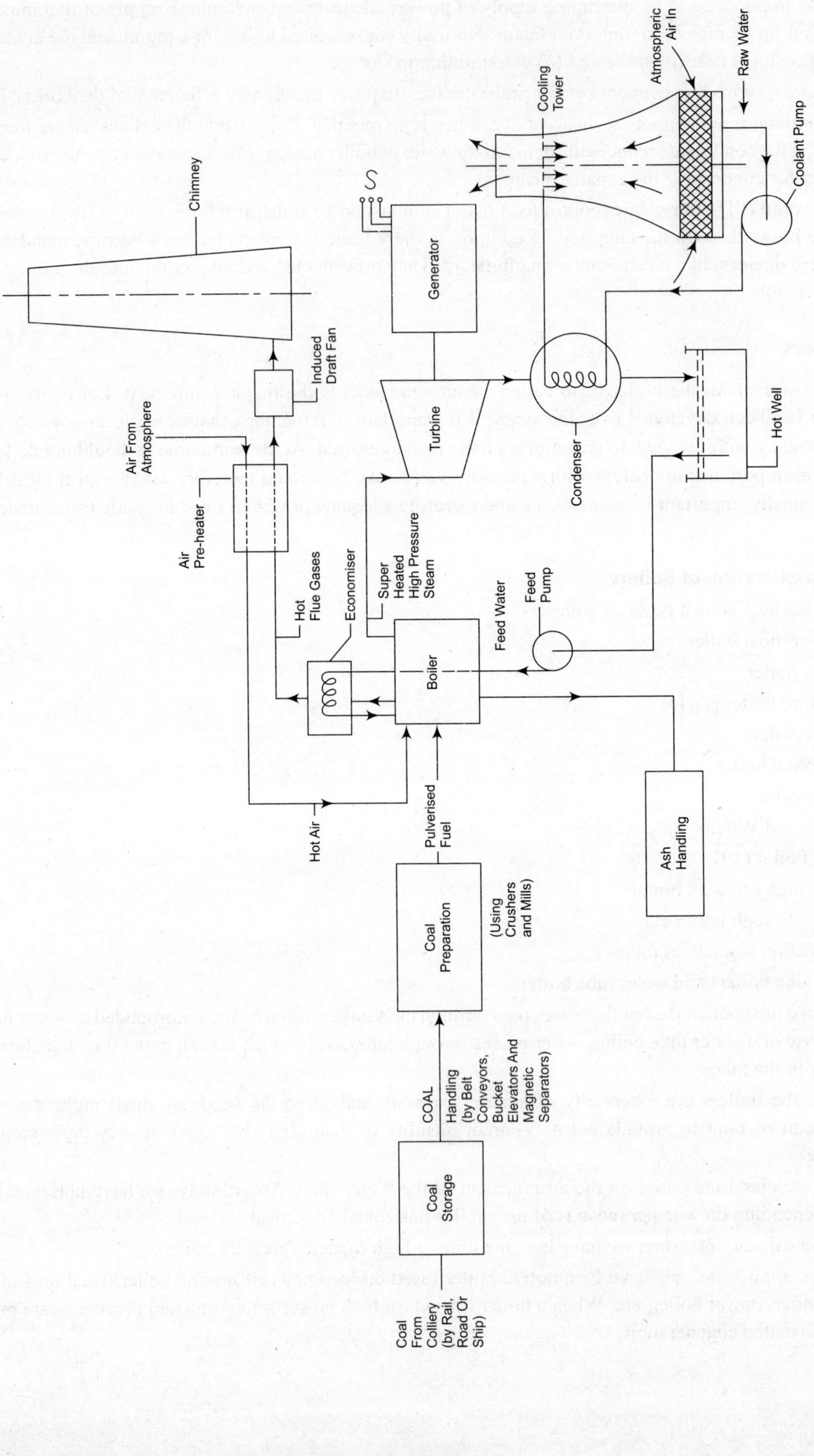

Fig. 2.2 Schematics of Steam Power Plant Using Pulverised Fuel

2.1.2 Simple Vertical Boiler (Fig. 2.3)

Description: It consists of a cylindrical shell made of low carbon steel. It is placed with its axis vertical and hence gets the name vertical boiler. It has a central flue, connected to the fire place, and the hot gases of combustion pass through this. Water surrounding the flue and the fire place gets heated and steam is formed. Steam is collected in the space above the water, and is taken out from the main steam stop valve when required.

The fuel (coal, charcoal or wood) is burnt on the grating and the ash is collected at the bottom.

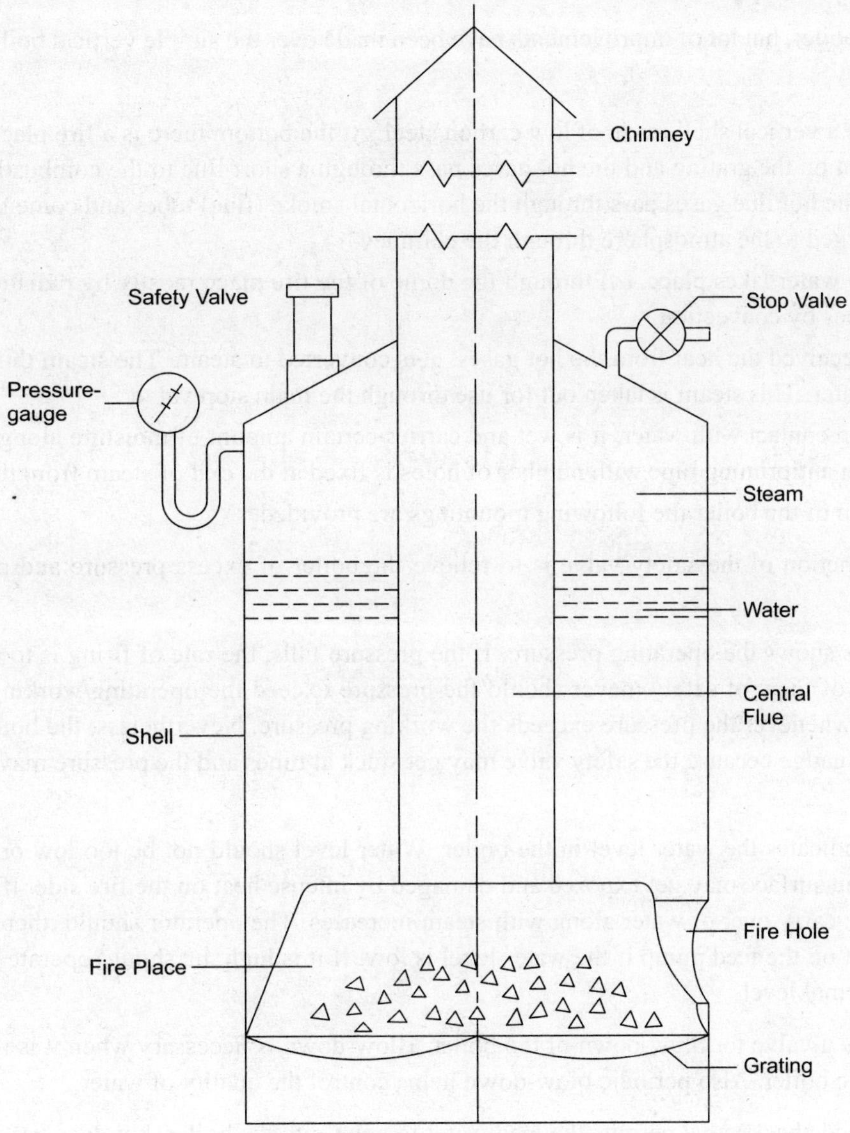

Fig. 2.3 Simple Vertical Boiler

Discussion: The boiler is simple in construction but it has very low thermal efficiency. (Efficiency indicates how much of the available energy in fuel is transferred to steam). Its rate of steam generation is also low. Hence it is superseded by more efficient boilers.

The equation of heat transfer to water may be written as

$$Q = Ah\,(t_f - t_w)$$

Where Q = Heat transferred (Joules/hr)

A = Heating surface area (m^2)

h = Heat transfer coefficient (J/m^2hC)

t_f = Temperature of flue gases (C)

t_w = Water temperature (C)

It is seen that for a given t_f and t_w, Q depends on the area A of the heating surface. "A" could be increased by having large number of smaller diameter tubes. Smaller diameter tubes are more efficient in taking out the heat from the flue gases. This concept is used in Cochran boiler.

2.1.3 Cochran Boiler (Fig. 2.4)

This is also a vertical boiler, but lot of improvements have been made over the simple vertical boiler and is therefore able to give better efficiency.

Description: This has a vertical shell made of low carbon steel. At the bottom there is a fire place with a hemispherical dome. The fuel is burnt on the grating and the hot gases pass through a short flue to the combustion chamber. From the combustion chamber the hot flue gases pass through the horizontal smoke (flue) tubes and come to the smoke box, from where they are discharged to the atmosphere through the chimney.

The heat transfer to water takes place, (*a*) through the dome of the fire place mostly by radiation and (*b*) through the walls of the smoke tubes by convection.

The water having received the heat from the hot gases, gets converted to steam. The steam thus generated collects in the space above the water. This steam is taken out for use through the main stop valve.

Since the steam is in contact with water, it is wet and carries certain amount of moisture along with it. To reduce the moisture carry over, an antipriming pipe with number of holes is fixed at the exit of steam from the boiler.

For proper operation of the boiler the following mountings are provided.

Safety valve: The function of the safety valve is to relieve the boiler of excess pressure and protect the boiler from explosion.

Pressure gauge: This shows the operating pressure. If the pressure falls, the rate of firing is to be increased and vice-versa. From the point of view of safety, never should the pressure exceed the operating/working pressure. The safety valve no doubt opens whenever the pressure exceeds the working pressure. Nevertheless, the boiler operator must have an eye on the pressure gauge because the safety valve may get stuck at times and the pressure may increase to dangerous limits.

Water gauge: This indicates the water level in the boiler. Water level should not be too low or too high. If the water level is too low, heating surface may get exposed and damaged by intense heat on the fire side. If on the other hand, the water level is high, the carry over of water along with steam increases. The operator should, therefore, keep a watch on the water level and put on the feed pump if the water level is low. If it is high, he should operate the blow-down till the water comes to the normal level.

Blow-off cock: This is a valve for blow-down of the boiler. Blow-down is necessary when it is required to bring down the level of water in the boiler. Also periodic blow-down helps control the quality of water.

Feed check valve: Feed check valve permits the feed water to enter into the boiler, but does not allow it to flow back.

The boiler has the following provisions for cleaning, maintenance and inspection.

Manhole: This opening is intended for the boiler inspector to get into the boiler for examining the inside of the boiler for possible corrosion, deposits and unsound conditions.

Mudholes: These are provided at the bottom of the boiler to help remove the sediments and dirt settling at the bottom. When cleaning of boiler is undertaken the mud hole covers are removed and the dirt, taken out.

Chimney: Helps to suck out and leave the flue gases into atmosphere and when this happens air from the atmosphere is drawn into the furnace to help combustion of fuel. The arrangement of providing combustion air in this manner is called natural draft.

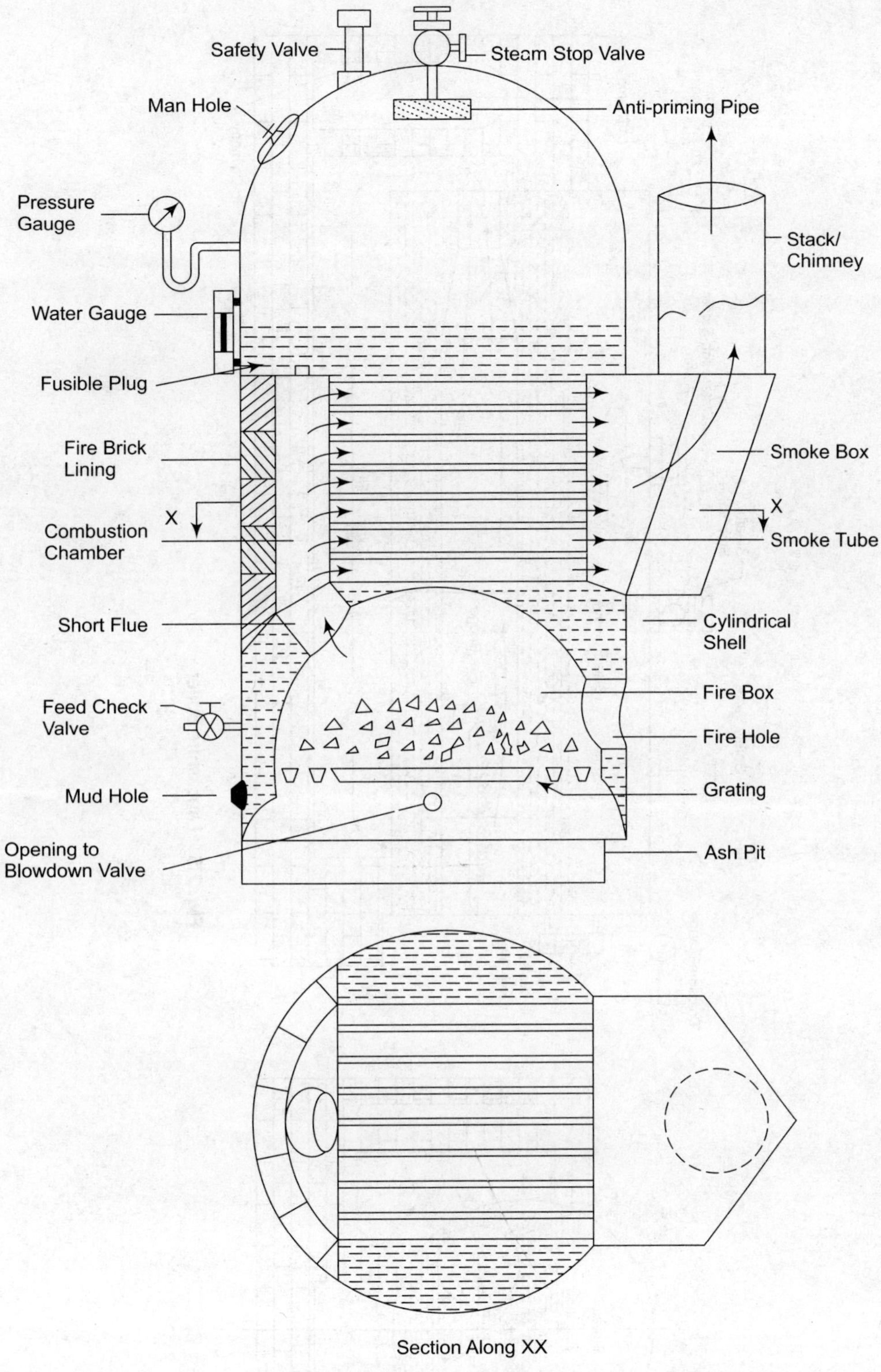

Fig. 2.4 Cochran Boiler

Feed pump: One of the important accessories of a boiler is feed pump. It is required for pumping water into the boiler.

Fusible plug (Refer Fig. 2.18): In the event of water level going too low and exposing the heating surface to combustion gases with no water on the water side, the heating surface may burn out. A fusible plug, when provided, melts and allows the steam to rush in to the fire place and quench the fire. Thus the damage to heating surface is averted.

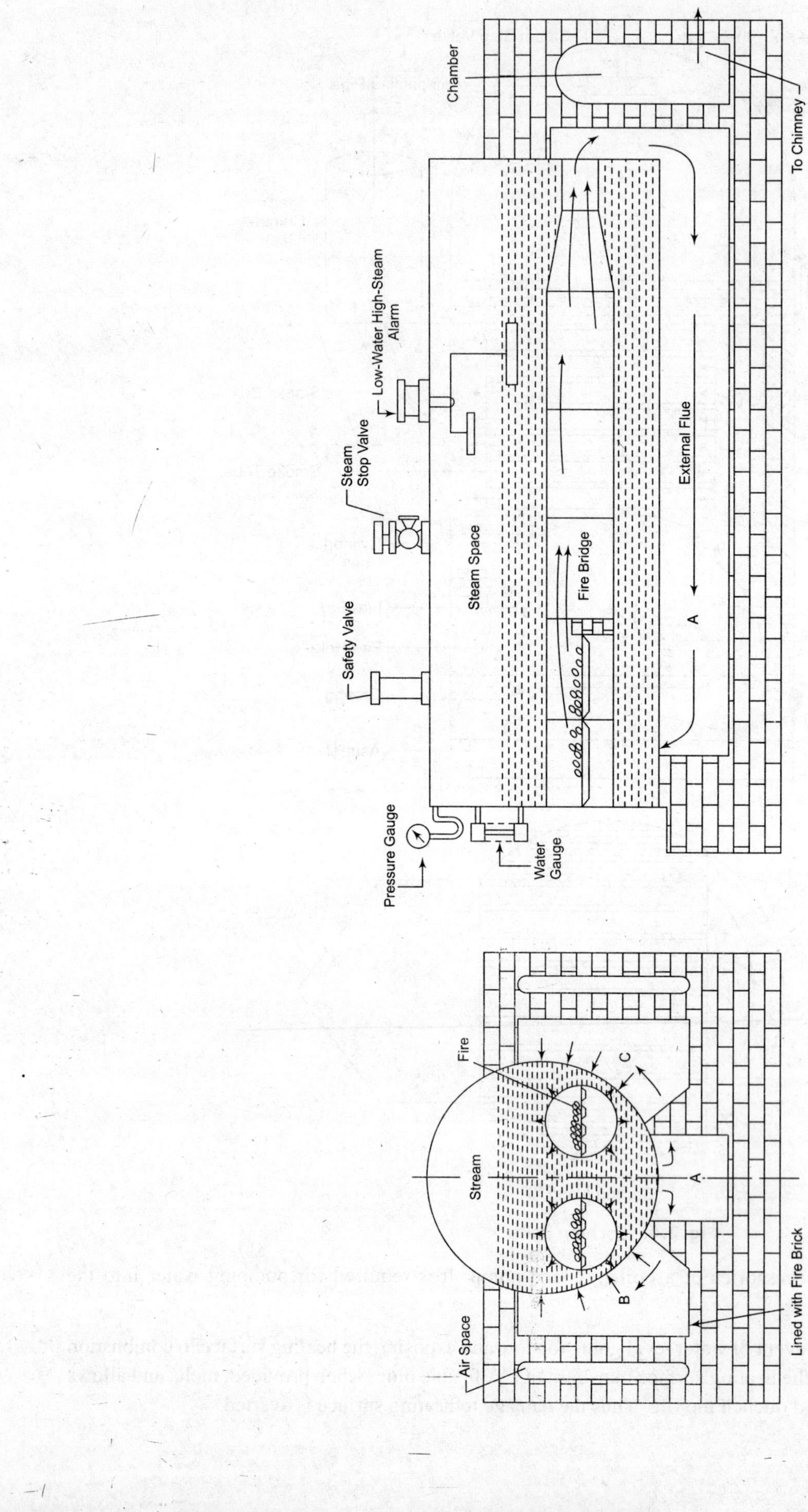

Fig. 2.5 Lancashire Boiler

Cochran boiler, with its number of horizontal smoke tubes and dome —like structure of the fire place has more heating surface. This feature increases the heat transfer and therefore the efficiency of the boiler.

2.1.4 Lancashire Boiler (Fig. 2.5)

This is a horizontal smoke tube boiler with two big internal flue tubes running the entire length of the boiler. The fire place is arranged at the front of the flue tubes. The fire bridge confines the fuel to burn over the grate. The boiler is set in brick work in which flue gas passages are formed. A is a flue gas passage below the boiler shell and B and C are the passages at the side.

Fuel is burnt on the gratings and the hot gases travel along the internal flue tubes and at the end, dip and pass into the bottom flue passage A. Coming out of passage A, the gases travel backwards through the side flue passages B and C as indicated by arrows.

Passages B and C are connected to a chamber at the end of the boiler from where the gases are led to the atmosphere through a chimney. The three 'passes' (*a*) through the internal flue tubes, (*b*) through the bottom flue and (*c*) through the side flues result in more effective heat transfer.

The boiler is provided with the usual mountings like pressure gauge, water gauge, blow-off cock, main stop valve and safety valve. In addition, there is also a low-water-high-steam alarm, which gives an audio-signal for the conditions of low water level or high steam pressure.

A typical Lancashire boiler has a drum about 2 m diameter and length 7 m. It is a good stationary boiler, highly reliable and easy to operate.

Superheater may be added at the end of the main flue tubes to superheat the steam and thus improve the efficiency of the boiler.

Cornish Boiler: This resembles a Lancashire boiler, but has only one inner flue tube instead of two. The capacity is smaller.

Typical Size: 1 m in diameter and 4 m in length.

2.1.5 Locomotive Boiler

This is also a smoke tube boiler. It is used mostly on locomotive for providing steam to drive steam engines which provide the motive power.

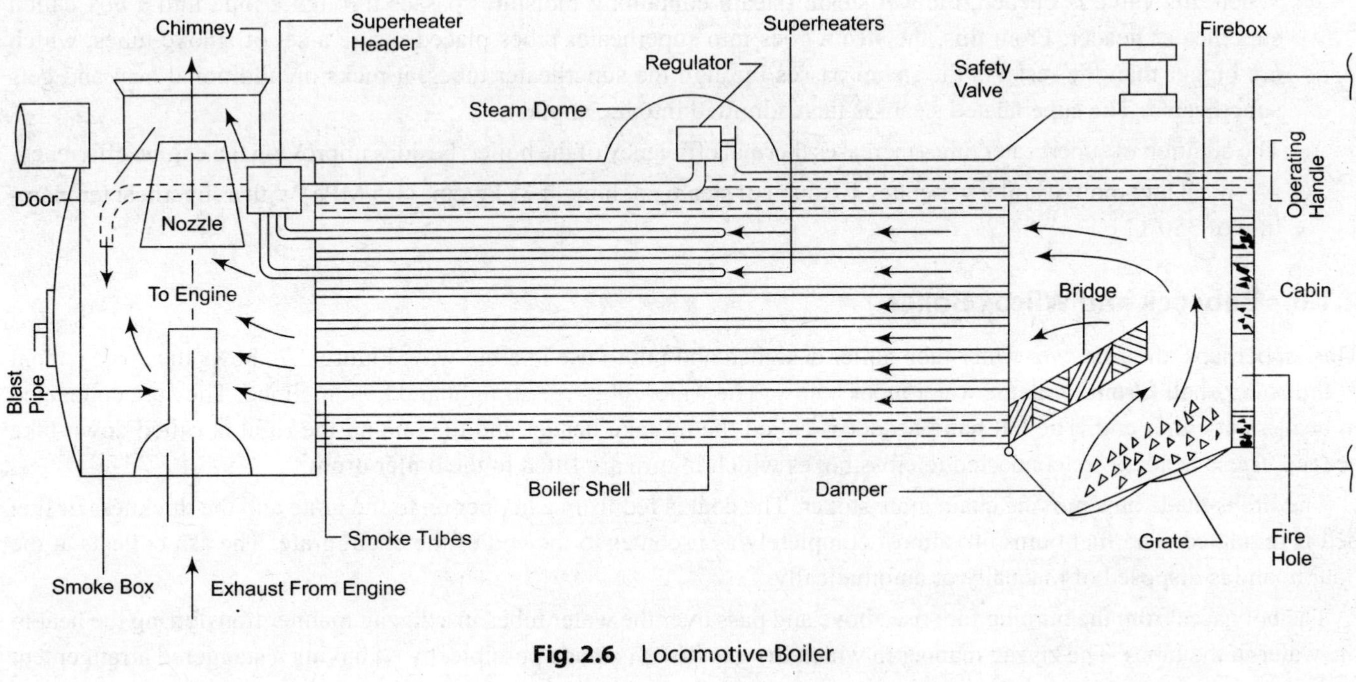

Fig. 2.6 Locomotive Boiler

Figure 2.6 shows a diagrammatic view of a Locomotive boiler.

The main parts of this boiler are:

1. Boiler shell (kept in a horizontal position)
2. Fire box with water jacket
3. Smoketubes
4. Superheater tubes
5. Smoke box
6. Chimney
7. Blast pipe
8. Damper
9. Steam dome with regulator
10. Safety valve and other usual mountings

Working

(*i*) **Flow of gases:** Coal in small lumps is used as fuel. The fuel is fed into the firebox through the fire hole. The fuel burns on the grating and the hot gases rise up, get deflected by the bridge at the top, pass through the smoke tubes and finally get discharged into the atmosphere through a short chimney.

(*ii*) **Heat Transfer:** The heat transfer takes place, in the first instance, through the walls of the fire box to the water surrounding the fire box. Here the mode of heat transfer is mostly radiation. Further heat transfer, by convection, takes place through the walls of the smoke tubes and also the superheater.

(*iii*) **Draft:** The air required for combustion is drawn through the grate from the atmosphere. The chimney, being short, is not able to create the required suction effect (called natural draft). Therefore, a blast pipe is arranged below the chimney, through which (blast pipe) is discharged the exhaust steam from the steam engine. The steam, passing through a nozzle at the top of the blast pipe creates enough suction to draw out the flue gases from the smoke tubes. This, in turn, results in the air rushing into the fire box.

The amount of air admitted into the fire box is regulated by a damper and thus the rate of firing is controlled.

(*iv*) **Superheater:** The steam generated in the shell, collects in the space above the water and in the steam dome. The regulator (stop valve) is enclosed in the steam dome and is operated through a long rod extending into the cabin. When this valve is opened, the wet steam (steam containing moisture) passes through a pipe into a box called superheater header. From this, the steam goes into superheater tubes placed inside a set of smoke tubes, which are bigger than the rest. As the steam passes through the superheater tubes, it picks up additional heat and gets superheated. The superheated steam is then admitted into the engine.

The addition of superheater tubes increases thermal efficiency of the boiler, besides improving the engine efficiency. A typical locomotive boiler generates 10 tonne of steam per hour at 15 kg/cm^2 (1.5 MPa) with a superheat temperature of 350°C.

2.1.6 Babcock and Wilcox Boiler

This is, perhaps, the first ever water tube boiler designed and put to use in a big way. Figure 2.7 shows the arrangement of the boiler shell (drum) with the water tubes below. The water tubes are straight and kept inclined. They are connected to headers at either end. The left hand header is called the up-take header and the one on the right is called down-take header. These headers are connected to cross boxes which in turn are fitted to the boiler drum.

The fire is made on a moving chain grate stoker. The coal is fed from a hopper on to the grate and the thickness of fuel bed is regulated. The fuel burns out almost completely as it comes to the end of the chain grate. The ash collects at the bottom and is disposed of manually or automatically.

The hot gases from the burning fuel rise above and pass over the water tubes in a zigzag manner transferring the heat to the water in the tubes. The zigzag manner in which the gas flows is made possible, by (*i*) having a staggered arrangement

[Figure 2.7(*a*)] of water tubes and by (*ii*) providing baffles B as shows in Fig. 2.7. Super heater tubes kept in the passage of flue gases get the required heat for superheating the steam. The gases, after cooling down, are sucked through a chimney and led out into the atmosphere. The damper at the entrance of the chimney regulates the air for combustion and thus the rate of firing is controlled. A chain connected to the damper passes over pulleys and hangs in front of the boiler at a convenient height to facilitate boiler operator to operate the damper.

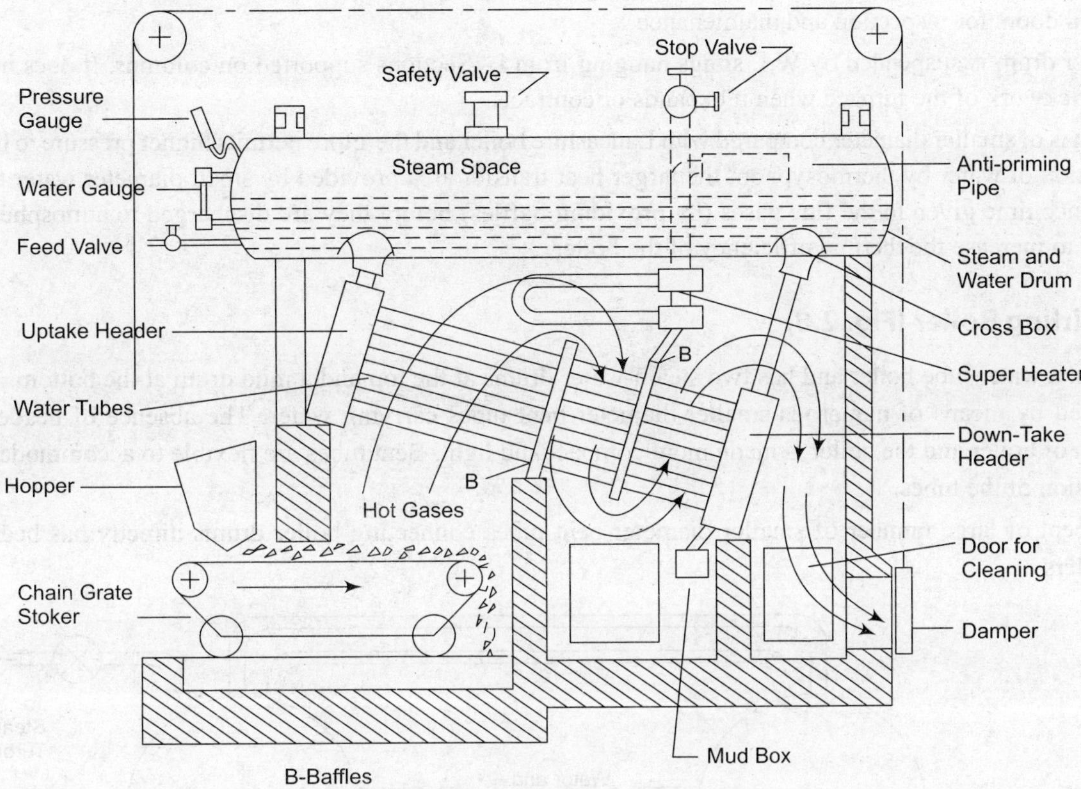

Fig. 2.7 Babcock and Wilcox Boiler

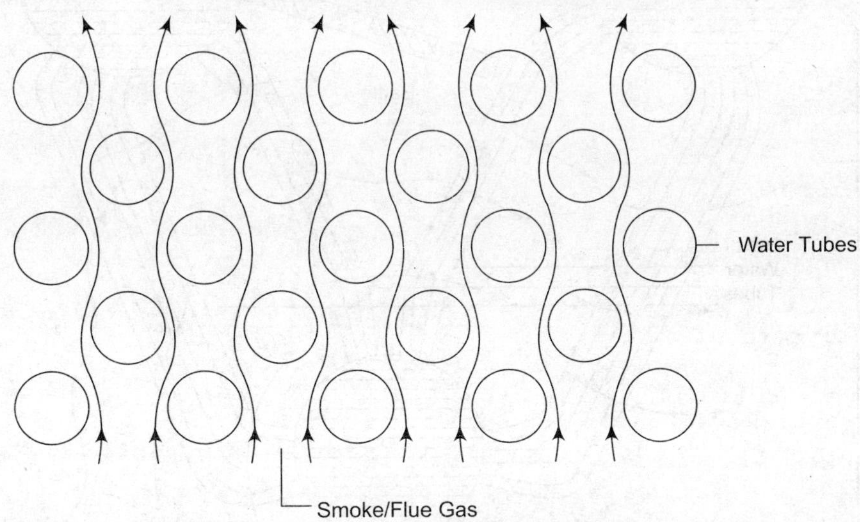

Fig. 2.7 (*a*) A Staggered Arrangement of Water Tubes Help Better Heat Transfer

The water in the water tubes gets heated and becomes lighter. Because of the lower density the hot water moves up through the uptake header and enters the boiler drum. The comparatively cold water from the boiler drum comes down through the downtake header and enters the water tubes for getting heated further. This circulation of water, which is brought about by the differential temperature, is called 'Thermosyphon'.

The feed water entering through the feed check valve is directed backwards to help thermosyphon.

Water getting heated thus, gets converted into steam and collects in the space above the water. The steam then passes through the antipriming pipe and gets into the superheater, where it is superheated. The superheated steam is then available for use.

The boiler has the usual mountings like pressure gauge, water gauge, safety valve and main stop valve. It has man-hole, mudbox, and doors for inspection and maintenance.

The boiler drum is suspended by W.I. slings hanging from I - Sections supported on columns. It does not, therefore, strain the brickwork of the furnace when it expands or contracts.

The drum is of smaller diameter compared with Lancashire boiler and therefore permits higher pressure to be developed. The circulation of water by thermosyphon, the larger heat transfer area provided by small diameter water tubes and the good residence time given to the flue gases (by providing baffles) before they are discharged to atmosphere—all these factors help to increase the thermal efficiency of the boiler.

2.1.7 Stirling Boiler (Fig. 2.8)

This is again a water tube boiler and has two steam-water drums at the top and a mud drum at the bottom. These drums are connected by means of numerous smaller diameter bent tubes carrying water. The absence of headers simplifies construction of boiler and the boiler is made more compact and light. Bent tubes are flexible to accommodate expansion and contraction of the tubes.

The concept of large number of smaller diameter bent tubes connecting boiler drums directly has been used in all modern boilers.

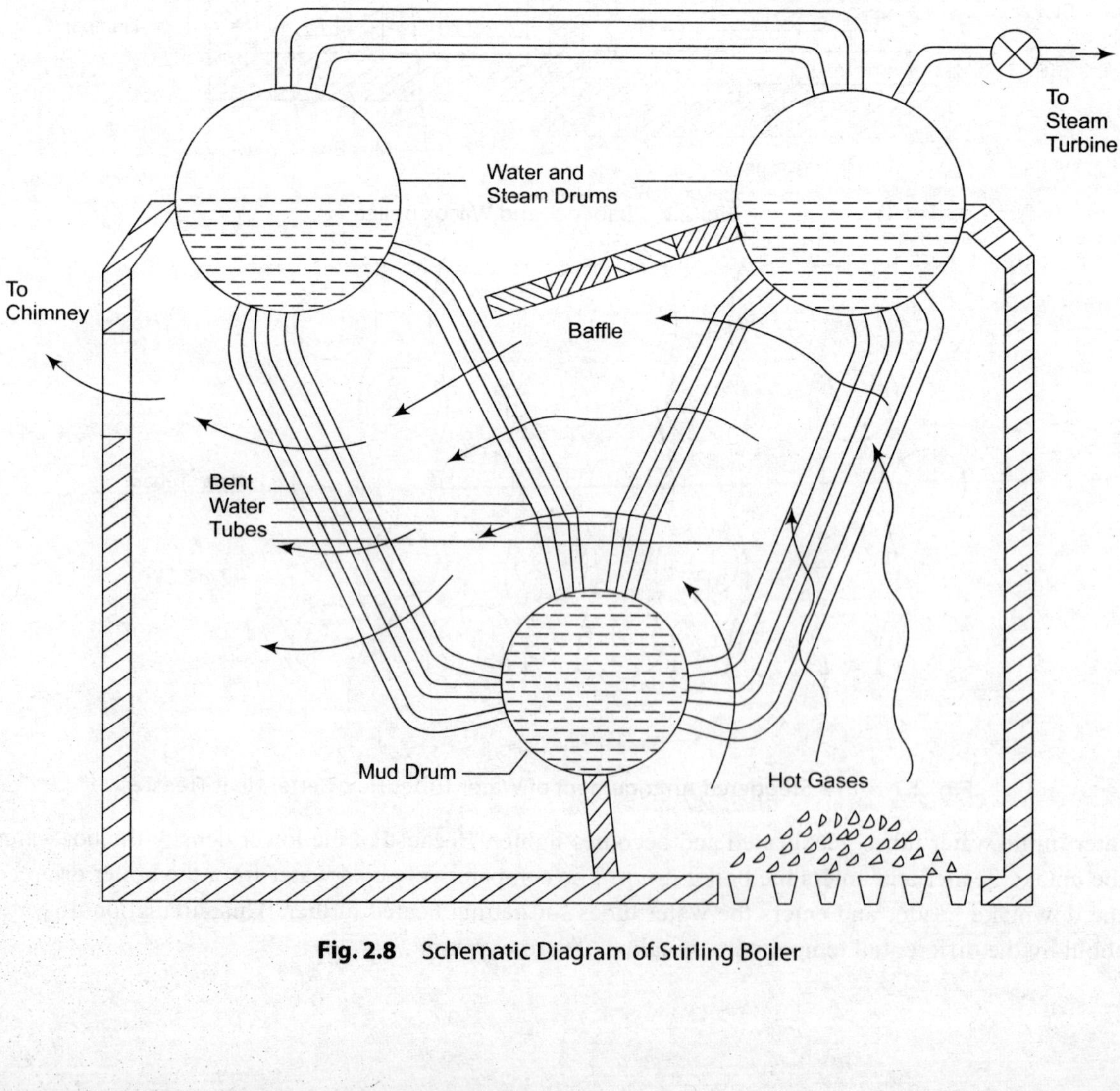

Fig. 2.8 Schematic Diagram of Stirling Boiler

2.1.8 Modern High Pressure Boilers

The requirements of a good steam boiler are:

1. Boiler should have good thermal efficiency.
2. Rate of evaporation should be rapid.
3. Boiler should respond readily to varying loads.
4. Maintainability must be good.
5. Safety and reliability must be ensured.

With the above objectives in view modern high pressure boilers have been developed. Modern boilers operate at very high pressures, even critical pressure (225 atm) and above. Higher pressure means higher energy availability. Higher temperature associated with higher pressure enables to realise better thermodynamic efficiency, which means conversion of large quantity of thermal energy in steam to mechanical energy.

Modern boilers are essentially water tube boilers, for they offer the following advantages.

1. They can have many long, straight or curved water tubes providing plenty of heating surface. This is in contrast with the smoke tube boilers in which the smoke tubes must always be straight and short to facilitate cleaning operation.
2. For a given capacity, the drum of a water tube boiler is smaller in size. This permits generating steam at higher pressures.
3. The heating surface, being provided mostly by water tubes, heat transfer is more effective and faster.
4. Forced circulation of water used in these boilers increases heat transfer and thus the rate of evaporation.

Some of the high pressure boilers developed are: (*a*) LaMont boiler, (*b*) Benson boiler, (*c*) Schmidt Hartmann boiler, (*d*) Loeffler boiler and (*e*) Velox boiler.

The present day boilers have much in common with LaMont and Benson boilers. We shall therefore examine these two boilers only.

LaMont boiler: Figure 2.9. shows schematically the arrangements in the LaMont Boiler.

The feed pump forces the water through the economiser, which, lying in the path of flue gases, transfers some heat into feed water. The heated water enters the steam-water drum, from where the water is drawn by the circulating pump to be sent through evaporator. The evaporator adds enough heat to the water so that as the water comes back to the drum, part of it evaporates and becomes steam. The unevaporated water together with the heated feed water goes for recirculation.

The steam formed in the drum, goes through the superheater and gets superheated.

The flue gases from the furnace pass first through the evaporator section, then through the super-heater section and finally through the economiser and airheater banks. By this process, most of the heat from the flue gases is extracted before the gases are led out to the atmosphere.

The preheating of air together with arrangement of burning fuel on mechanical stokers greatly improves combustion and very high temperatures are realised.

The evaporator section is formed by water tubes lining the walls of the combustion chamber. This type of wall is called 'water wall'. The water wall not only absorbs the heat, but also protects the brick work of the furnace from damage by the intense heat of fire. The heat transfer takes place largely by radiation in this section.

A typical boiler develops as high a pressure as 160 bars and generates 700 tonne of steam per hour.

The important features to be noted are:

1. Forced circulation of water, which improves heat transfer manifold and increases the evaporative capacity.
2. Water level in the steam-water drum is maintained by automatic regulation of the feed water pump. This arrangement is necessary because of the smaller reserve of water held in the steam-water drum, which is now reduced in size.
3. The smaller drum together with smaller tubes result in reduction of weight of the boiler.

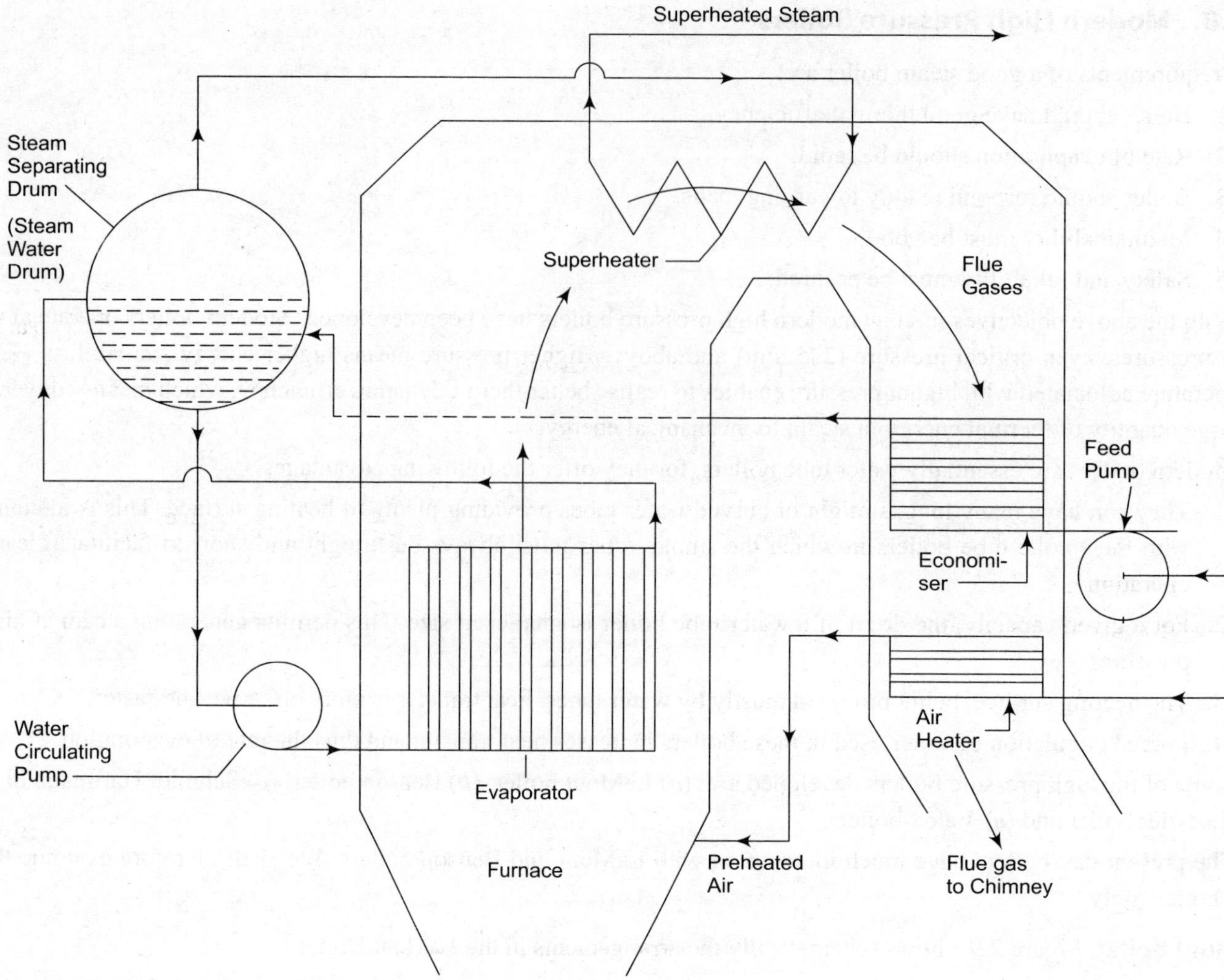

Fig. 2.9 LaMont Boiler

4. The evaporative section is made in the form of water wall, which in addition to helping heat transfer, protects the furnace walls from the high temperature flame.

Benson boiler: Figure 2.10. shows a schematic diagram of a Benson boiler. The boiler has no drum and is designed to operate at critical pressure of 225 atm at which the steam and water have the same density.

This is a once-through boiler with many parallel circuits, each of which consists of a long single tube functioning differently as economiser, evaporator and superheater in different sections of the tube.

The efficiency is almost 90% and the boiler generates steam very fast.

The unit is very compact and light.

The disadvantages are (*i*) the formation of heavy deposits on the heating surfaces when impure water is used; (*ii*) the boiler control for variable loads is difficult.

Modern high pressure boiler using pulverised coal: World's coal reserve is quite substantial and naturally every effort should be made to use coal as fuel and conserve the much scarcer and much precious petroleum products.

Coal, in lump form, burnt on variety of mechanical stokers cannot release heat as fast as pulverised fuel (*i.e.*, fuel ground to fine powder almost as fine as face powder). The reason is simple. Pulverised fuel has much larger surface area to come into contact with the oxygen of the air and therefore burns vigorously. This also results in very high temperatures in the furnace.

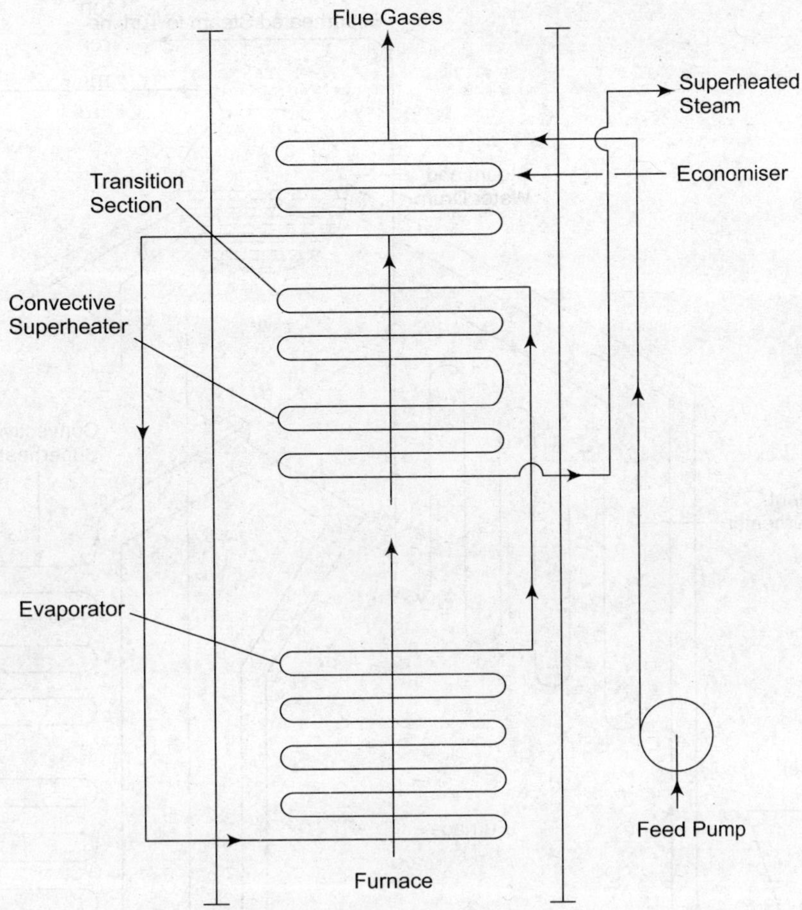

Fig. 2.10 Benson Boiler Schematics

The modern boiler shown schematically in Fig. 2.11 has a pulversing unit. Pulverised coal is carried into the boiler by a stream of air called primary air and burnt like a jet of oil or gas using special fuel burners.

Water walls protect the furnace refractory by absorbing radiant heat of the flame emanating from the burners. A circulating pump keeps a rapid flow of water in these water tubes enabling a fast heat transfer. The heated water goes into the steam separating drum (steam and water drum) and a part of it becomes steam.

The steam thus generated, passes through convective superheater placed in the passage of the flue gases. It then enters the radiant superheater before it is used in the steam turbine.

Auxiliaries like economiser and air preheater help to extract maximum possible heat from the flue gases.

Induced draft or forced draft (sometimes both combined) is mostly arranged to assist the natural draft produced by the chimney.

A typical 50 MW boiler develops 140 atm and delivers 375 tonne/hr at a super-heat temperature of 540°C.

2.1.9 Boiler Mountings

Boiler mountings are part and parcel of the boiler and are mounted directly on the boiler. These are required for proper functioning of the boiler.

The important boiler mountings are the following:

1. Safety valve
2. Pressure gauge
3. Water gauge
4. Steam stop valve
5. Feed check valve
6. Blow-off cock.

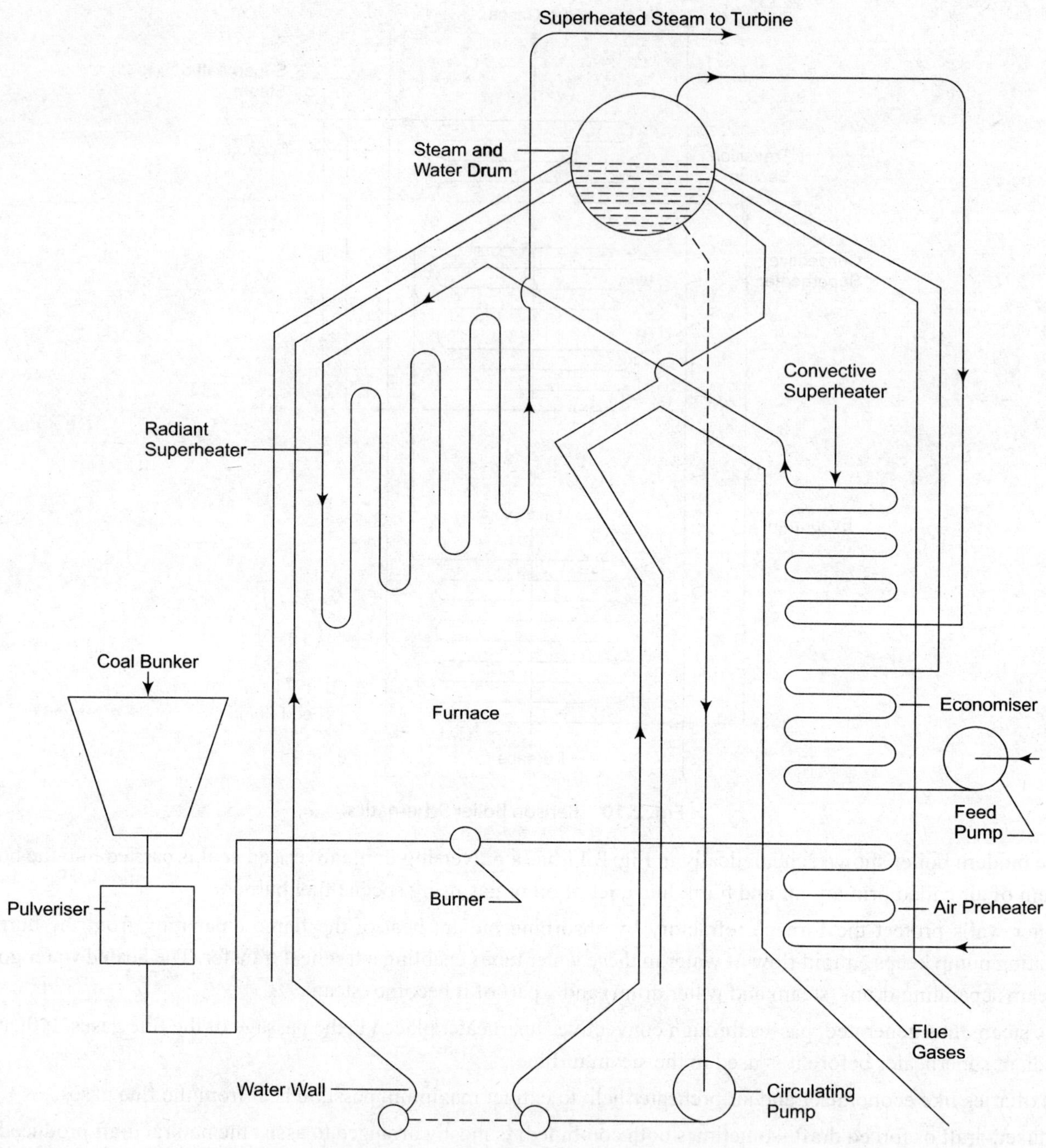

Fig. 2.11 Modern Boiler Using Pulverised Coal

Safety valve: Three types are available: (*a*) Dead weight, (*b*) Lever loaded and (*c*) Spring loaded type.

Dead weight safety valve: Figure 2.12 shows diagrammatically the arrangement of a dead weight safety valve. The dead weight is calculated based on the working pressure (p) and the area of cross section of the valve (a) acted upon by the pressure. Dead weight may be reduced, for a given pressure, by reducing the area 'a'. But remember that this area must be large enough to release the steam quickly and bring back the pressure to normal level in the event of rapid build-up of excess pressure.

Lever loaded safety valve: Figure 2.13 shows the arrangement.

Here the governing equation is

$$W(M+N) = paM$$

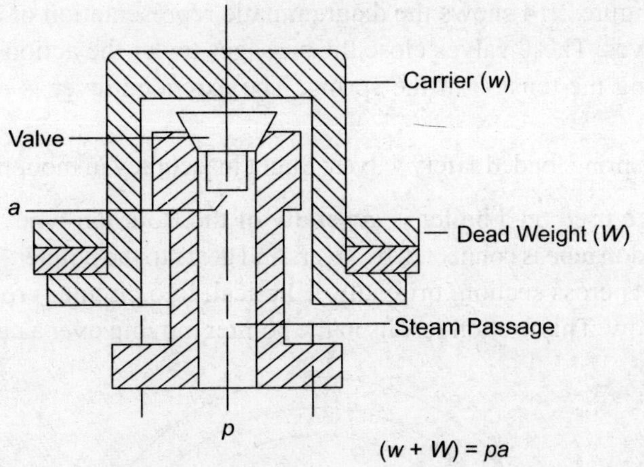

Fig. 2.12 Dead Weight Safety Valve

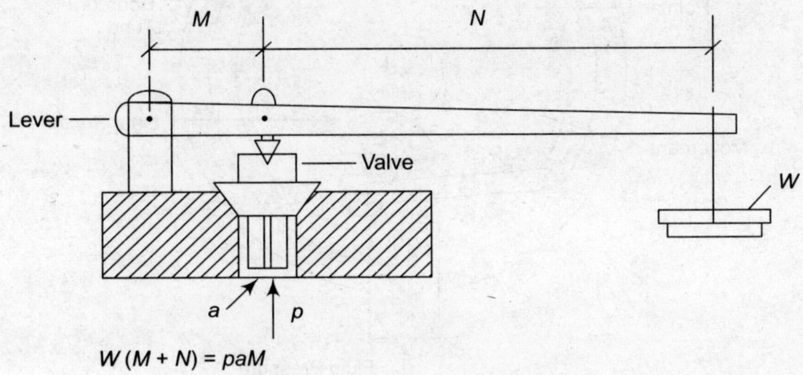

Fig. 2.13 Lever Loaded Safety Valve

By suitably designing the lever arms, the dead weight 'W' can be reduced. The working pressure may be adjusted by varying the quantum of dead weight.

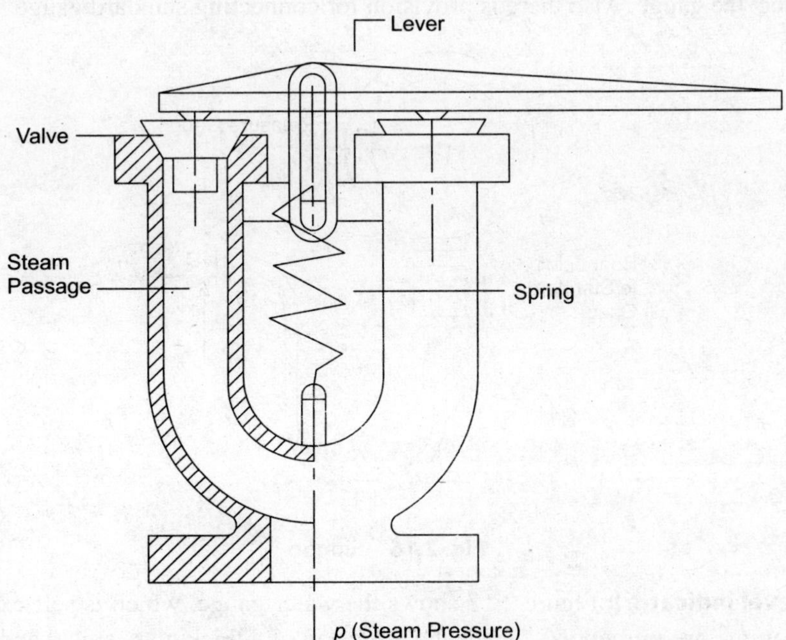

Fig. 2.14 Spring Loaded Safety Valve

Spring loaded safety valve: Figure 2.14 shows the diagrammatic representation of a spring loaded safety valve. It has two steam passages and two valves. These valves close the passages under the action of the central spring. The pressure adjustment is possible by varying the tension in the spring. The extended lever is operated occasionally to check the functioning of the valves.

All high pressure boilers have spring loaded safety valves. Slight inaccuracy in mounting does not affect the performance.

Pressure gauge: Pressure gauge used on a boiler is generally of the Bourdon type. Figure 2.15 shows the details of a pressure gauge in which the Bourdon tube is connected by levers and gears to the pointer. The fluid pressure, when transferred to Bourdon tube, tries to make its cross section, originally elliptical, into circular cross section. In the process, the tube moves out as indicated by the arrow. This motion results in the pointer moving over a circular scale indicating the pressure.

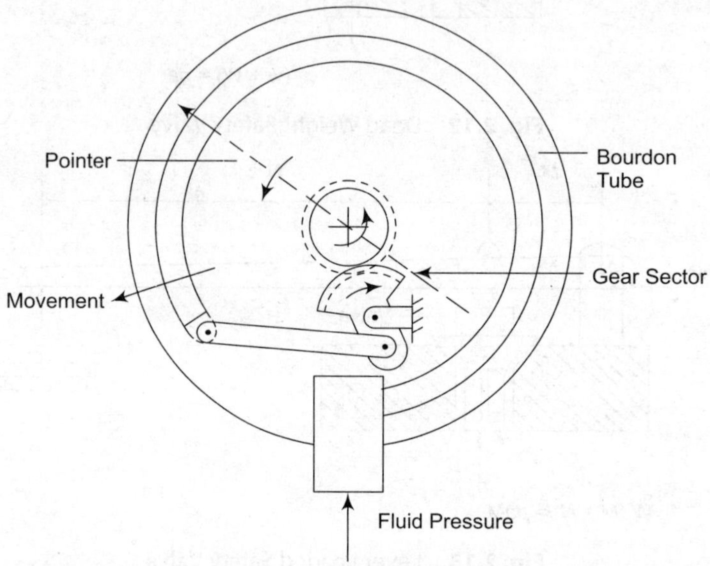

Fig. 2.15 Bourdon Pressure Gauge (Back Cover Removed)

The pressure gauge is connected to the boiler through a siphon (Fig. 2.16) filled with water so that the high temperature of the steam does not affect the gauge. Also there is provision for connecting standard gauge for inspection purposes.

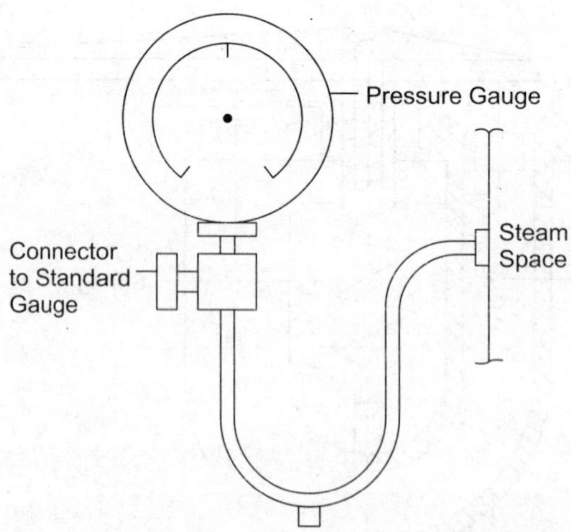

Fig. 2.16 Siphon

Water gauge (Water level indicator): Figure 2.17 shows the water gauge, which is self explanatory. From the point of view of safety, two features, are introduced. One is the provision of a thick glass shield and the other is the automatic closing of the steam and water passages in the event of breakage of the glass tube.

Figure 2.18 shows the steam stop valve, the feed check valve and the fusible plug.

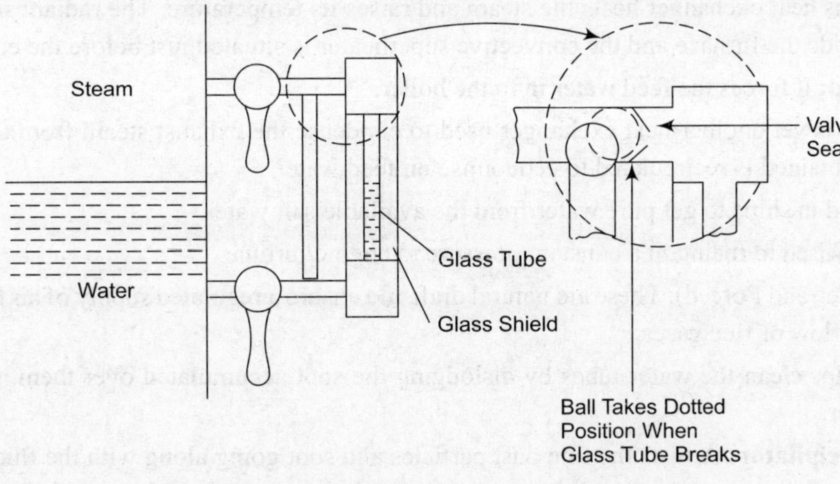

Fig. 2.17 Water Gauge

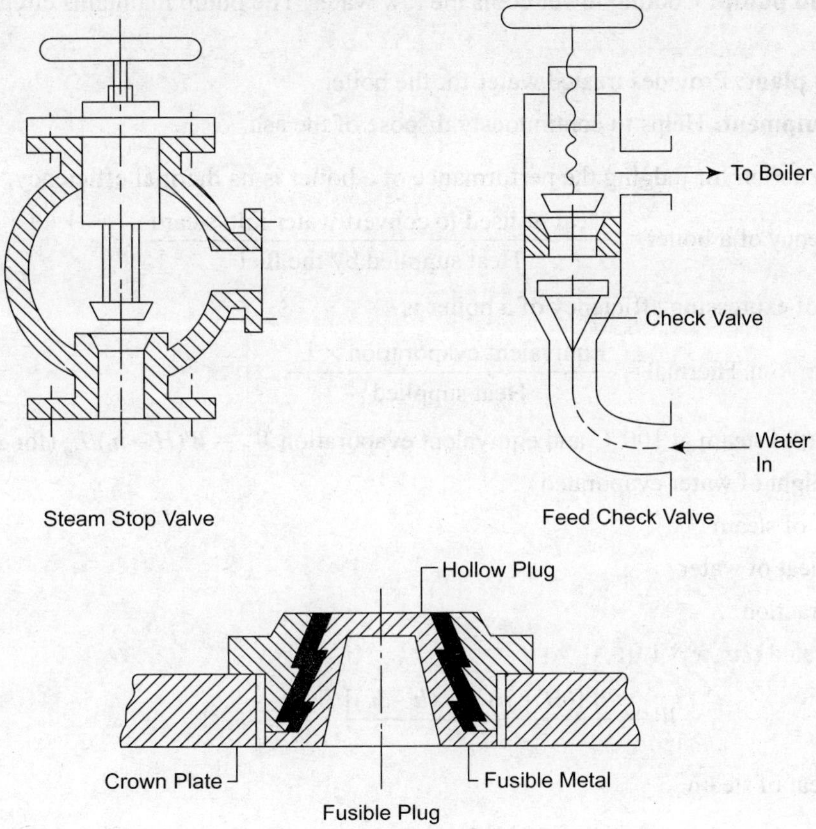

Fig. 2.18

2.1.10 Boiler Accessories

Some of the accessories like air preheater and economiser are used to improve the thermal efficiency of the boiler. Some are used to help maintenance and others for pollution control leading to preservation of safe environment.

Following is a brief description of accessories used on a modern boiler.

1. **Air Preheater:** It is a heat exchanger used to heat the combustion air utilising the flue gases. Combustion with heated air results in higher furnace temperature. Heat recovered improves the thermal efficiency of the boiler.
2. **Economiser:** This is also a heat exchanger but used to heat feed water, using again flue gases. It is placed in the path of flue gases just before the air-preheater.

3. **Superheater:** This heat exchanger heats the steam and raises its temperature. The radiant superheater taking the radiant heat is inside the furnace and the convective superheater is situated just before the economiser.
4. **Feed water pump:** It forces the feed water in to the boiler.
5. **Condenser:** This is yet another heat exchanger used to condense the exhaust steam from the engine or turbine. The condensate obtained is recirculated to economise on feed water.
6. **Evaporator:** Used in ships to get pure water from the available salt water.
7. **Reducing valve:** Used to maintain a constant pressure to engine/turbine
8. **Draft fans** (Induced and Forced): These aid natural draft and ensure a regulated supply of air for combustion. They also help smooth flow of flue gases.
9. **Soot blower:** Helps clean the water tubes by dislodging the soot accumulated over them. Clean surface means better heat transfer.
10. **Electrostatic precipitator:** Arrests the fine dust particles and soot going along with the flue gases. This helps to maintain a cleaner and safer environment.
11. **Cooling tower and pump:** Cooling tower cools the raw water. The pump maintains circulation of cooling water in the condenser.
12. **Water treatment plant:** Provides treated water for the boiler.
13. **Ash handling equipment:** Helps to continuously dispose of the ash.

Boiler efficiency: The criterion for judging the performance of a boiler is its thermal efficiency.

$$\text{Thermal efficiency of a boiler} = \frac{\text{Heat utilised to convert water into steam}}{\text{Heat supplied by the fuel}}$$

The standard method of expressing efficiency of a boiler is

$$\eta\ \text{Thermal} = \frac{\text{Equivalent evaporation} \times L_o}{\text{Heat supplied}}$$

Where L_o = Latent heat of steam at 100°C and equivalent evaporation $W_1 = W(H - h_1)/L_o$ (for $x = 1$)

In this W = Actual weight of water evaporated

H = Total heat of steam

h_1 = Sensible heat of water

x = Dryness fraction.

If the steam is unsaturated (*i.e.*, $x < 1.0$)

$$W_1 = \frac{W[(H - h)\,x + (h - h_1)]}{L_o}$$

Where h = Sensible heat of steam.

Indian boiler regulation: Lays down certain rules and regulations which must be followed in the design, construction, operation and maintenance of a boiler or any pressure vessel. It also gives test procedures and schedule to ensure that the system works satisfactorily and is safe.

2.1.11 Steam Turbine

Steam turbine is a prime mover, which converts the high pressure energy in steam to mechanical energy in the form of rotation of the turbine shaft.

Working principle: The high pressure steam, when expanded in a nozzle, results in a steam jet of high velocity (Fig. 2.19). When this steam jet impinges on curved blades fixed to the rotor of the turbine, the kinetic energy of the jet is transferred to the blades and thus the rotor is set in motion.

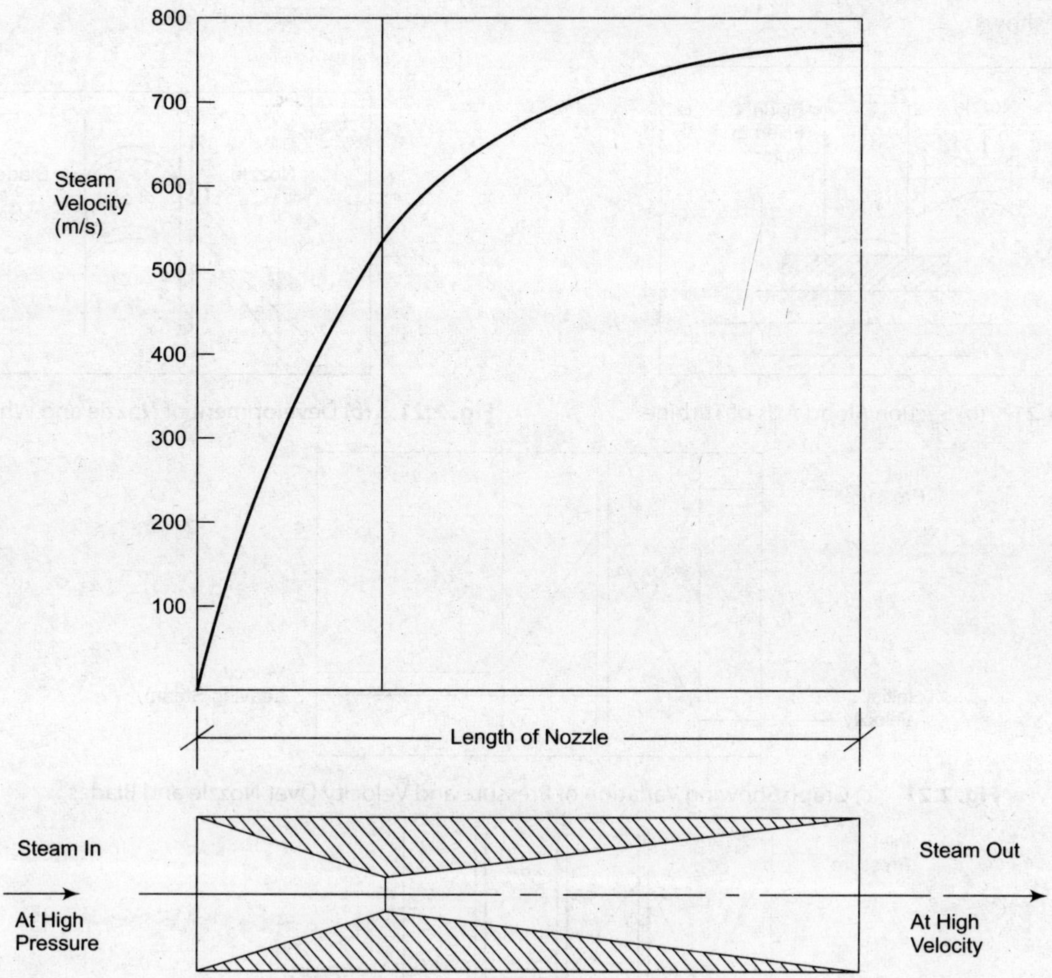

Fig. 2.19 Converging–Diverging Nozzle

Impulse Steam Turbine: The earliest impulse steam turbine was of De Laval (Fig. 2.20). This is a single rotor turbine and the speed attained is about 30,000 rpm. The high rpm is not only dangerous, but also not practically usable unless speed reduction is brought about using a gear box.

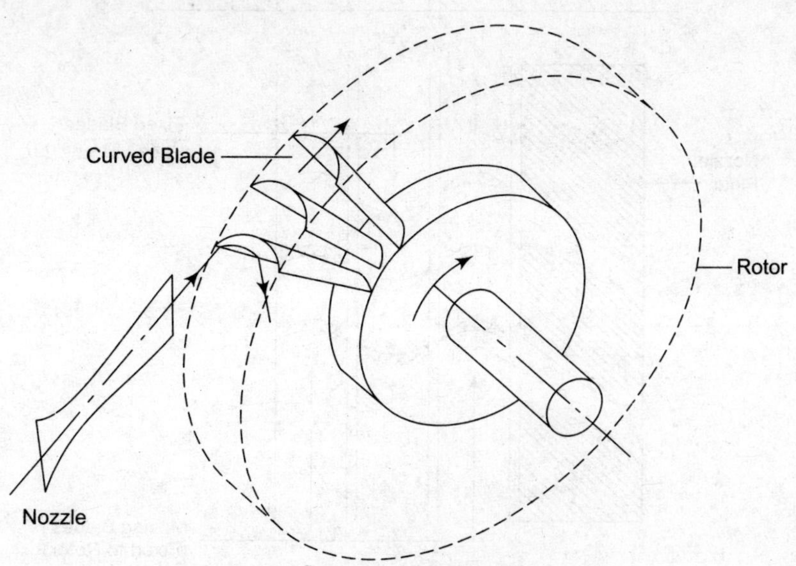

Fig. 2.20 Delaval Turbine

Figure 2.21 shows

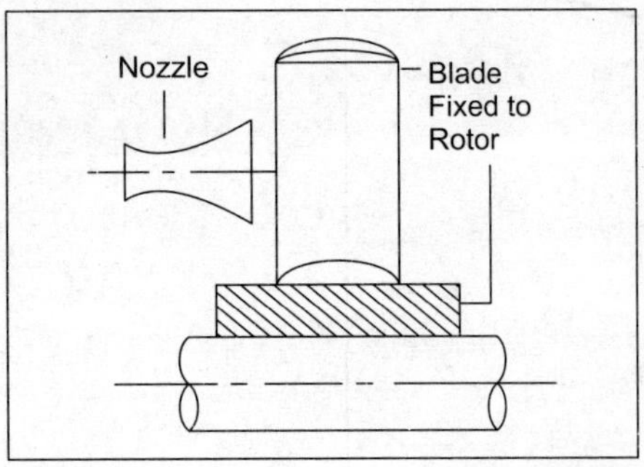

Fig. 2.21 (*a*) Section Along Axis of Turbine

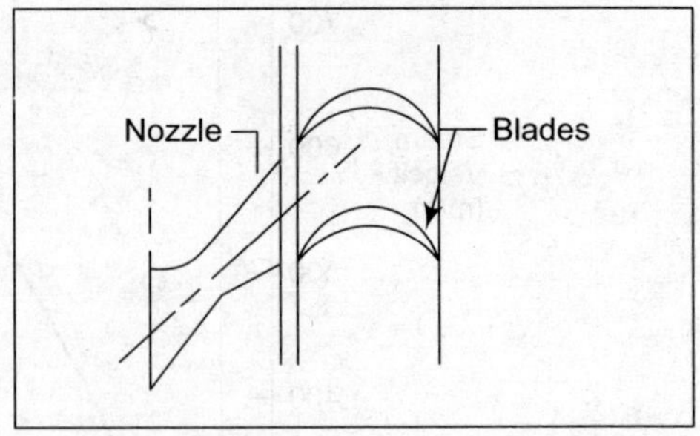

Fig. 2.21 (*b*) Development of Nozzle and Wheel/Rotor

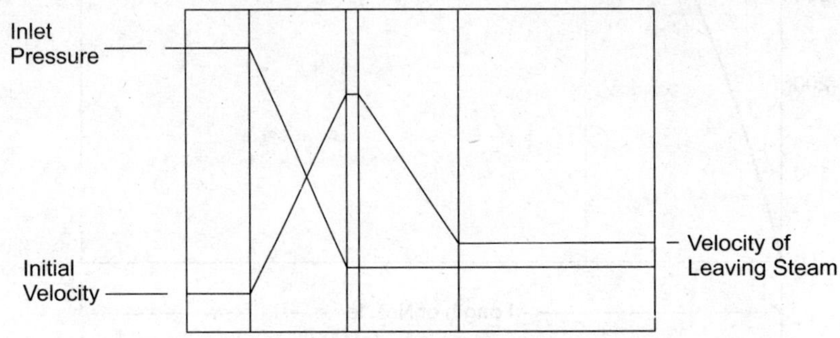

Fig. 2.21 (*c*) Graph Showing Variation of Pressure and Velocity Over Nozzle and Blades

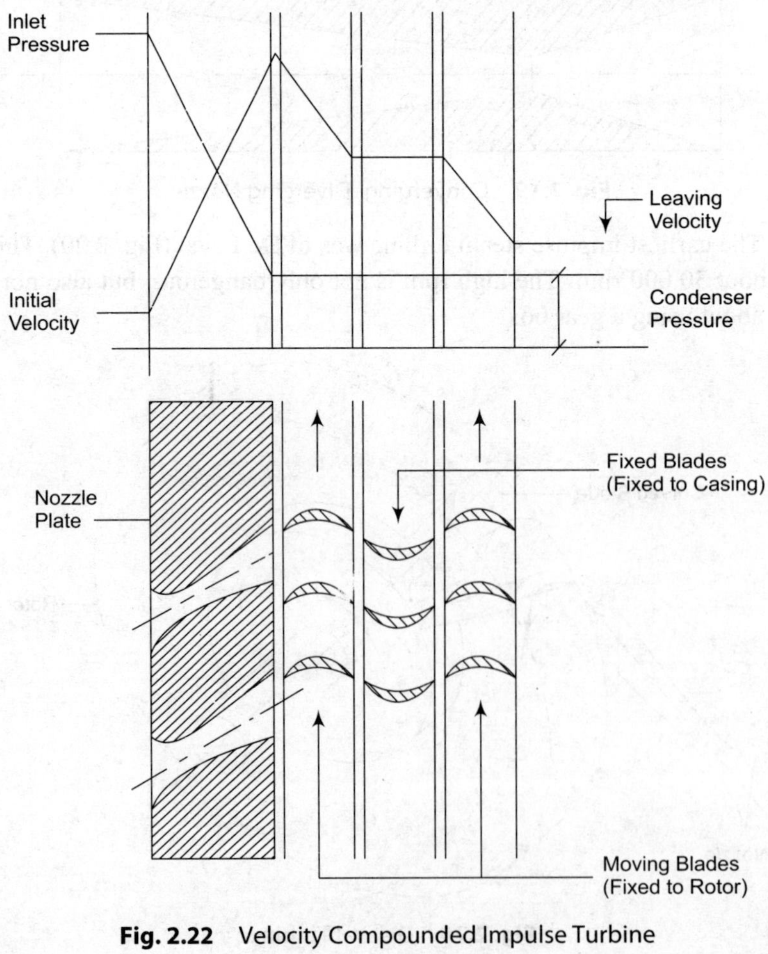

Fig. 2.22 Velocity Compounded Impulse Turbine

(*a*) Section along the axis of the Delaval turbine

(*b*) The development of nozzle and wheel/rotor

(*c*) The graph of variation of pressure and velocity over nozzle and blades.

In order to overcome the difficulty of high speeds, the present day steam turbines are all of multi-rotor design. Depending on how the energy of the steam is absorbed by the blades, we have (*a*) Velocity compounded, (*b*) Pressure compounded, (*c*) Velocity compounded and (*d*) Reaction turbines. Here we examine only (*i*) Velocity compounded and (*ii*) Reaction turbines.

Velocity Compounded Impulse Turbine: Figure 2.22 shows the developed view of a velocity compounded impulse turbine. The steam passing through the nozzle plate attains the maximum possible velocity. The velocity is then absorbed in several stages. In the figure only two stages are shown for the sake of simplicity.

The fixed blades (fixed to casing) change the direction of steam without causing loss of energy.

The absorption of the energy in several stages brings down the speed of turbine.

Reaction Turbine: Figure 2.23 shows the development of a reaction turbine. Here there are no nozzles as seen in an impulse turbine. But the passages between moving blades and fixed blades are so formed that nozzle action is obtained as the steam passes, through them.

There is a continuous reduction in pressure of steam from inlet to outlet, pressure drop occuring over fixed and moving blades. The velocity available at the outlet of fixed blades is absorbed in the moving blades immediately next to them.

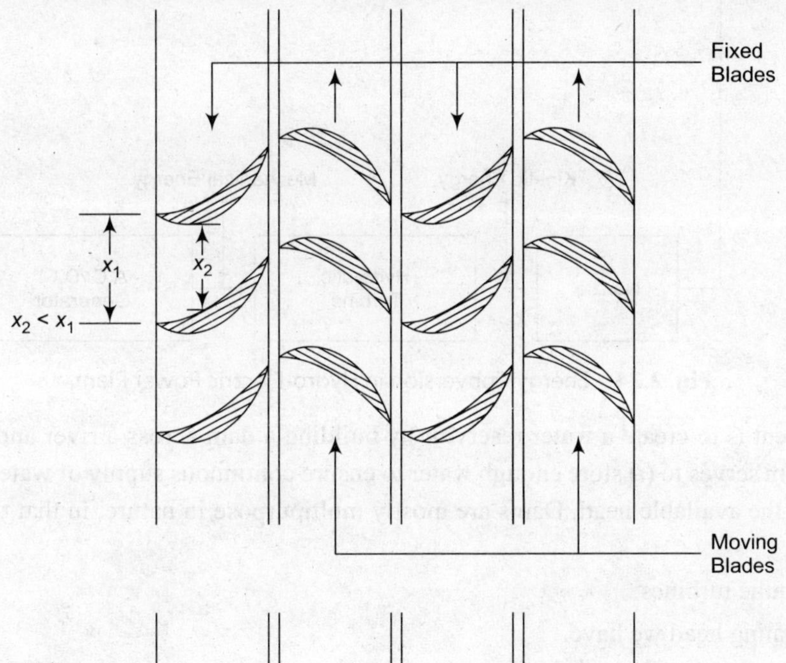

Fig. 2.23 Reaction Turbine

The fixed blades not only help in changing the direction of steam but also act as nozzles converting pressure energy into velocity energy. Since the moving blade passages also act as nozzles there is increase in velocity and hence increase in kinetic energy.

The kinetic energy made available is in two forms: (*a*) By way of change in momentum of steam occuring over the moving blades and (*b*) By the reaction produced by the expanding steam in the passages of the moving blades. It is for this reason that this turbine is called impulse reaction or simply reaction turbine.

2.2 HYDRO-ELECTRIC POWER PLANT

In a hydro-electric power plant a hydraulic turbine is used as a prime mover. The hydraulic turbine uses the potential energy of water available at high elevations as in water falls. The energy conversion scheme in a hydro-electric power plant is shown in Fig. 2.24.

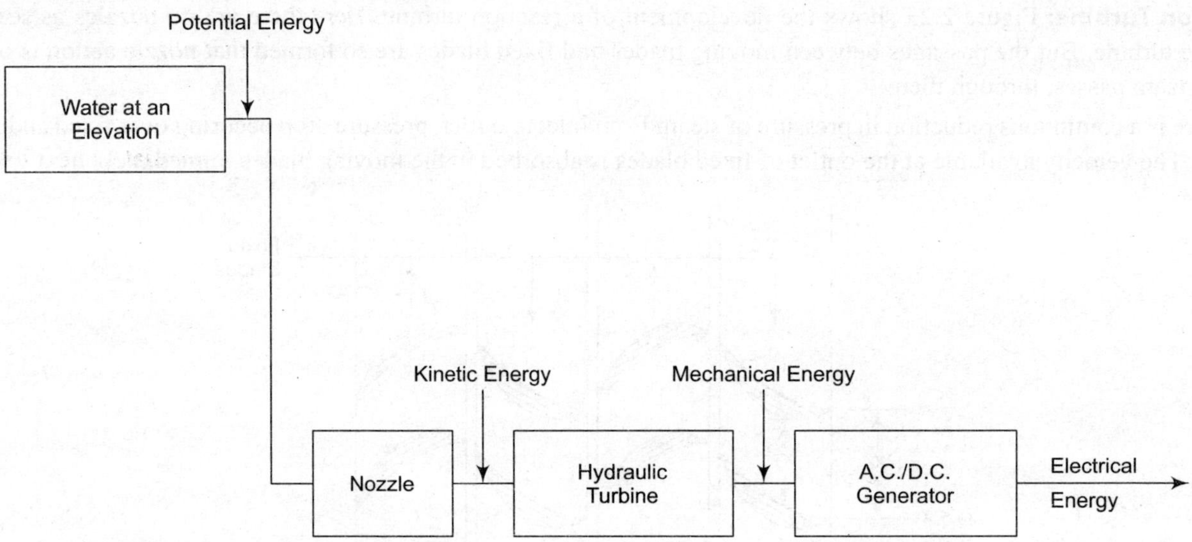

Fig. 2.24 Energy Conversion in Hydro-Electric Power Plant

The general arrangement is to create a water reservoir by building a dam across a river and use this water to drive a hydraulic turbine. The dam serves to (*i*) store enough water to ensure continuous supply of water through the turbine year through and (*ii*) increase the available head. Dams are mostly multipurpose in nature, in that they help in irrigation and sometimes flood control.

Classification of hydraulic turbines:

(*a*) Based on the operating head we have,

(*i*) High head turbine (for heads greater than 300 m)

Example: Pelton wheel

(*ii*) Medium head turbine (30 to 300 m)

Example: Francis turbine

(*iii*) Low head turbine (for 3 to 30 m)

Example: Kaplan turbine

(*b*) Based on the direction in which water flows in to the wheel,

(*i*) Tangential flow: Water impinges the wheel tangentially as in Pelton wheel.

(*ii*) Axial flow: Water flows in a direction parallel to the axis of the wheel as in a Kaplan turbine.

(*iii*) Mixed flow: Water flows radially into the wheel and leaves axially as in Francis turbine.

(*c*) Based on how the potential energy of water gets converted to mechanical energy of wheel,

(*i*) **Impulse turbine:** Pelton wheel is an example (Refer Fig. 2.25). Here the nozzle converts the potential energy of water into kinetic energy of the jet. The jet issuing from the nozzle with high velocity impinges on the pelton buckets and the kinetic energy is absorbed by the wheel.

(*ii*) **Reaction turbine:** Kaplan turbine is an example (Refer Fig. 2.27). Here most of the potential energy of the water is absorbed by the wheel as the water passes over the blades. The blade passages provide the nozzle action of converting potential energy to kinetic energy, and as the water leaves the wheel it produces a reaction on the wheel causing it to rotate.

We shall examine in some detail the different types of hydraulic turbines mentioned earlier.

2.2.1 Pelton Wheel

The turbine has a rotor (wheel) with a number of pelton buckets fixed round the rotor (Fig. 2.25). The water is brought down from the reservoir (Refer Fig. 2.26) by means of pipes called 'PENSTOCK' and is made to pass through a nozzle at the end. The nozzle converts the potential energy of water into kinetic energy and therefore water issues out of the nozzle with a high velocity. The jet impinges at the centre of the buckets in a direction tangential to the wheel and sets the wheel in motion. Each bucket has a splitter and the jet striking it gets split and flows out in opposite directions as shown in the inset of Fig. 2.25. By this arrangement the axial thrust is balanced.

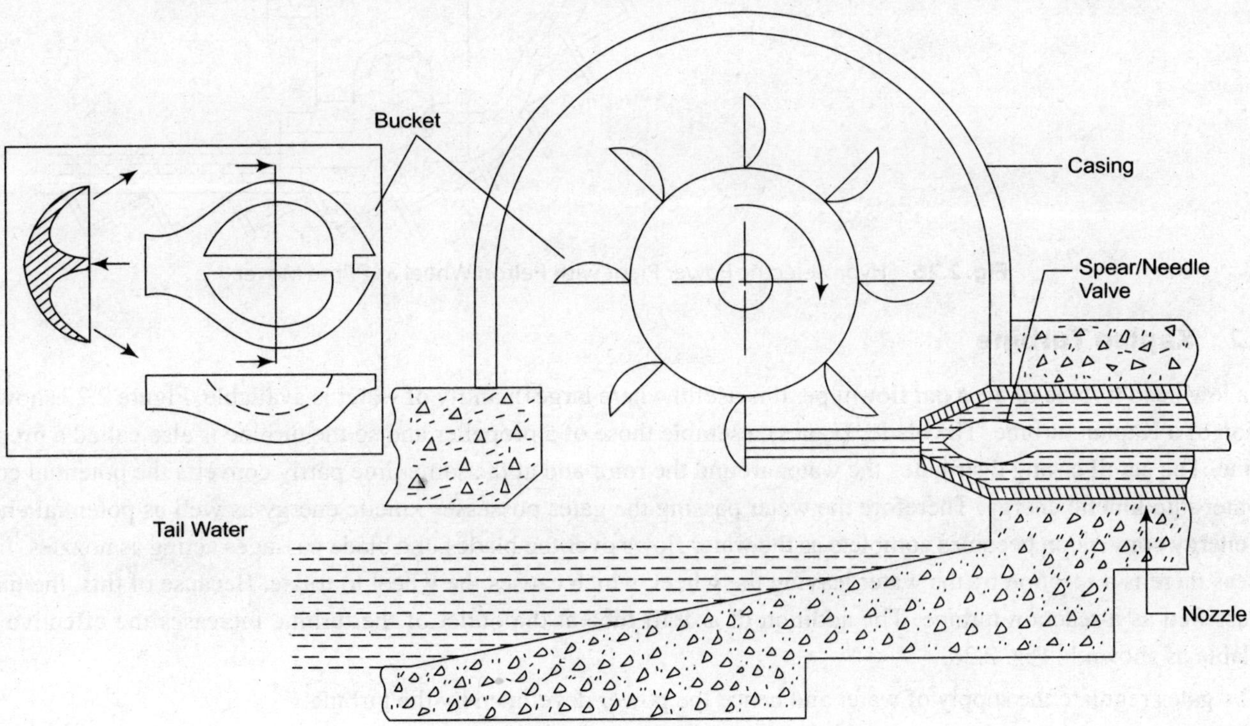

Fig. 2.25 Pelton Wheel (High Head Turbine)

This turbine is called an impulse turbine because the potential energy of water is completely converted into kinetic energy in the form of high velocity jet using a nozzle.

The needle valve helps to control the rate of flow of water and hence the power output.

Layout of a hydro-electric plant using pelton wheel is shown in Fig. 2.26. The figure is self-explanatory. In case the distance between the reservoir and the turbine is considerable, a surge tank (not shown) is introduced in the supply line. This relieves pressure build up in the event of sudden stoppage of water flow and thus prevents damage to pipeline.

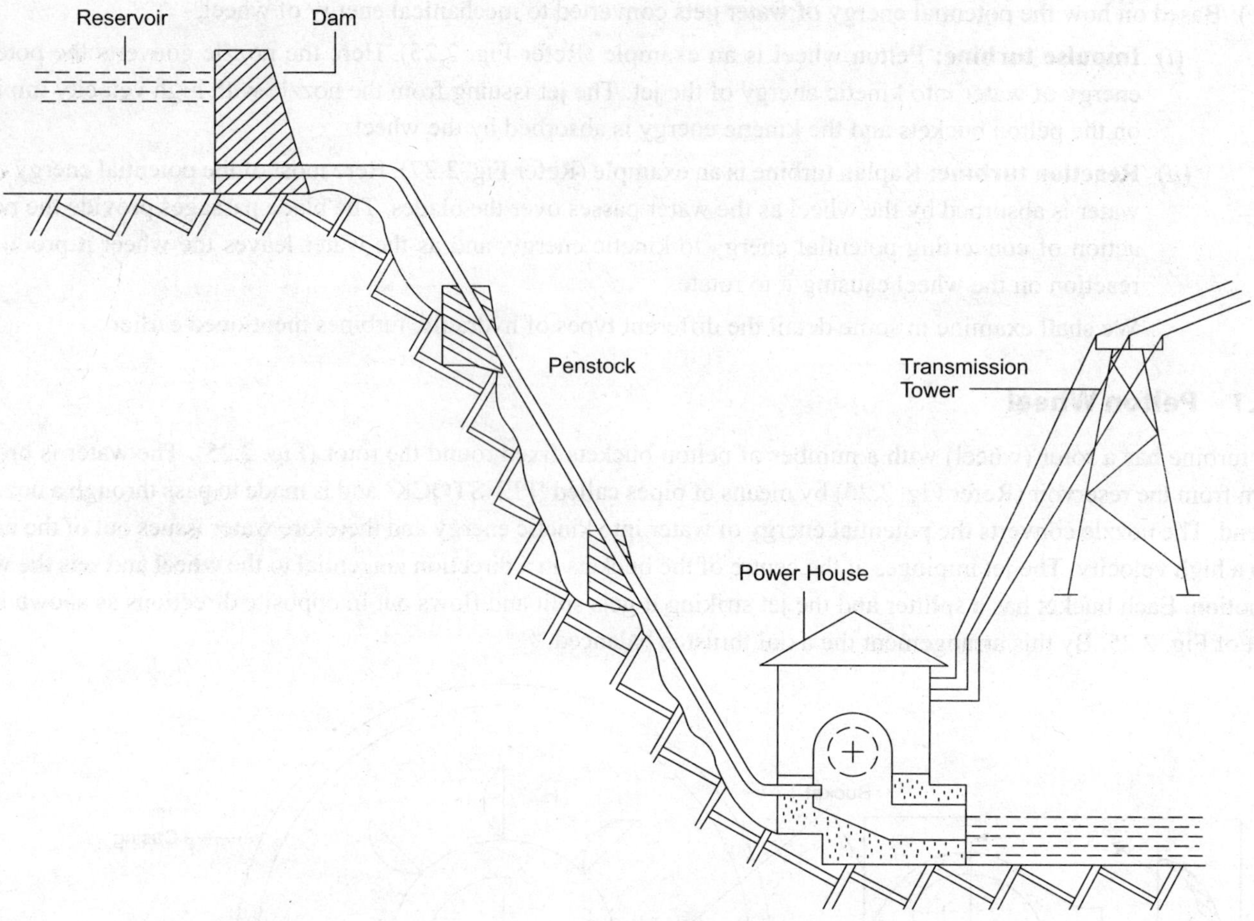

Fig. 2.26 Hydro-electric Power Plant with Pelton Wheel as Prime Mover

2.2.2 Kaplan Turbine

It is a low head turbine of the axial flow type. It is useful where large quantity of water is available. Figure 2.27 shows the section of a Kaplan turbine. The blades (vanes) resemble those of a propeller and so the turbine is also called a propeller turbine. The scroll casing distributes the water around the rotor and at the same time partly converts the potential energy of water into kinetic energy. Therefore the water passing the gates possesses kinetic energy as well as potential energy. The energy conversion becomes complete as the water flows over the blades, the blade passages acting as nozzles. In this process there is a reaction by the water leaving the wheel, which causes the wheel to rotate. Because of this, the turbine is classified as a reaction turbine. The addition of a draft tube at the outlet of the turbine increases the effective head available as shown in Fig. 2.28.

The gates regulate the supply of water and hence the power developed by the turbine.

Figure 2.28 shows the layout of a hydro-electric power plant using a Kaplan turbine.

2.2.3 Francis Turbine

This is a mixed flow, medium-head turbine. Figure 2.29 shows a section of the turbine with the regulating gates (wicket gates), the rotor and the draft tube. This also has a spiral casing and the working of the turbine is similar to that of Kaplan turbine, the conversion of energy taking place as the water passes through the gates and flows over the blades. This turbine differs from the Kaplan turbine in that (*i*) it has 16 to 24 blades as against 3 to 6 in Kaplan turbine; (*ii*) water enters the wheel radially and leaves axially (mixed flow) as against a purely axial flow in Kaplan turbine.

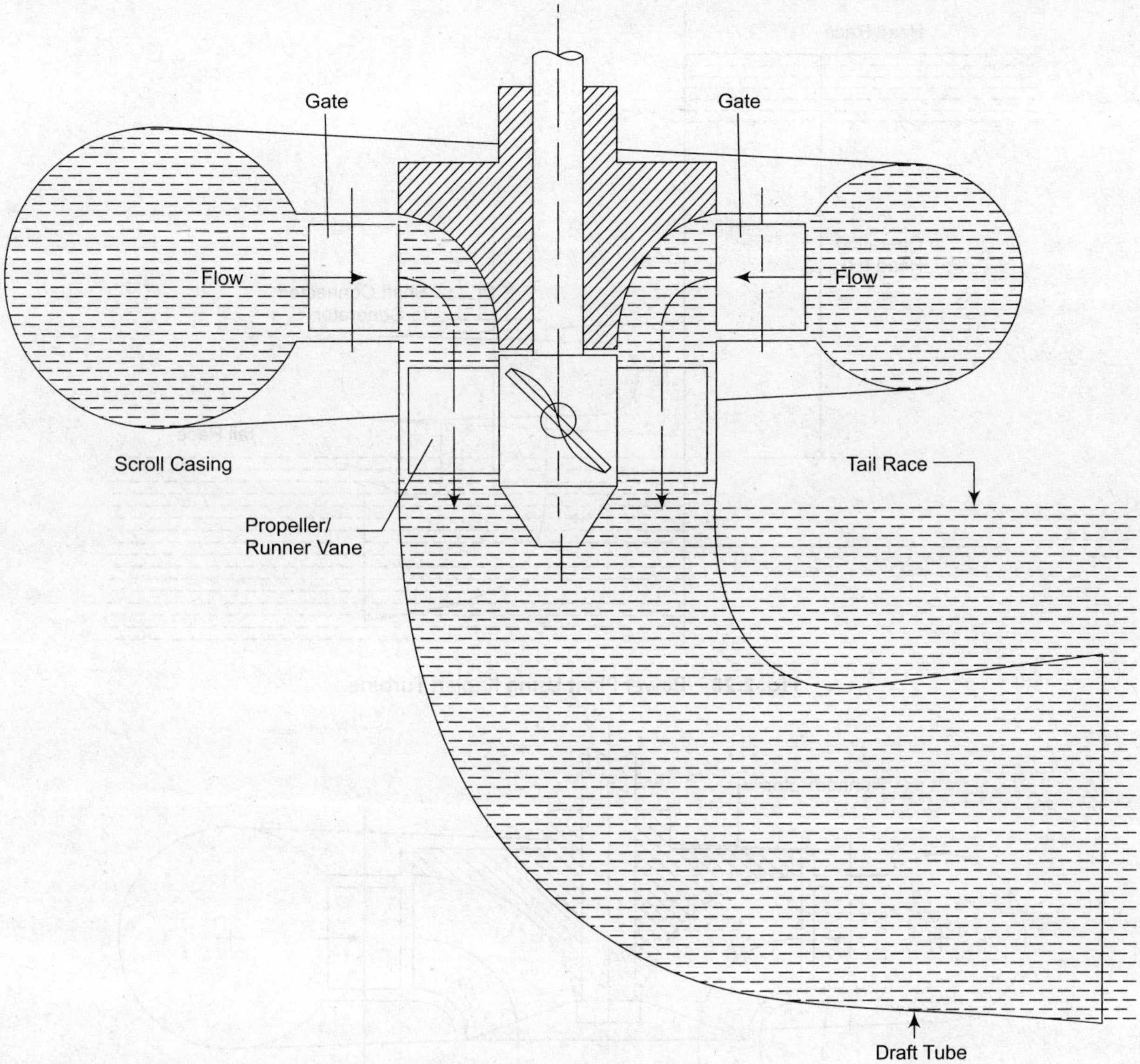

Fig. 2.27 Kaplan Turbine

Advantages of hydro-electric power plant

1. Hydel energy is renewable and is non-depletable.
2. It is made available by nature free of cost and hence running cost is the least compared to other power plants.
3. The cost per unit of electrical energy generated in hydro-electric plant is therefore the lowest.
4. There is no pollution problem.
5. The operation and maintenance of the plant is simple.
6. The reliability is excellent.
7. It is capable of meeting the variable load requirements very well.
8. Life of plant is longer.
9. Offers irrigation facility and flood control.

Disadvantages

1. The cost of constructing the dam is high and therefore the initial investment is very huge.
2. The power availability is subject to vagaries of nature.

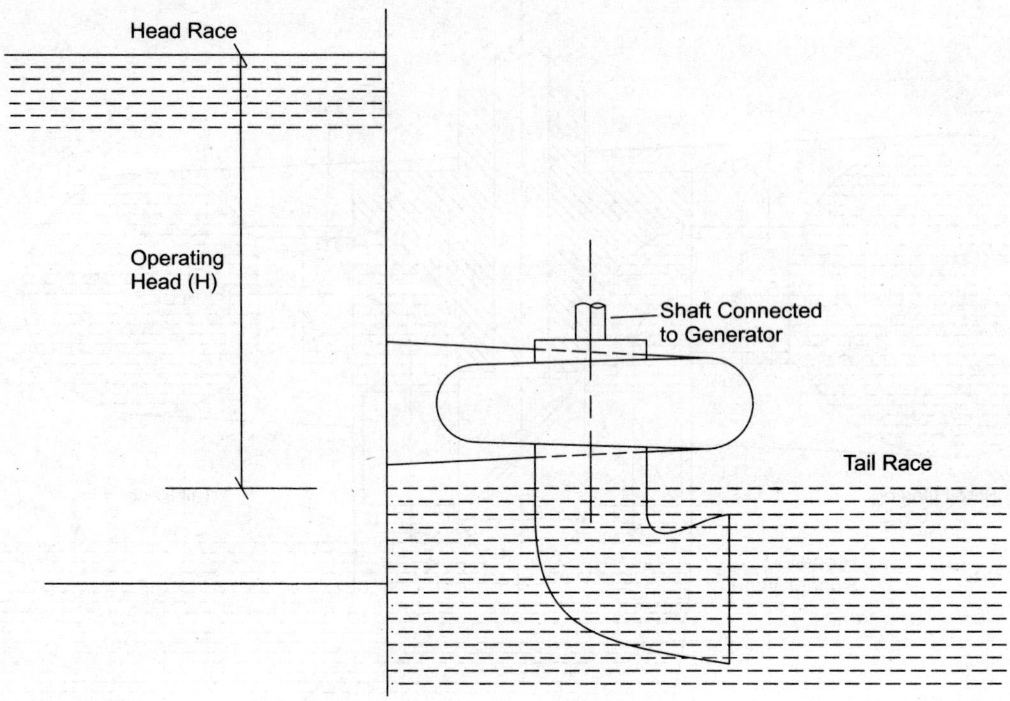

Fig. 2.28 Power Plant Using Kaplan Turbine

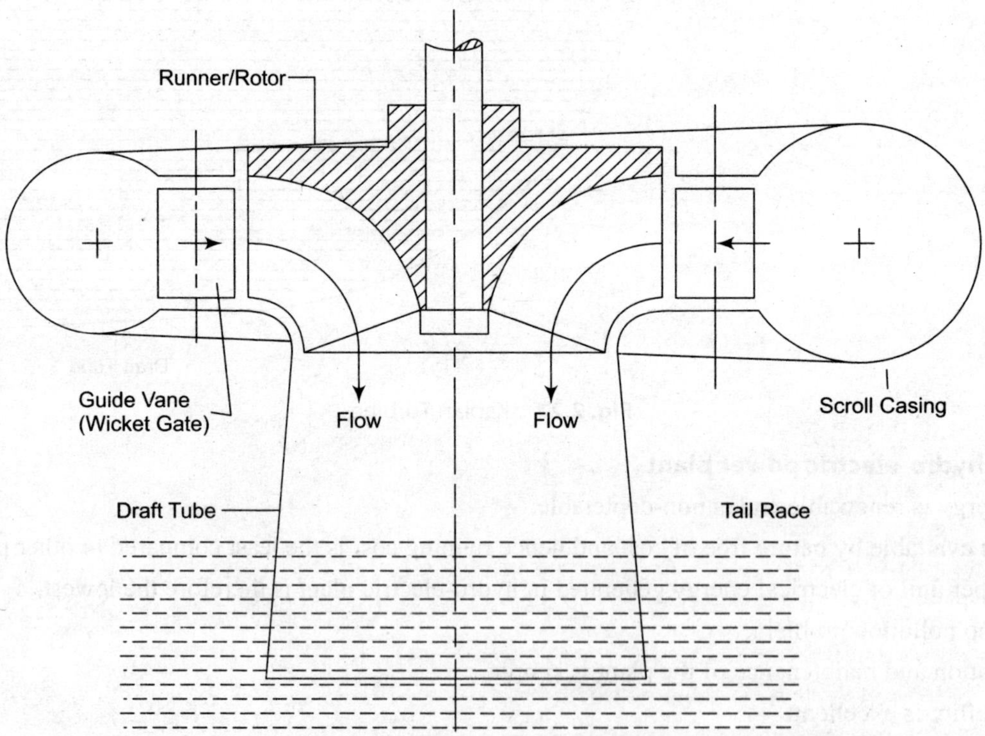

Fig. 2.29 Francis Turbine (Mixed Flow)

3. It also takes considerable time for erecting the plant.
4. Often situated away from load centre, transmission cost is more. It also means waste of power.

Power developed by a hydro-electric turbine

$$P_t = \frac{wQHg}{1000} \eta_t$$

where

P_t = Power output of the turbine (kW)

w = Density of water (kg/m^3)

Q = Rate of flow of water (m^3/s)

H = Net head of water (m)

η_t = Efficiency of the turbine (usually 85 to 90%)

g = Acceleration due to gravity

2.3 NUCLEAR POWER PLANT

Power plant which uses nuclear energy for power production is called nuclear power plant.

Nuclear energy is derived from splitting or fissioning of the nucleus of fissionable materials like uranium U-235. Uranium has several isotopes; Ex: U-234, U-235 and U-238 (isotopes are atoms of the same element having different atomic masses)

Of the several isotopes, U-235 is the most unstable isotope, which is easily fissionable and hence used as the fuel in an atomic reactor.

In 1802 John Dalton, an English Chemist, laid the foundation for understanding the structure of matter. He showed that an element is made up of extremely small particles called atoms. (In Greek atom means cannot be cut). Atom is defined as the smallest particle of an element which retains the same properties as the element itself. Similarly the smallest particle of a compound which still retains the properties of the compound is called a molecule.

Dalton assumed that atoms of any one element were identical in all respects. This is only partially true. Again his postulate of 'atom cannot be split' is no longer valid. It is only because atoms can be smashed, we are now able to derive enormous quantity of energy called atomic energy/nuclear energy.

Joule brought out the equivalence of thermal and mechanical energy and the law of conservation of energy was established.

With Einstein propounding the theory of relativity, mass is also now included in the concept of energy and we have the equivalence of energy and mass established in the famous but simple equation

$$E = M * c^2$$

Where E = Energy (J)

M = Mass (kg)

c = Velocity of light (3×10^8 m/s)

This simple, but powerful equation forms the basis for atomic energy. It is seen that just 1 kg of U-235 gives out as much thermal energy as 4500 tonne of good quality coal.

2.3.1 Fission Energy

Chain Reaction: Let us see how the reaction is set up in a nuclear reactor (a chamber or an enclosure in which the splitting or fissioning of atoms takes place) and thermal energy is released (Refer Fig. 2.30).

When a neutron is captured by a nucleus of an atom U-235, the atom splits up into two fragments (Krypton and Barium) and about 2.5 neutrons are released and with it a large amount of energy is also released. This is called the fission process. The neutrons so produced are very fast and can be made to fission other nuclei of U-235, thus setting up a chain reaction. When a large number of fission occurs, enormous amount of heat is generated.

Out of 2.5 neutrons released in fission of each nucleus of U-235, one neutron is used to sustain the chain reaction, about 0.9 neutron is captured by U-238, which gets converted into fissionable material Pu-239 and about 0.6 neutron is partly absorbed by control rod material, coolant, moderator and a part escapes from the reactor.

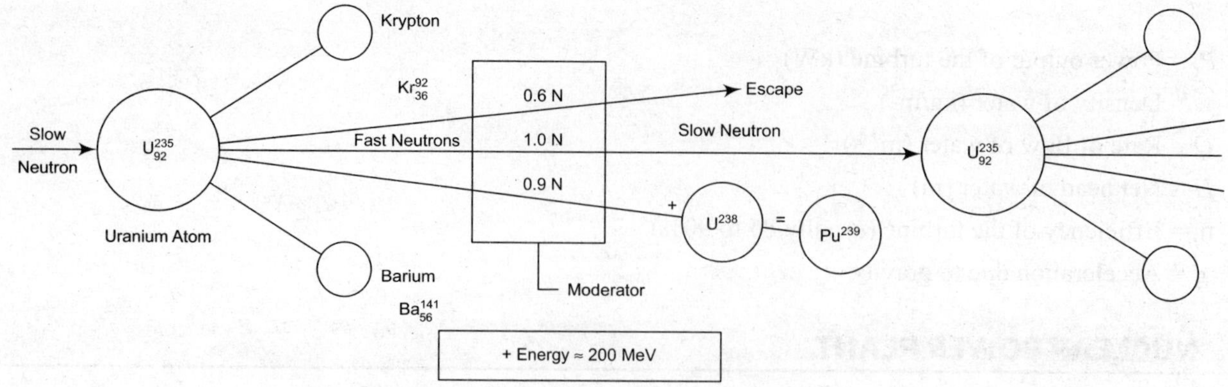

Fig. 2.30 Chain Reaction in the Nuclear Reactor.

If Thorium is used in the reactor core it gets converted to fissionable material U-233.

Th-232 + Neutron → U-233

Pu-239 and U-233 so produced are fissionable materials and are called secondary fuels. They can be used as nuclear fuels. U-238 and Th-232 are called fertile materials.

2.3.2 Nuclear Reactor: Major Parts and Their Functions (Fig. 2.31)

Fuel Rods: Contain the nuclear fuel. Nuclear fuel is a fissionable material, whose nuclei undergo fission in the reactor. The fission chain reaction, once set up, releases a large quantity of energy. U- 235 and Pu-239 are examples of fissionable material.

Moderator: The function of a moderator is to reduce the speed of fast moving neutrons. Fast moving neutrons collide with the nuclei of the moderator material and get slowed down. The slow neutrons (thermal neutrons) so formed are easily captured by the nuclear fuel enabling the chain reaction to proceed smoothly.

Graphite, heavy water and beryllium are generally used as moderators.

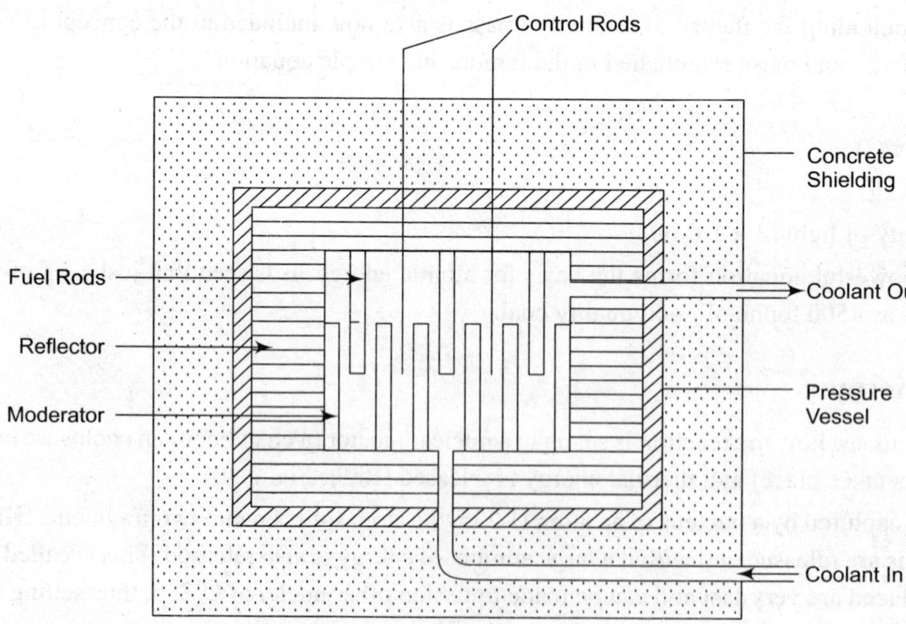

Fig. 2.31 Parts of a Nuclear Reactor

Control rods: Nuclear reactor is normally loaded with enough fuel to operate for a few months. The consumption of this fuel and power level of the reactor depends upon the neutron flux in the reactor core. The lowering of the control rods results in large quantity of neutrons getting absorbed by the control rods and hence the neutron flux is reduced causing the nuclear reactor to slow down. On the other hand the withdrawal or raising of the control rods results in greater neutron, movement and consequently accelerates the fission process.

Control rods are made of boron or cadmium which are capable of absorbing neutrons.

Coolant: The purpose of the coolant is to transfer the heat from the nuclear reactor to a boiler for raising steam. Coolants used are:

1. Gaseous: (Air, CO_2, Helium)
2. Water: Ordinary and heavy
3. Liquid Metals: Sodium, potassium or a mixture of the two
4. Inorganic and organic fluids

Pressure Vessel: It is a strong walled container, which houses the core of the power reactor containing fuel, moderator, reflector and control rods.

Reflector: Reflector surrounds the core and sends back the neutrons that tend to escape. The returned neutrons cause more fission and improve the performance of the reactor. The neutron economy of the reactor is thus improved.

Reflector is made of graphite or beryllium.

Biological Shielding: Protects the operating personnel from radiation hazards. During fission of nuclear fuel, alpha, beta and gamma rays are emitted. To prevent escape of these rays endangering the operating personnel heavy shielding is provided in the form of thick layer of concrete (about 2 m).

2.3.3 Nuclear Power Plant Layout

Figure 2.32 shows the schematics of a nuclear power plant. Hot coolant from the reactor passes through the steam generator (a heat exchanger) and is recirculated using a coolant pump.

The steam generated is used in the steam turbine for conversion into mechanical energy and then to electrical energy as in the conventional steam power plant.

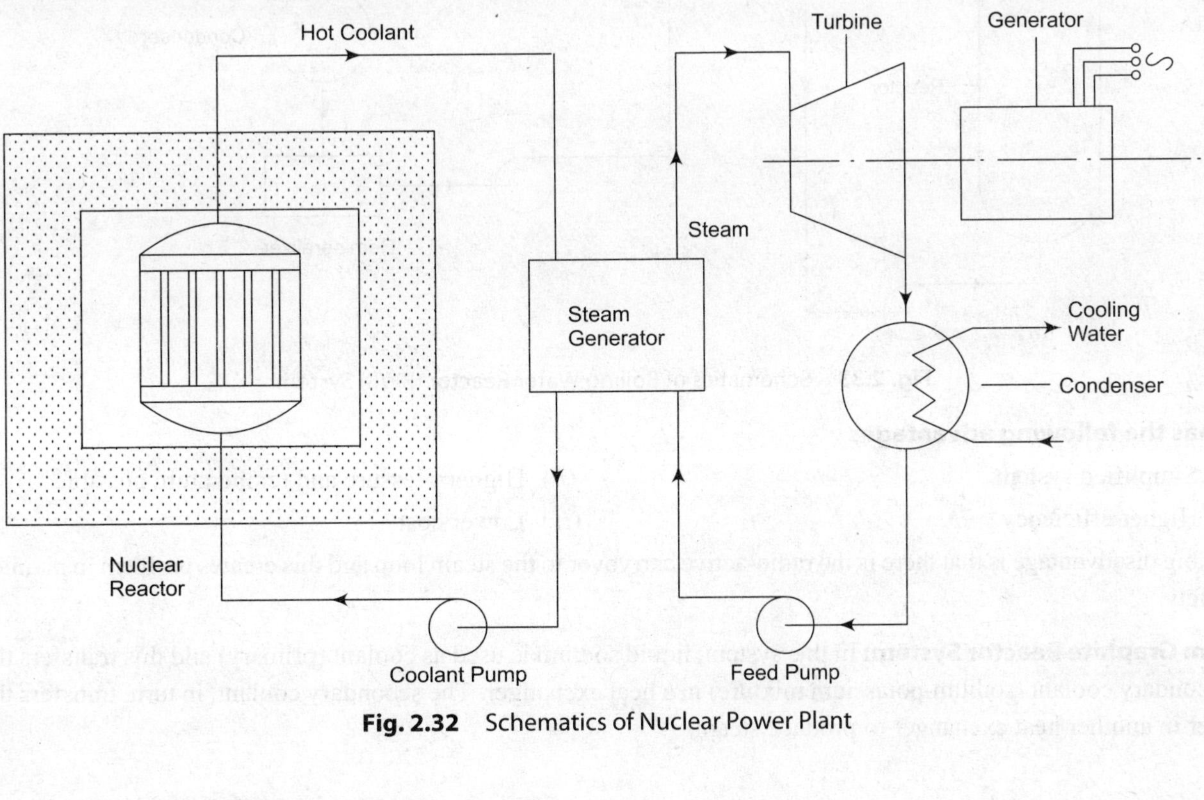

Fig. 2.32 Schematics of Nuclear Power Plant

2.3.4 Classification of Reactors

Reactors are classified based

(*i*) Neutron energy: Ex. (1) Fast reactor (works without moderator) (2) Thermal reactor (works with moderator)

(*ii*) Types of coolant: Ex. Gas-cooled, liquid metal cooled, water-cooled.

(*iii*) Type of core:

(*a*) Homogeneous reactor: In this the fuel and moderator are in a uniform mix.

(*b*) Heterogeneous reactor: Here the fuel rods are inserted in moderator to form a uniform lattice.

(*iv*) Application: Research reactor, power reactor, etc.

2.3.5 Reactors for Power Generation

We shall examine here a few of the common power reactor systems.

Boiling water reactor: (BWR) (Fig. 2.33): This is a thermal reactor and uses water (ordinary or heavy) as coolant, and also as moderator. Saturated steam is generated in the reactor core itself. This steam is separated in the steam-water drum and sent directly to the steam turbine.

India's first nuclear power plant at Tarapur has 2 BWRs each of 200 MW capacity.

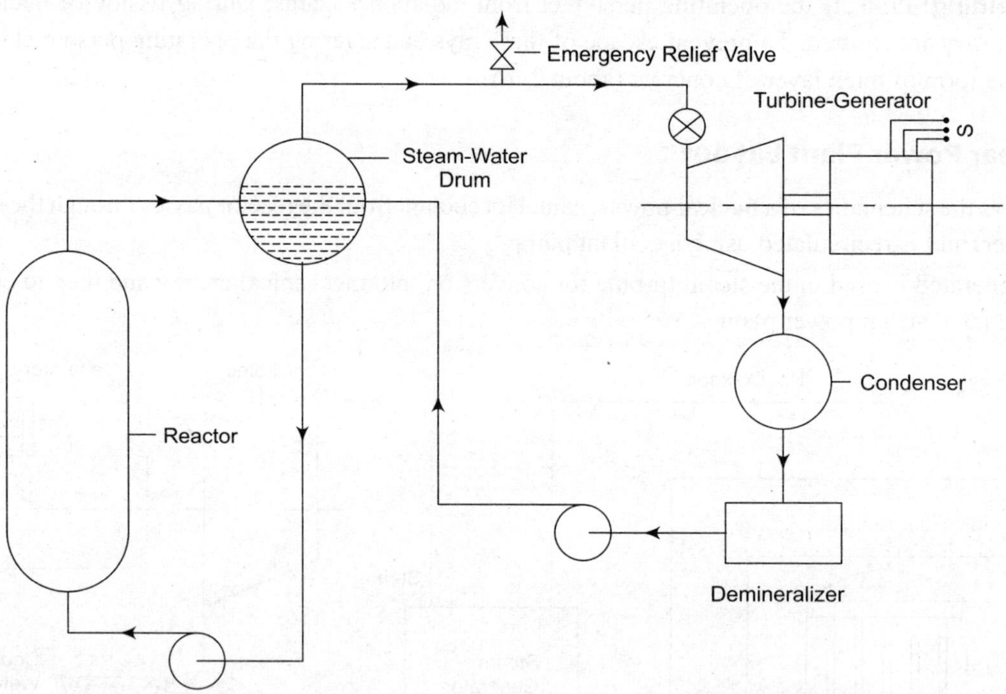

Fig. 2.33 Schematics of Boiling Water Reactor (BWR) System

BWR has the following advantages

(*i*) Simplified system

(*ii*) Higher pressure and temperature possible

(*iii*) Higher efficiency

(*iv*) Lower cost

The big disadvantage is that there is the radio-active carryover to the steam loop and this creates problem in maintenance and safety.

Sodium Graphite Reactor System: In this system, liquid sodium is used as coolant (primary) and this transfers the heat to a secondary coolant (sodium-potassium mixture) in a heat exchanger. The secondary coolant, in turn, transfers the heat to water in another heat exchanger to produce steam.

Graphite is used as moderator and reflector.

It has the advantage of attaining high temperature without pressurising the reactor.

The addition of the heat exchangers adds to the cost. But there is no problem of radio-active carry over as in BWR.

Fast Breeder Reactor (F.B.R.): In this reactor the core containing U-235 is surrounded by a blanket of fertile material U-238. No moderator is used. The fast moving neutrons liberated in the fission of U-235 are absorbed by U-238 which gets converted to fissionable Pu-239.

Here also two liquid metal circuits are used for heat transport.

In addition to its use as power reactor breeder reactor produces more fissionable material than it consumes.

2.3.6 Advantages and Disadvantages of Nuclear Plant

Advantages

1. Space requirement is less
2. Consumes small quantity of fuel
3. Excellent reliability of operation
4. Not affected by weather conditions
5. Well suited to meet large power demands
6. Does not require large quantity of water

Disadvantages

1. Initial cost is higher than hydro or steam power plants
2. Not well suited for varying load
3. Problem with disposal of radioactive wastes
4. Maintenance cost is high
5. Requires trained personnel

2.3.7 Nuclear Fusion

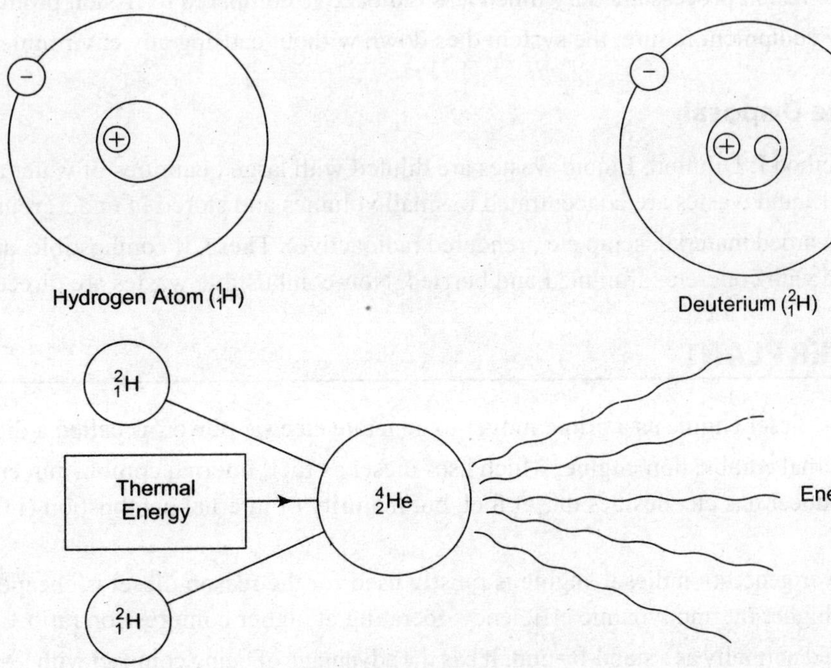

Fig. 2.34 Fusion Process

Another way of getting nuclear energy is by thermal fusion of heavy hydrogen to form helium (Fig. 2.34). From two heavy hydrogen atoms with atomic weight $2 \times 2.01625 = 4.03250$, one atom of helium is obtained with atomic weight 4.00389. The balance of mass 0.02861 is converted into energy according to the equation,

$$e = m * c^2$$
$$= 0.02861 \times (3 \times 10^{10})^2 \text{ erg}$$

It is on this principle, thermonuclear or "hydrogen" bomb has been developed. Unlike the fission of uranium, it has not been possible so far to let this fusion proceed in a controlled manner.

Two major requirements for the success of fusion reaction system are: (*a*) attaining extremely high temperature of the order of 10^8 to 10^9 K and (*b*) confining plasma (high temperature ionised gas) and maintaining the temperature.

Fission reaction provides the required level of temperature but containment of plasma has not been a total success. Several methods are being tried. One of them employs a powerful magnetic field as shown in Fig. 2.35.

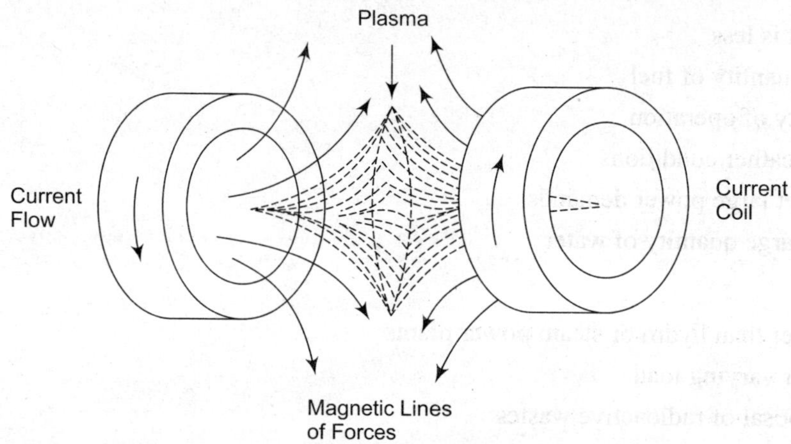

Fig. 2.35 Containment of Plasma by Magnetic Field

Advantages

1. The availability of the most favoured fuel isotope, deuterium (hydrogen-2) is almost unlimited.
2. The products of the fusion process are very much less radioactive compared to fission products.
3. In the event of any equipment failure, the system dies down without causing any environmental damage.

2.3.8 Nuclear Waste Disposal

(*i*) Liquid Wastes: Method I: Dilution. Liquid wastes are diluted with large quantities of water and released. Method II: Concentration. Liquid wastes are concentrated to small volumes and stored in underground tanks.

(*ii*) Solid Wastes: (Discarded material, scrap etc., rendered radioactive). These, if combustible, are burnt and the radio active ash is mixed with concrete, drummed and burried. Non-combustible wastes are directly burried.

2.4 DIESEL POWER PLANT

A power plant, which uses diesel engine as a prime mover to generate electric power, is called a diesel power plant.

Diesel engine is an internal combustion engine, which uses diesel as fuel. Internal combustion engine may use petrol, kerosene, natural gas, producer gas etc. besides diesel fuel, but majority of internal combustion (I.C.) engines use either petrol or diesel fuel.

For the purpose of power generation diesel engine is mostly used for the reason diesel is cheaper and therefore more economical. It also gives higher thermodynamic efficiency operating at higher compression ratio.

Diesel power plant is used normally as a stand-by unit. It has the advantage of being compact with less auxiliary equipment and also being able to start quickly and deliver the required power. The scheme of diesel power plant is shown in Fig. 2.36.

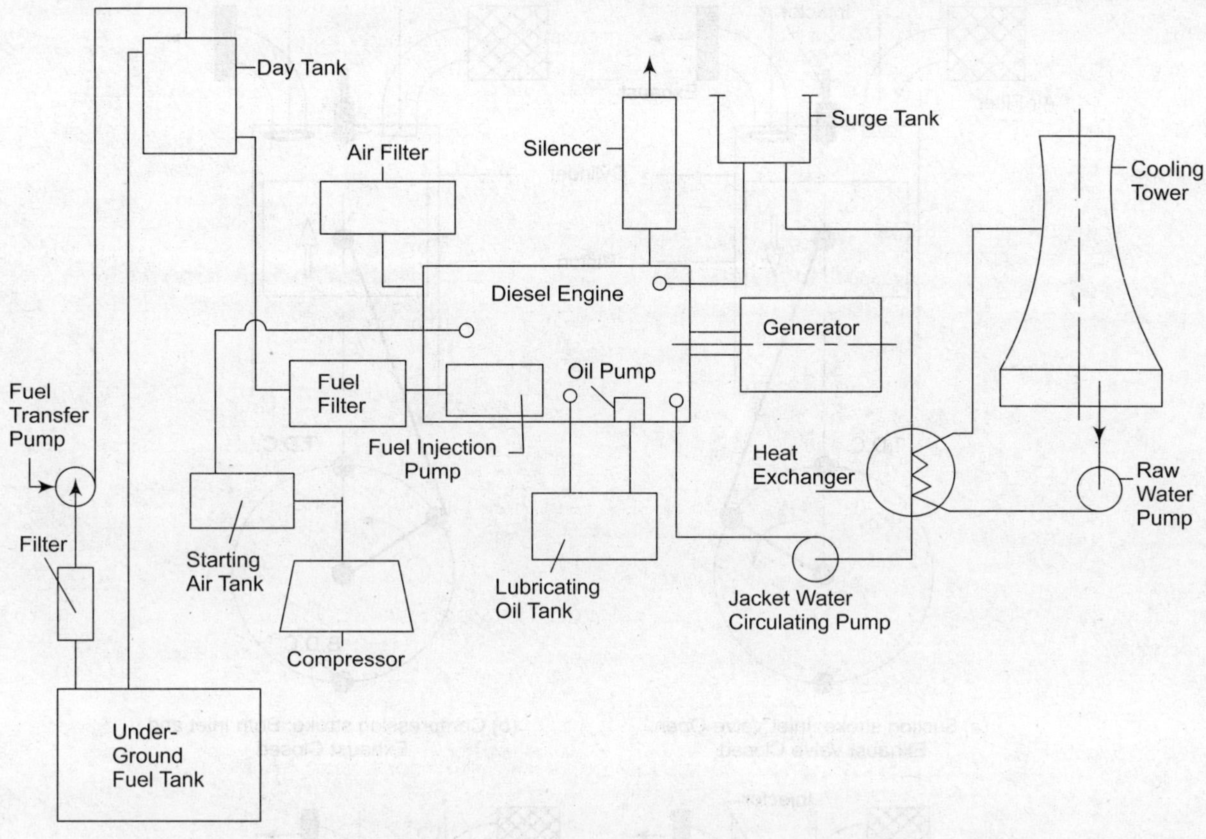

Fig. 2.36 Schematics of Diesel Engine Power Plant

Diesel fuel is stored in an under-ground tank which serves as a large reservoir. From this, fuel is pumped into the day tank. The day tank supplies the fuel to the engine. There is a cooling circuit which helps maintain the engine cylinders at optimum temperature. Lubricating oil tank and pump supply lubricating oil to the various moving parts, such as piston and cylinder, crank pin bearings, main bearings, etc. Generator with control unit produces electrical energy at the desired voltage and frequency.

Starting of the diesel engine is done using compressed air.

2.4.1 Classification of I.C. Engines

I.C. engines are classified based on type of fuel used. Ex.Petrol engine, diesel engine etc. Classification is also made on the number of strokes needed to complete one cycle of operations. Accordingly, we have (*i*) Four stroke engines and (*ii*) Two stroke engines.

Again we may classify engines based on type of cooling arrangement (air cooling/water cooling), number of cylinders and their arrangement and speed of engine.

2.4.2 Working of Four Stroke Diesel Engine

Figure 2.37 shows the working of a 4-stroke diesel engine.

Suction stroke [Refer Fig. 2.37 (*a*)]: As the piston moves down and crank rotates clockwise from TDC (Top Dead Centre), air from atmosphere is drawn into the cylinder through the inlet valve, which is kept open by the action of a cam (not shown). During this stroke the exhaust valve is kept closed.

Compression stroke [Refer Fig. 2.37 (*b*)]: Piston having reached the end of the suction stroke moves upwards with inlet and exhaust valves remaining closed. The air in the cylinder gets compressed (compression ratio being about 22:1) and its temperature rises above the auto-ignition temperature of fuel.

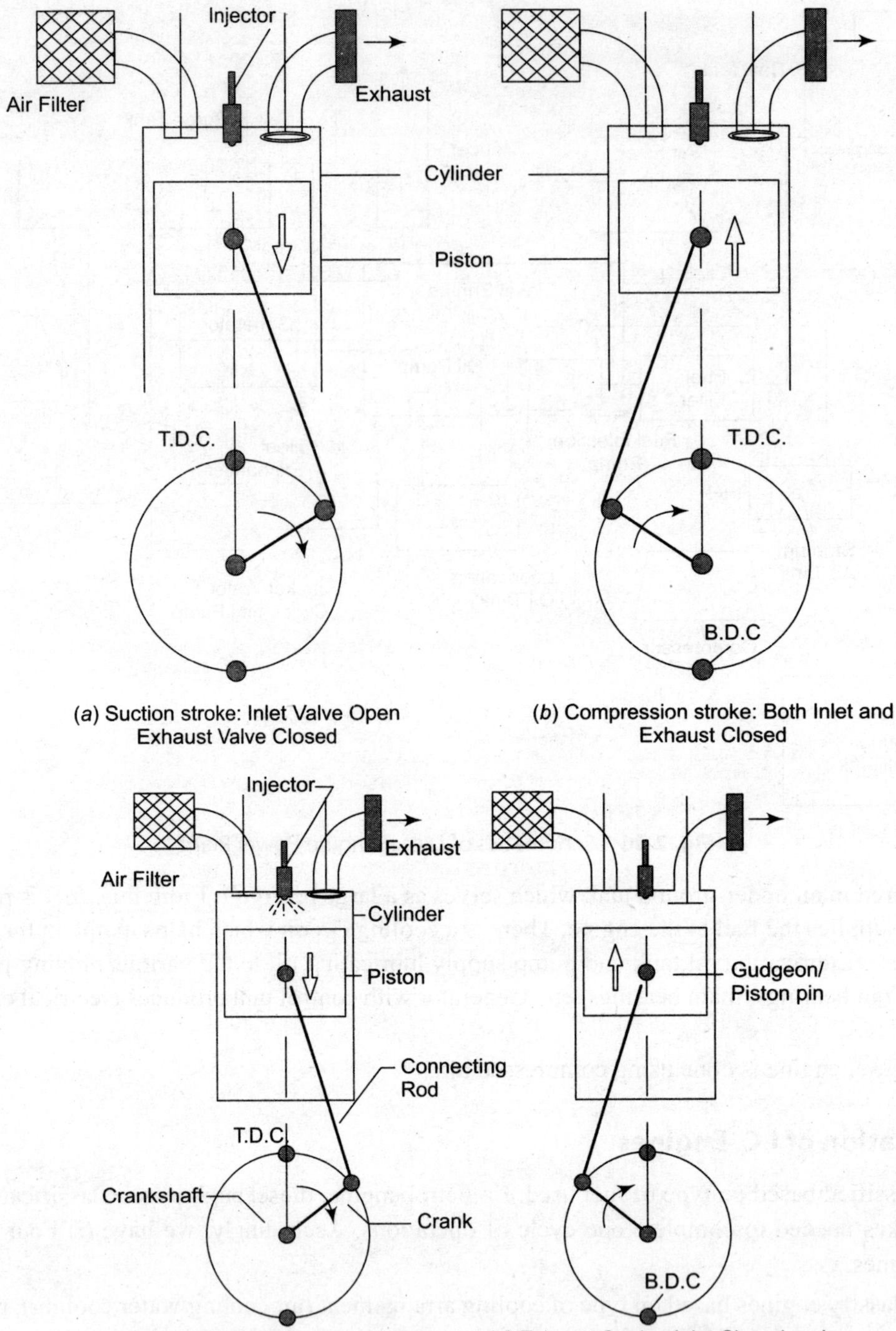

(*a*) Suction stroke: Inlet Valve Open Exhaust Valve Closed

(*b*) Compression stroke: Both Inlet and Exhaust Closed

(*c*) Power Stroke: Both Inlet and Exhaust Closed

(*d*) Exhaust Stroke: Inlet Closed and Exhaust Open

Fig. 2.37 Schematics of Four-Stroke Diesel Engine—Working of

At this point diesel fuel is injected. The atomised fuel getting mixed with the high temperature compressed air, burns almost instantaneously and the products of combustion expand.

Power stroke [Refer Fig. 2.37 (*c*)]: The expanding gases push down the piston delivering the power to the crank shaft through the crank and connecting rod mechanism.

Exhaust stroke [Refer Fig. 2.37 (*d*)]: Piston, having descended to the lowest position corresponding to the crank in the bottom dead centre (BDC), moves up again pushing the burnt gases through the exhaust valve. During this stroke, the inlet valve remains closed and the exhaust valve is kept open by another cam (not shown).

The four strokes of the piston-suction, compression, power and exhaust — constitute a cycle of operations, which repeats itself. Part of power delivered during power stroke is used up in executing the other three strokes.

In the preceding paragraphs a simplified explanation of the working of a 4-stroke diesel engine is given. This does not take into account the dynamics of the valve motion and the momentum of incoming air/outgoing gases into and out of the engine cylinder. In practice, the valves take some time to operate and therefore, are timed to open and close over a certain degrees of crank rotation [Refer Fig. 2.37(*e*)].

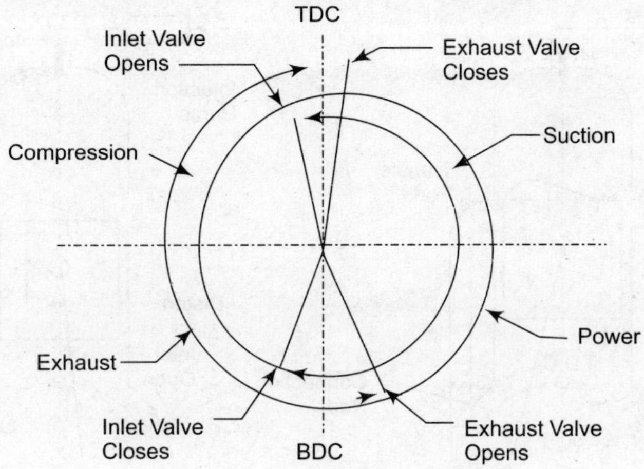

Fig. 2.37 (*e*) Valve Timing Diagram

Referring to inlet valve, it opens slightly ahead of TDC, *i.e.*, even before the piston descends on its suction stroke and closes a few degrees after BDC, *i.e.*, even after suction stroke is complete. The latter arrangement (keeping inlet valve open beyond BDC) helps use the momentum of the incoming air to get more air inducted into the cylinder.

Referring to the exhaust valve, it opens a few degrees earlier to BDC, *i.e.*, before the end of power stroke and remains open a few degrees after TDC, *i.e.*, after the completion of exhaust stroke, This arrangement coupled with earlier opening of inlet valve on suction stroke helps scavenge thoroughly the cylinder of all burnt gases.

Ignition delay: With reference to ignition there is a time lag between the injection of fuel and the ignition/start of combustion. This period is called ignition delay. To enable complete combustion and build up of maximum pressure at the right time injection is timed slightly ahead of TDC, *i.e.*, it takes place a few degrees before the end of compression stroke.

2.4.3 Working of a Two-Stroke Diesel Engine

Figure 2.38 shows one particular arrangement of a two stroke diesel engine. It has a cylinder, piston, crank and connecting rod mechanism with an injector for injecting the diesel fuel into the cylinder. The crank case, in addition to supporting the bearings of the crank shaft, holds lubricating oil for splash or pressure lubrication of the cylinder, piston and other moving parts.

As the piston comes down due to expanding hot gases with the crank rotating clockwise from the top dead centre (TDC), it uncovers the exhaust port first [Ref. Fig. 2.38(*a*)]. A little later it uncovers the transfer port. When this happens, air contained in the crank case is pushed out through the transfer port into the cylinder. The incoming air flushes out the exhaust gases and this is helped by the deflector integral with the piston crown. Thus, towards the end of piston travel due to expanding gases, we have exhaust of burnt gases, followed by filling in of cylinder with air from the crank case.

Now, the movement of piston upwards, [Ref. Fig. 2.38(*b*)] with crank rotating further clockwise form BDC to TDC results in first closing of the transfer port and then the exhaust port. Further movement results in compressing of the air. As in the case of 4 stroke diesel engine, fuel is injected at the end of compression and the atomised fuel burns almost instantly. The resulting hot gases expand providing the power to the crank shaft through the piston-connecting rod-crank mechanism.

During the upward movement of piston compressing the air, the inlet valve gets opened due to partial vacuum in the crank case and air from atmosphere is drawn in. When the piston moves down the inlet valve closes and the air in the crank case gets compressed. As soon as transfer port opens the compressed air is transferred into the cylinder.

Thus it is seen that during the two strokes of the piston, all the four operations — power, exhaust, suction and compression —take place. Therefore, this type of engine is called a two-stroke engine.

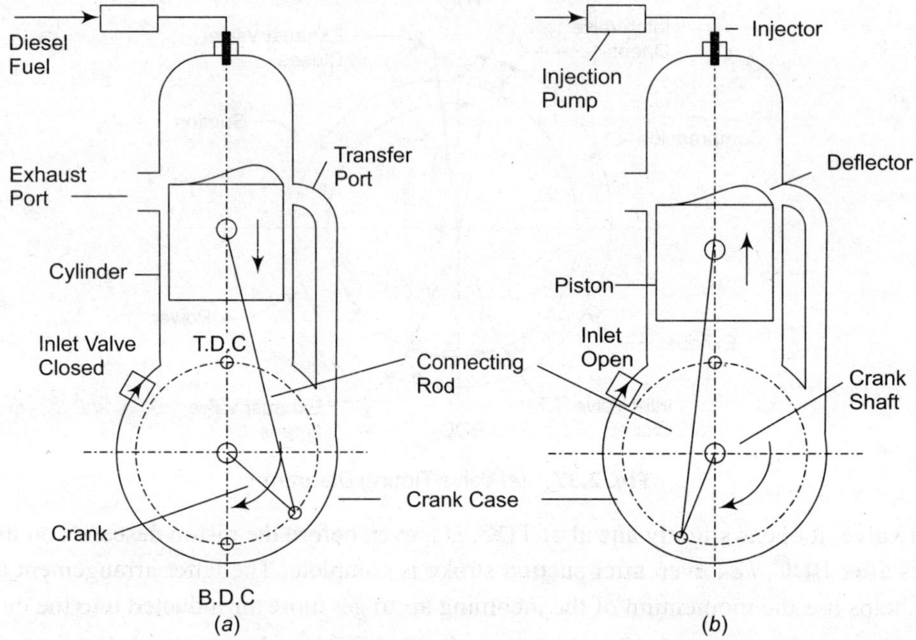

Fig. 2.38 Schematics of Two-Stroke Diesel Engine

2.4.4 (a) Internal Combustion Engine Components

Any internal combustion (I.C.) engine must have the following systems to help its proper running and maintenance.

1. Fuel system
2. Cooling system
3. Lubricating system
4. Inlet and exhaust system

Going in the reverse order, the inlet system takes care of the incoming air to the engine and the exhaust system takes care of the gases going out of the engine. Since the incoming air from the atmosphere may not be clean all the time, an air filter is introduced in the inlet passage of the engine. This prevents dust entering the engine cylinder and thus saves the cylinder and piston from early wear. This also reduces the noise of the air entering with high velocity. In the case of exhaust system, it is the noise that should be suppressed. For this a muffler/silencer is introduced in the exhaust outlet of the engine. These components — air filter and the silencer — may be seen in Fig. 2.36.

Lubricating system is provided to adequately lubricate the cylinder and piston, small end and big end bearings of the connecting rod, the main bearings and other moving parts. Components like oil filter and oil pump are used in this system.

The best method of lubricating is to use a pressurized system [Ref. Fig. 2.39(*a*)]. The lub oil pump draws oil from the sump (a portion of the crank case) and delivers it under pressure to the main oil gallery through a filter. From the main gallery, oil is branched off to the main bearings, the camshaft bearings etc. Lubricating of piston and cylinder is best done by splash-lubrication. The dip-rod attached to the big end of the connecting rod dips into the oil in the sump as the crank passes through the B.D.C. and the oil is splashed on to the piston and cylinder as the crank rotates. The oil after lubrication returns back to the sump. Incidentally, the splashing of the oil helps to cool the piston, which cannot be done otherwise.

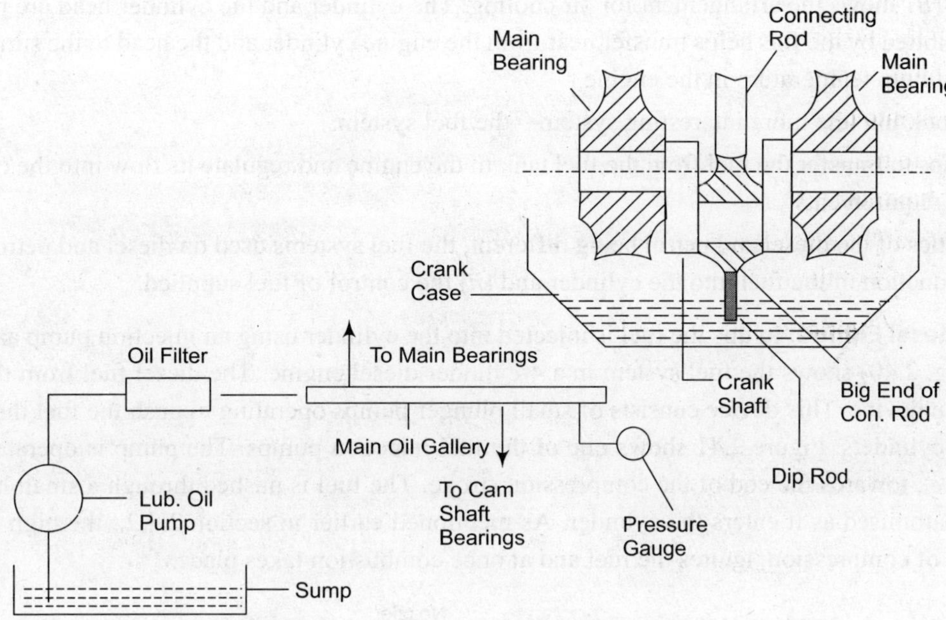

Fig. 2.39 (*a*) Lubricating System in I.C. Engine

Advantages of lubrication

1. Helps reduce the friction in the moving parts and thus enhances the mechanical efficiency.
2. Wear and tear of the engine is reduced.
3. Splash lubrication helps also cool the piston, which cannot be done otherwise.
4. Helps reduce gas leakage past the piston.

Next the cooling system. Cooling of the cylinder and cylinder head is a must because the very high temperature reached during combustion may damage them. Water cooling for larger engines and air cooling for smaller engines is generally adopted.

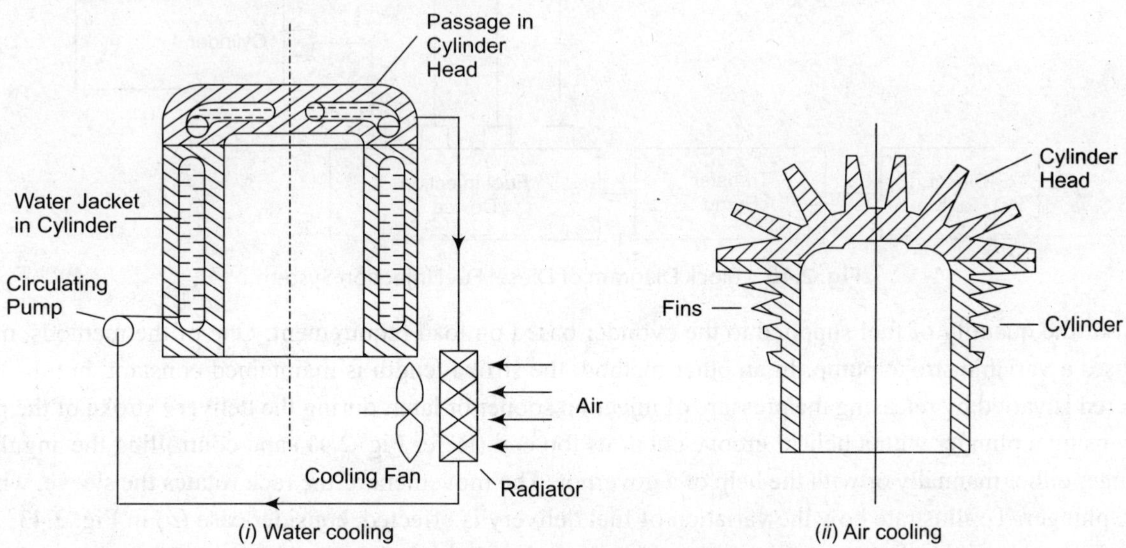

Fig. 2.39 (*b*) Cooling System in I.C. Engines

Figure 2.39 (*b*) (*i*) shows diagramatically the water cooling system. The cooling water is circulated through the water jacket in the cylinder and the passages in the cylinder head using a pump. A cooling fan and radiator combination is used to cool the heated water before it is pumped into the cylinder jacket. Alternately larger power engines use cooling stacks or a cooling tower to cool the heated water.

Figure 2.39 (*b*) (*ii*) shows the arrangement for air cooling. The cylinder and the cylinder head are provided with fins. The large area exhibited by the fins helps transfer heat from the engine cylinder and the head to the surroundings and thus helps maintain optimum temperature in the engine.

Now we shall look into the more interesting system—the fuel system.

Fuel system helps to transfer the fuel from the fuel tank to the engine and regulate its flow into the cylinder depending on speed and load requirements.

The characteristics of the diesel and petrol being different, the fuel systems used on diesel and petrol engines differ in respect of (*i*) introduction of the fuel into the cylinder and (*ii*) the control of fuel supplied.

Fuel System in Diesel Engine: In this the fuel is injected into the cylinder using an injection pump and an injector. The block diagram (Fig. 2.40) shows the fuel system in a 4-cylinder diesel engine. The diesel fuel from the tank is pumped to the fuel injection device. This device consists of small plunger pumps operating to push the fuel through the injectors (nozzles) into the cylinders. Figure 2.41 shows one of the fuel injection pumps. The pump is operated by a cam at the appropriate time *i.e.*, towards the end of the compression stroke. The fuel is pushed through a small hole in the injector (Fig. 2.42) and is atomised as it enters the cylinder. As mentioned earlier in section 2.4.2., the high temperature of air reached at the end of compression, ignites the fuel and at once combustion takes place.

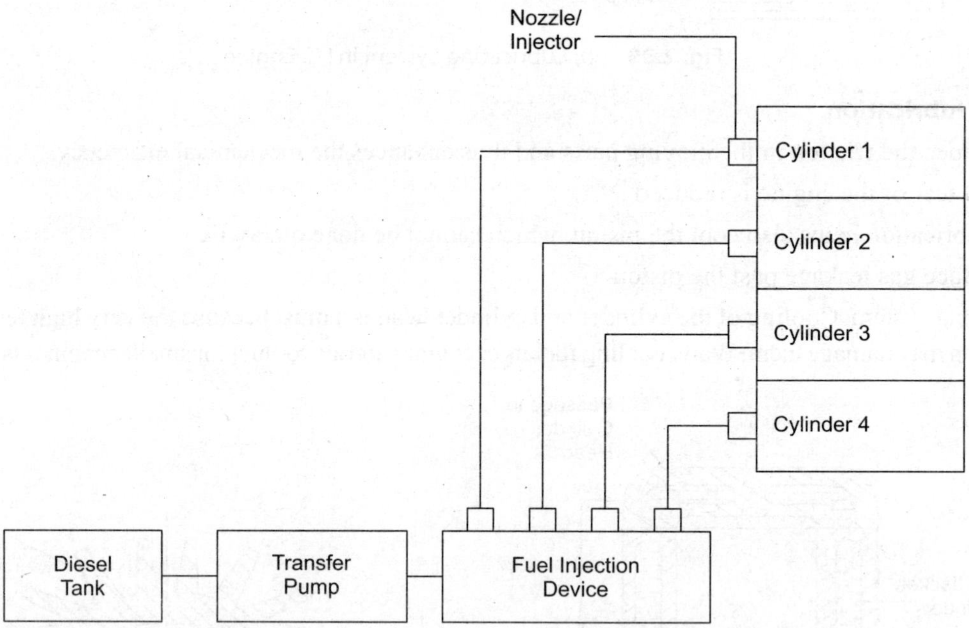

Fig. 2.40 Block Diagram of Diesel Fuel Injection System

Control of the quantity of fuel supplied to the cylinder based on load requirement: One of the methods, used for the purpose, uses a variable stroke pump. In an other method, the stroke length is maintained constant, but the quantity of fuel delivered is varied by releasing the pressure of injection sooner or later, during the delivery stroke of the pump. This is done by using a plunger with a helical groove cut at its top end (Refer Fig. 2.41) and controlling the angular position of the plunger either manually or with the help of a governor. The movement of the rack rotates the sleeve, which in turn rotates the plunger. To illustrate how the variation of fuel delivery is effected, consider case (*a*) in Fig. 2.43. As soon as the plunger is displaced by a distance x_1, the pressure is released with the fuel finding its way back to the suction chamber. Now consider case (*b*). In this the plunger's angular position is changed and as the upward displacement of the plunger reaches a value x_2, the pressures is released. Since $x_2 < x_1$, the pressure in the case (*b*) is released much earlier than in case (*a*) and therefore much less fuel is pushed into the cylinder.

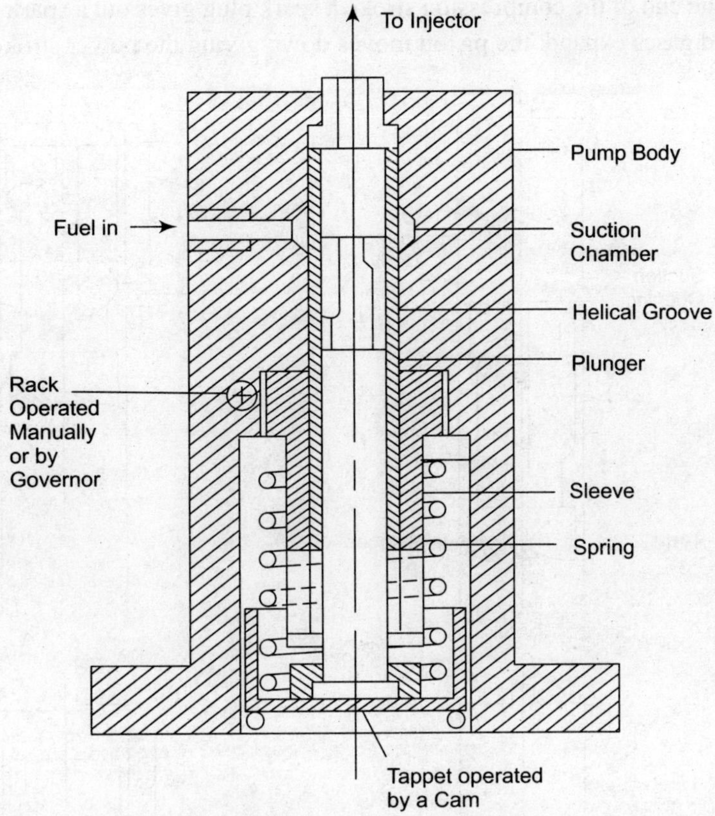

Fig. 2.41 Fuel Injection Pump

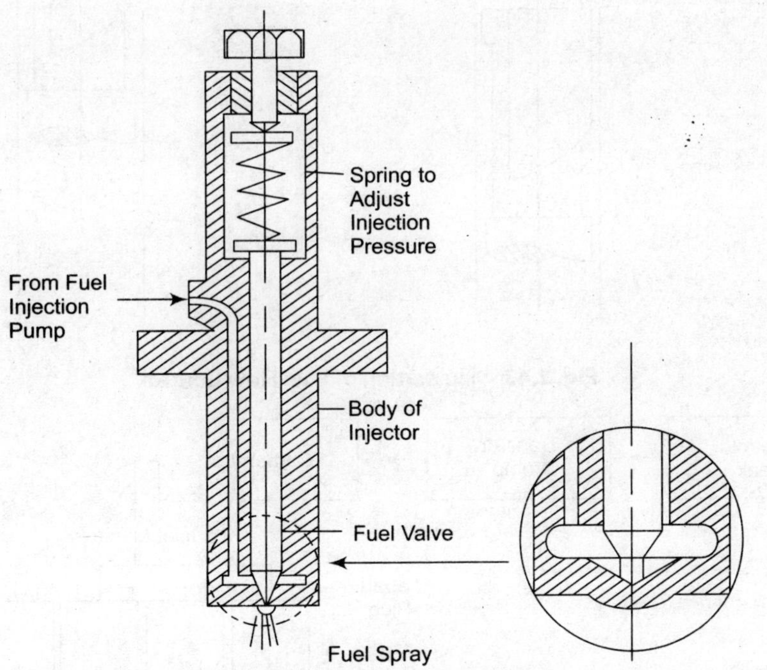

Fig. 2.42 Fuel Injector

Fuel system in a petrol engine: The block diagram of the fuel system in a petrol engine is shown in Fig. 2.44. The petrol is transferred from the tank and made available at the carburettor (Fig. 2.45). During the suction stroke, the carburettor vapourises the petrol and a mixture of petrol vapour and air enters the engine cylinder. (**Note:** In the diesel engine only air is drawn during suction stroke.) This petrol vapour-air mixture is compressed to about 1/7 its original volume during the compression stroke. At the end of the compression stroke a spark plug gives out a spark and the mixture is ignited. As the combustion proceeds and gases expand, the piston moves down giving the power stroke.

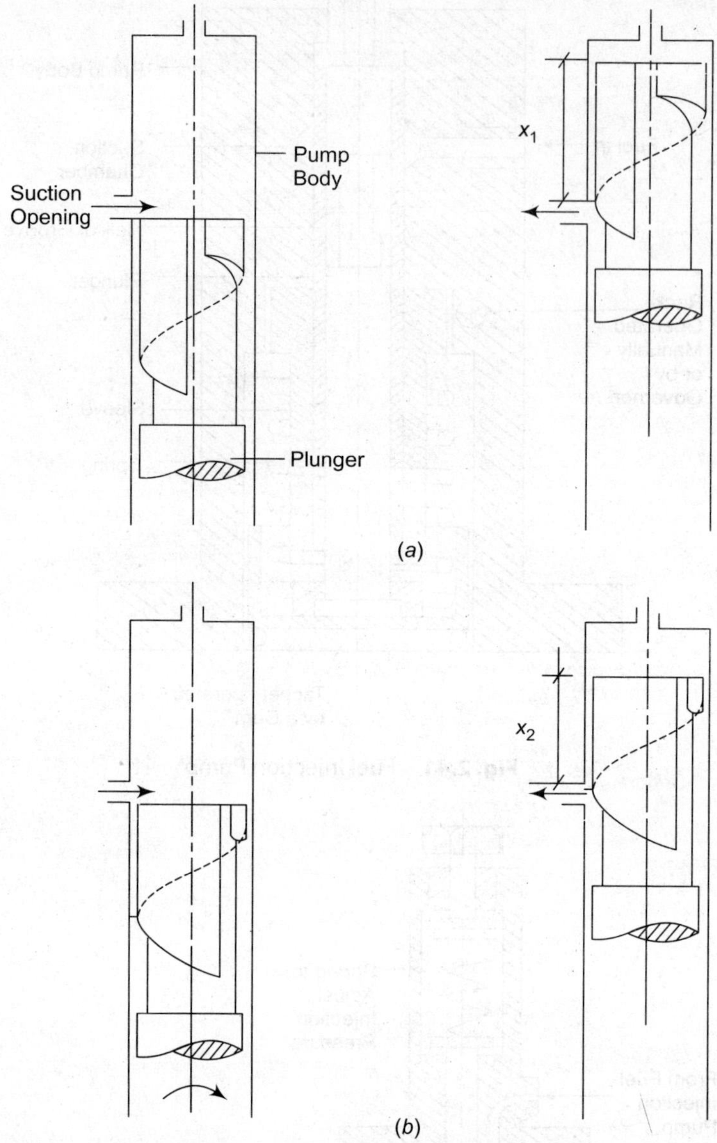

Fig. 2.43 Illustration of Fuel Flow Control

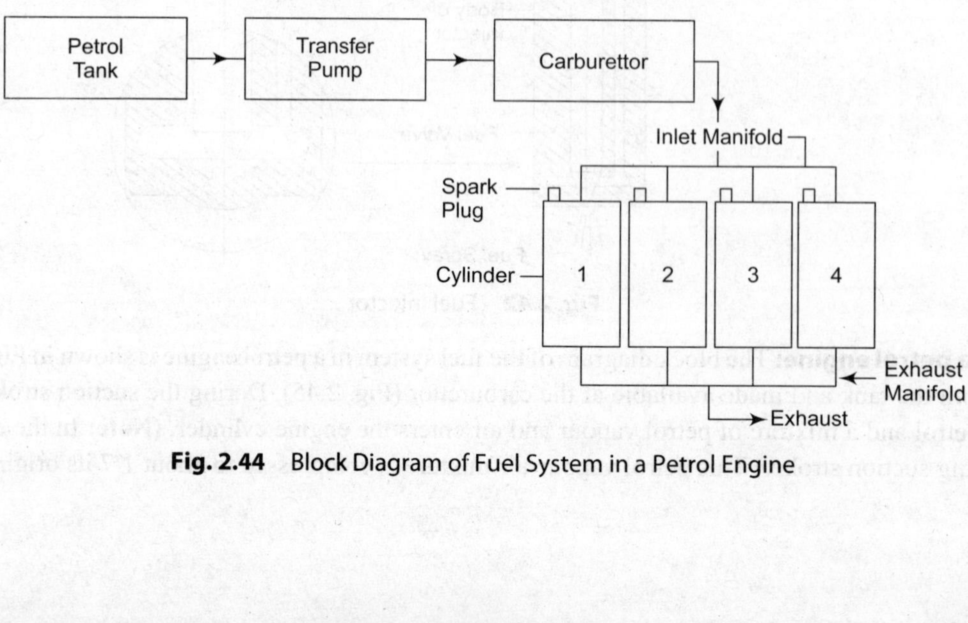

Fig. 2.44 Block Diagram of Fuel System in a Petrol Engine

Carburettor: The basic principle here is that when a volatile fuel is placed in the passage of a high velocity air, the fuel gets vapourised soon. Based on this principle, a fuel jet is placed in the throat of a venturi (a passage of varying cross section) of the carburettor (Fig. 2.45). During the suction stroke, the air is drawn through the venturi of the carburettor and since the pressure at the throat reduces and velocity increases, the petrol is drawn out of the jet and is soon vapourised. The petrol vapour together with the air enters the cylinder via the butterfly valve. The float chamber maintains a constant level of fuel in the jet and prevents the fuel from spilling. Only when the suction takes place the fuel is drawn out. Once the maximum level is reached in the float chamber, the float rises up and closes the fuel supply by means of a needle valve.

The butterfly valve connected to the governor or accelerator pedal helps to control the amount of fuel-air mixture admitted to the cylinder depending on the load conditions.

The carburettor discussed above gives the basic concept. Here the fuel flow is dependent on the pressure drop in the venturi and the pressure drop depends on the rate of flow of air. Therefore, the air-fuel ratio provided is theoretically constant. However, in practice it is not so and compensating devices are provided to maintain proper air-fuel ratio. One

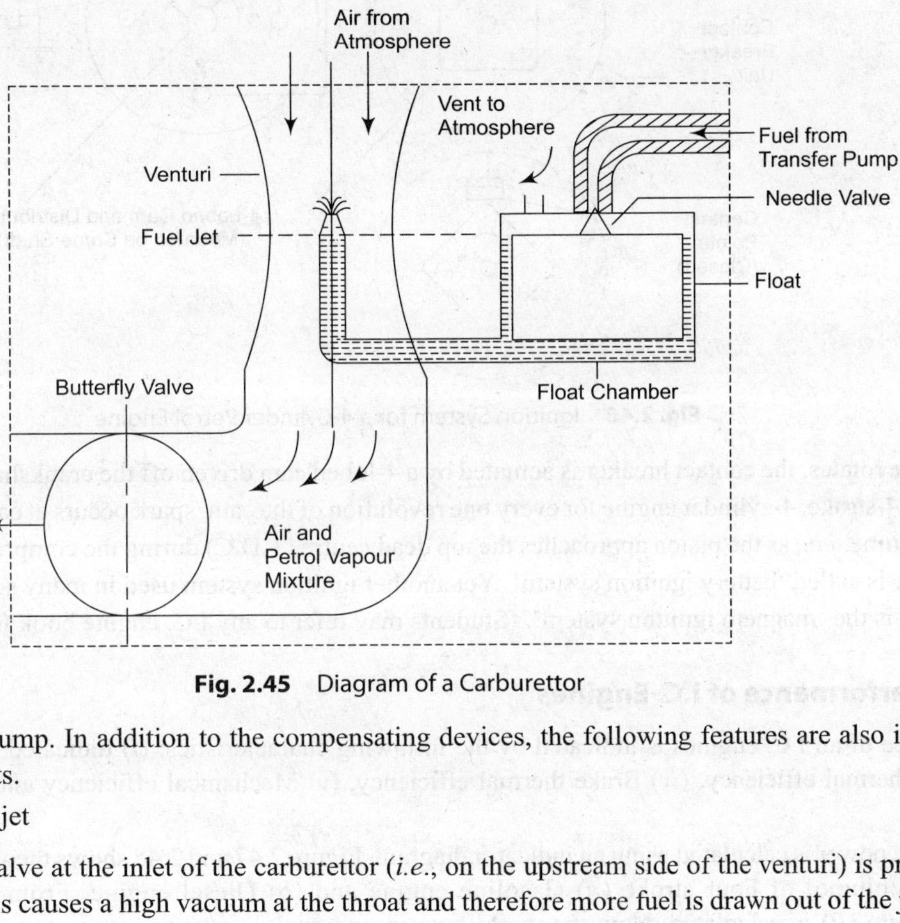

Fig. 2.45 Diagram of a Carburettor

such is the accelerating pump. In addition to the compensating devices, the following features are also incorporated to meet specific requirements.

1. Choke and 2. Idling jet

Choke: A choke/choke valve at the inlet of the carburettor (*i.e.*, on the upstream side of the venturi) is provided to help choke the air passage. This causes a high vacuum at the throat and therefore more fuel is drawn out of the jet. As a result, an extra rich mixture is made available, necessary for starting the engine from cold.

Idling jet: With the throttle (butterfly valve) fully closed petrol does not issue from the main fuel jet. An idling jet provided beyond the throttle releases a quantity of air-fuel mixture necessary to sustain running of the engine at no load when the throttle-is nearly closed.

Accelerating pump: When the throttle is suddenly opened for either speeding up the engine or for taking up more load air-fuel ratio becomes leaner momentarily. This will affect the performance of the engine. To overcome this difficulty an accelerating pump is used, which pushes in extra fuel at the required moment and the proper air-fuel ratio is restored.

Ignition system: In the fuel system using petrol, the petrol-air mixture does not get ignited on its own. It requires a sub-system —an ignition system to do the job. Figure 2.46 shows an ignition system for a 4 cylinder petrol engine. In this, each time the contact is made and broken by the contact breaker a very high voltage develops across the secondary winding of the ignition coil and a spark appears at the spark plug.

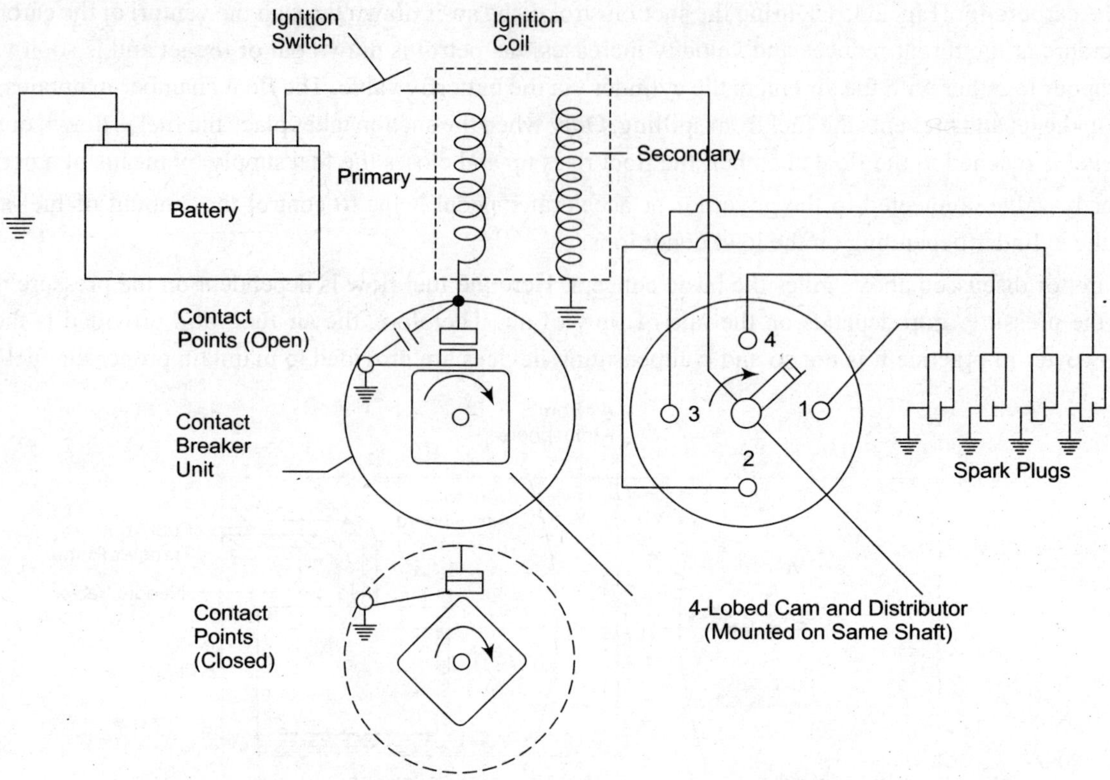

Fig. 2.46 Ignition System for a 4-Cylinder Petrol Engine

As the engine rotates, the contact breaker is actuated by a 4-lobed cam driven off the crankshaft at half the speed of the crankshaft. In a 4-stroke, 4-cylinder engine for every one revolution of the cam, spark occurs at each of the four cylinders at the appropriate time, *i.e.*, as the piston approaches the top dead centre (T.D.C) during the compression stroke. The system discussed above is called 'battery ignition system'. Yet another ignition system used in many small engines as well as in aircraft engines is the 'magneto ignition system'. (Students may refer to any I.C. engine book for information.)

2.4.4 (b) Performance of I.C. Engines

The performance of an I.C. engines is indicated by the following characteristics: (*i*) Indicated power, (*ii*) Brake power, (*iii*) Indicated thermal efficiency, (*iv*) Brake thermal efficiency, (*v*) Mechanical efficiency and (*vi*) Brake specific fuel consumption.

The indicated power is calculated using an indicator diagram. Figure 2.47 and 2.48 shows theoretical indicator diagrams (Pressure v/s Volume) of Four stroke (*a*) Gasoline engine and (*b*) Diesel engine. From the diagram, the Mean Effective Pressure (*P*) is calculated. Now the work done in one cycle – suction, compression, power and exhaust is $P_m \times A \times L$, where P_m = Mean effective pressure (N/m^2), A = Area of cross section of the cylinder (m^2) and L is the stroke length (m).

Therefore, Work done/min = $P_m \times L \times A \times N/2$, where, N is rpm of crank shaft.

(**Note:** A 4-stroke engine has one power stroke in 2 revolutions of crank shaft)

(*i*) Indicated Power (IP)

$$(\text{Work done/second}) = P_m \times L \times A \times N \text{ (N.m/s)}$$

With this understanding a general equation for IP could be written as.

$$\text{IP} = n \times P_m \times A \times L \times N \times K \text{ (10/6) kW},$$

where n = no. of cylinders, P is in bars and K = 1/2 for 4-stroke and = 1 for 2-stroke engines

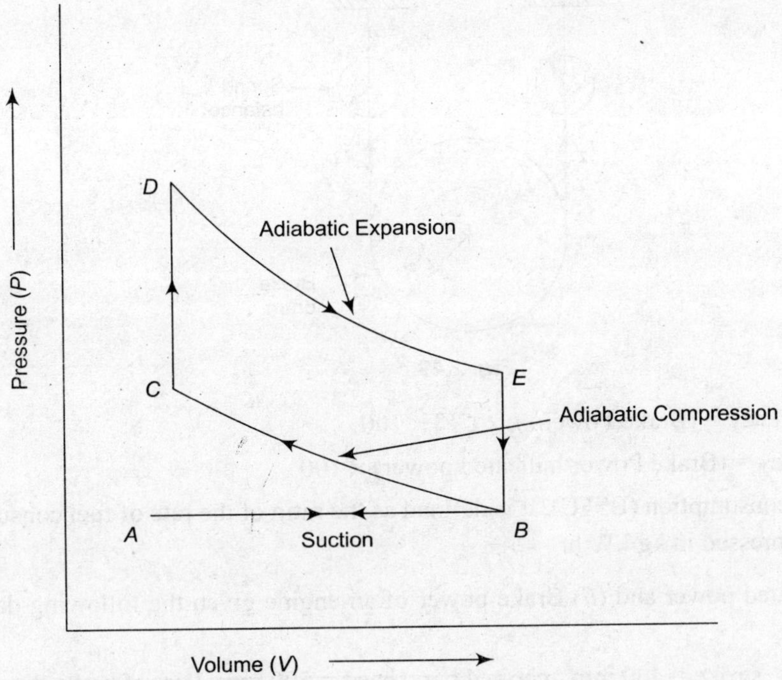

Fig. 2.47 Theoretical Otto Cycle (PV diagram) for Gasoline Engine

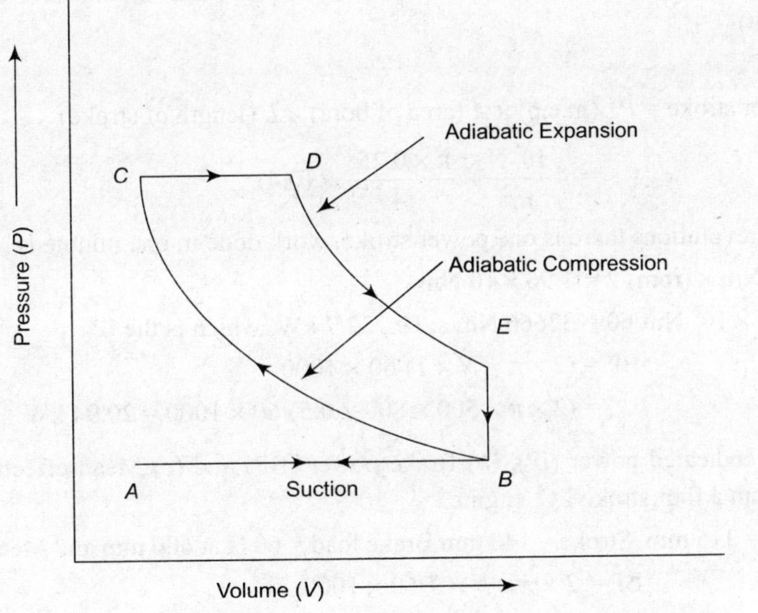

Fig. 2.48 Theoretical Diesel Cycle (PV diagram) for Diesel Engine

(*ii*) Brake Power = $\frac{2\pi NT}{60 \times 1000}$ where T = torque is in Nm. Different types of dynamometer have been developed to measure the torque. Taking the case of a belt dynamometer as shown below (Fig. 2.49).

$$T = (T_1 - T_2) \times R,$$

where T_1 and T_2 are the tensions in the tight and slack side of the belt and R is radius of brake drum (m).

Friction Power = Indicated power – Brake power

(*iii*) Indicated thermal efficiency = [Indicated Power/$m_f \times CV$] × 100

where m_f = mass of fuel consumed/hr. and CV is the calorific value of fuel in kJ/kg.

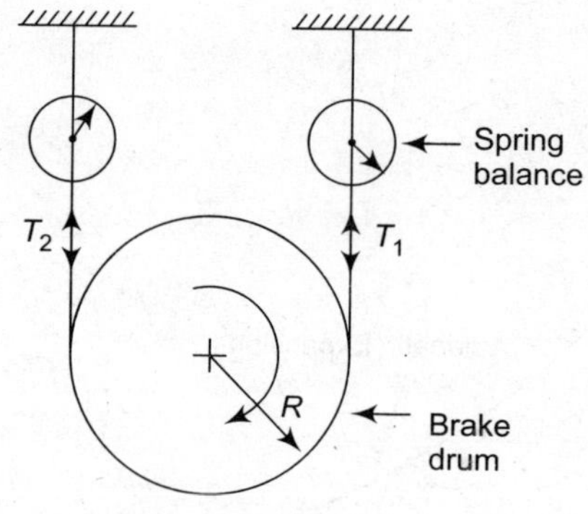

Fig. 2.49

(*iv*) Brake thermal efficiency = [Brake Power/m_f × *CV*] × 100

(*v*) Mechanical efficiency = (Brake Power/Indicated power) × 100

(*vi*) Brake specific fuel consumption (BSFC). It is defined as the ratio of the rate of fuel consumed to the brake power developed and is expressed in kg/kW hr.

Example 1: Find (*i*) Indicated power and (*ii*) Brake power of an engine given the following data. Engine type: Single cylinder four stroke engine.

Bore diameter = 250 mm, stroke = 400 mm, mep = 4 bar, speed = 500 rpm. Dia. of brake drum = 1000 mm. Effective brake load = 800 N.

(Caution. Keep track of the dimensions of the data. The calculated value should have the same dimension as the dimension of the end result).

(*i*) Indicated Power (IP)

Work done in power stroke = P_m (m.e.p) × *A* (area of bore) × *L* (length of stroke)

$$= 4\frac{10^5\text{ N}}{\text{m}^2} \times \frac{\pi \times 0.25^2}{4} \times (0.4) \qquad \left(\text{Since } 1b = \frac{10^5\text{ N}}{\text{m}^2}\right) \quad ...(1)$$

Since in every two revolutions there is one power stroke, work done in one minute is

$$7840 \text{ Nm} \times (\text{rpm})/2 = 1.96 \times 10 \text{ Nm}$$

and $\quad 1.96 \times 10^6$ Nm/60 = 32666 Nm/s. *i.e.*, 32.7 kW, which is the IP. ...(2)

(*ii*) Brake power (BP) $\quad$ BP = (2 × π × *N* × *T*)/60 × 1000

$$= (2 \times \pi \times 500 \times 800 \times 0.5)/60 \times 1000 = 20.94 \text{ kW}.$$

Example 2: Calculate (*a*) Indicated power (IP), (*b*) Brake power (BP) and (*c*) Mean effective pressure (*P*) given the following data obtained from a four stroke I.C. engine.

N = 1000 rpm, Bore dia = 115 mm, Stroke = 140 mm Brake load = 60 N at 600 mm and Mechanical efficiency = 80 %.

$$\text{BP} = 2 \times \pi \times N \times T/60 \times 1000$$

where $\quad T = 60 \times 0.6 = 36$ Nm

∴ $\quad$ BP = 3.7 kW.

$$\text{Mech. } \eta = (\text{BP/IP}) \times 100 \quad ...(3)$$

∴ $\quad \text{IP} = \dfrac{3.7}{0.8} = 4.71$ kW.

$$A = \frac{\pi \times 0.115^2}{4} = 0.0103 \text{ m}^2 \text{ and } L = 0.14 \text{ m}.$$

From equation of IP, $\quad 4.71 = P_m \times 0.0103 \times 0.14 \times 600 \times 10/6$

∴ $\quad P_m = 3.9\ b.$

Example 3: An experiment on a 4-stroke single cylinder gasoline engine gave the following data.

The indicator diagram measurements were (*i*) area of diagram (*a*) = 4 cm^2, (*ii*) length (*l*) = 6.5 cm, ...(4)

Spring constant (*S*) = 100 N/cm /cm. Engine speed = 450 rpm.

Stroke volume = 9.425×10^{-4} m^3.

Calculate mean effective pressure (P_m) and indicated power,

$$\text{m.e.p } (P_m) = (4/6.5) \times S \quad \text{...(5)}$$

where $S = [100 \text{ N/cm}^2]/\text{cm} = [10^6 \text{ N/m}^2]/\text{cm}$

$\therefore$ $\text{m.e.p} = 0.615 \text{ cm} \times [10^6 \text{ N/m}^2]/\text{cm}$

$$= 6.15 \times 10^5 \text{ N/m}^2. \quad \text{...(6)}$$

$$= 6.15 \text{ b}.$$

$$\text{IP} = 1 \times 6.15 \times 9.425 \times 10^{-4} \times \frac{450}{2} \times \frac{10}{6}$$

$$= 2.17 \text{ kW}.$$

Example 4: Given the following data find IP, BP, brake thermal efficiency and specific fuel consumption.

Single cylinder 4-stroke engine Bore = 180 mm, Stroke = 200 mm, *N* = 300 rpm, Torque obtained in the brake dynamometer = 200 Nm, mep = 6*b*. Fuel consumption = 4 kg/h. The calorific value of the fuel is 42000 kJ/kg.

$$\text{BP} = (2 \times \pi \times N \times T)/60 \times 1000$$

$$= (2 \times \pi \times 300 \times 200)/60 \times 1000 \quad \text{...(1)}$$

$$= 6.28 \text{ kW}.$$

$$\text{IP} = n \times P \times L \times A \times N \times K \text{ (10/6)}$$

$$= 1 \times 6 \times 0.2 \times \pi \times \frac{(0.18)^2}{4} \times 300 \times \frac{1}{2} \times \frac{10}{6} \quad \text{since } n = 1, \text{ and } K = 1/2$$

$$= 7.63 \text{ kW}$$

Brake Thermal Efficiency = (BP/m × CV) × 100, where m is the mass of fuel consumed in one hour and CV is the calorific value of fuel. ...(2)

$$= (\text{Brake power output/Rate of heat input}) \times 100.$$

$$= [6.28 \times 1000 \text{ (J/s)} \times 3600 \text{ (s/hr)}]/4\text{(kg/hr)} \times 42000 \text{ (kJ/kg)}$$

= 13.47 %. (It is the % of heat converted to BP)

Brake Specific Fuel Consumption (BSFC) defined as the mass of fuel consumed in one hour in developing 1 kW brake power.

BSFC = m_f/BP, where m_f = fuel consumption/hr, and BP is in kW.

$$= 4/6.28$$

$$= 0.636 \text{ kg/kW h}.$$

Example 5: Following are the data obtained from a single cylinder diesel engine.

BP = 70 kW, Mech. Efficiency = 85 %, Calorific value = 40,000 kJ/kg. Brake thermal efficiency = 35%

Calculate IP, FP and brake specific fuel consumption.

$$\text{IP} = \text{BP}/\eta_{mech} = 70/0.85 = 82.35 \text{ kW}$$

$$\text{FP} = \text{IP} - \text{BP} = 82.35 - 70 = 12.35 \text{ kW}.$$

Brake thermal efficiency (η_{BTH}) = BP/m_f × CV

$\therefore$ $m_f = 70 \text{ k(J/s)}/(0.35 \times 40000 \text{ kJ/kg})$

$$= 0.005 \text{ kg/s}$$

$$= 18 \text{ kg/h}.$$

and $\text{BSFC} = (m_f/\text{BP}) = 18/70$ ...(3)

$$= 0.257 \text{ kg/kWh}.$$

2.4.5 Gas Turbine

Turbine, in which hot air or burnt gases are expanded to obtain motive power, is called a gas turbine.

Gas turbine, as a prime mover, finds wide application in aircrafts. It is also used in railways and water transport system. In recent years, it has made its way into the field of power generation as well.

In a power plant, gas turbine-generator is used generally as a topping plant. The exhaust of gas turbine is then employed in a steam generator to produce steam, for either power generation/process steam/combined use. When steam is used for power generation, the gas turbine plant in combination with steam power plant is called a combined cycle power plant and has much higher efficiency than what is obtainable in individual plants.

Figure 2.50 shows the schematics of a simple gas turbine plant. The gas turbine portion of the plant consists of (*i*) Gas turbine, (*ii*) Rotary compressor driven by the turbine, (*iii*) Combustion chamber and (*iv*) Fuel injection system. The rotary compressor continuously draws in atmospheric air, compresses it to 3–4 atm. and sends it to the combustion chamber. The fuel is also injected into the combustion chamber continuously. The fuel once ignited continues to burn adding heat to the gases. The hot gases then enter the turbine, where they expand doing work. A part of the power developed by the turbine is used up by the compressor and the rest is available as useful power.

Gas turbine systems: Gas turbine systems are of two types, (*i*) Constant volume system and (*ii*) Constant pressure system.

Despite higher theoretical efficiency attainable in a constant volume system, the constant pressure system is preferred because of its mechanical simplicity. Constant volume system requires the introduction of valves to separate combustion chamber from the compressor and the turbine and is marked by intermittent combustion. On the other hand, in a constant pressure system, the combustion is continuous and this helps smooth operation of the turbine. Figure 2.50 depicts such a system.

Open and Closed Cycle Systems: The system examined earlier (Fig. 2.49) is of the open cycle type. This means, the working fluid — atmospheric air, after doing work in the system, goes out with products of combustion back to atmosphere. In a closed system, however, (Ref. Fig. 2.50) the working fluid is confined to the system and is recirculated. The heat is added to the fluid earlier to its entering the turbine using the heat exchanger and adding heat by external combustion of solid, liquid or gaseous fuel. After the fluid leaves the turbine, it is cooled in a cooler (another heat exchanger) before it is sent back to the compressor.

Advantages and Disadvantages of Closed Cycle System

Advantages

1. Possibility of using high pressure through out the cycle, reduces the size of plant for a given output.
2. Varying pressure level in the circuit enables accommodating wide range of loads, without alteration of the maximum cycle temperature and hence with little variation in the overall efficiency.

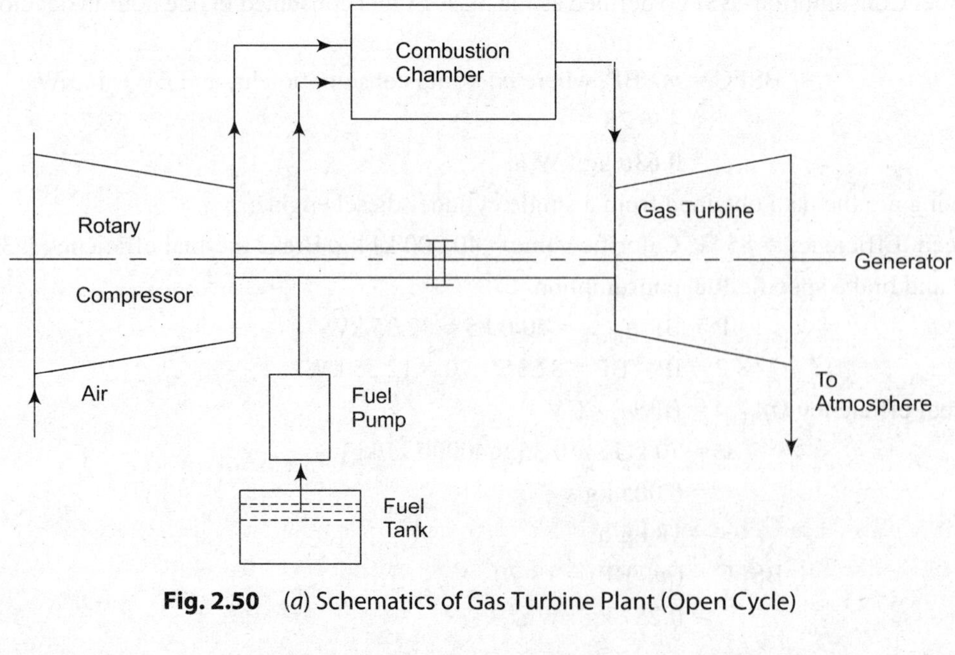

Fig. 2.50 (*a*) Schematics of Gas Turbine Plant (Open Cycle)

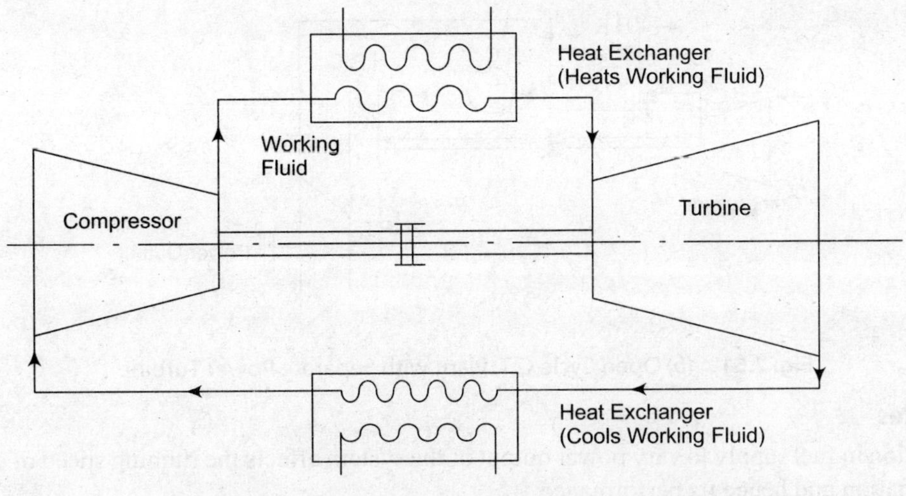

Fig. 2.50 *(b)* Schematics of Gas Turbine (Closed Cycle)

3. Wear and tear of the machinery is very much reduced, especially the erosion of turbine blades.
4. Higher density of working fluid improves heat transfer.
5. Working fluid could be other than air, having more desirable thermal properties.

Disadvantages

1. Auxiliary systems are necessary for heating and cooling of the working fluid and hence the initial cost goes up.
2. Maximum operating temperature is limited, which affects the overall efficiency.

Improving efficiency and power output of gas turbine: Some of the gas turbine architecture are given below. Students are advised to look for other possible arrangements for efficiency, economy and meeting specific requirements.

(a) Open cycle single shaft arrangement [Ref. Fig. 2.51(*a*)].

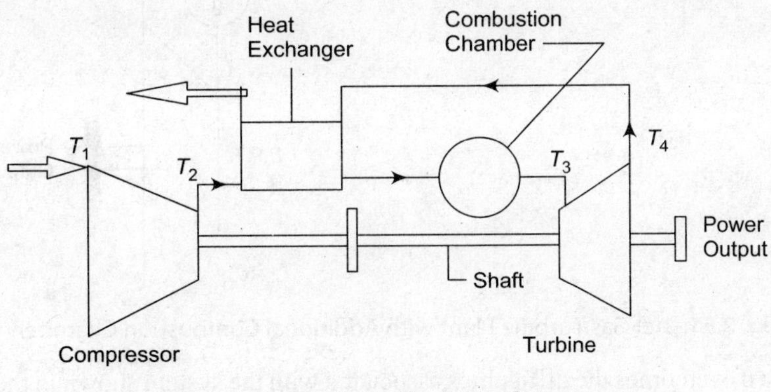

Fig. 2.51 *(a)* Open Cycle Single Shaft Arrangement

When the turbine outlet temperature is greater than the outlet temperature of compressor ($T_4 > T_2$) the exhaust of the turbine can be used to heat the compressed air. This reduces the fuel consumption, which in turn enhances the overall efficiency. It should, however be noted that there would be a definite pressure drop in the working fluid as it passes through the heat exchanger and this in turn would affect the efficiency to some extent.

This system is suited only for fixed load conditions, such as for peak load power generation.

(b) Open cycle gas turbine with separate power turbine [Ref. Fig. 2.51(*b*)].

This arrangement is helpful where flexibility in operation is essential as in road, rail and marine applications.

Here the high pressure (H.P.) turbine and the compressor together act as a gas generator for the low pressure (L.P.) turbine. The L.P.T. is mechanically independent, which is easy to control. The power output is controlled by controlling fuel supply.

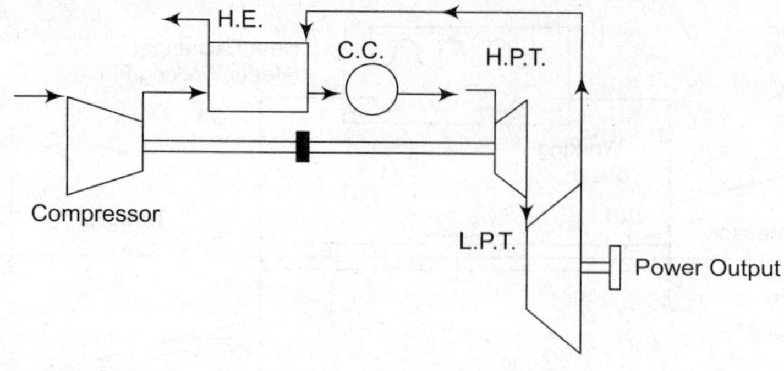

Fig. 2.51 (*b*) Open Cycle G.T. Plant with Separate Power Turbine

Disadvantages

1. Alteration in fuel supply to vary power output of the system affects the running speed of compressor-turbine combination and hence its performance.
2. Also at part load, reduction of fuel lowers maximum temperature of the cycle affecting the overall efficiency.

(*c*) Gas turbine plant with additional combustion chamber [Ref. Fig. 2.51(*c*)].

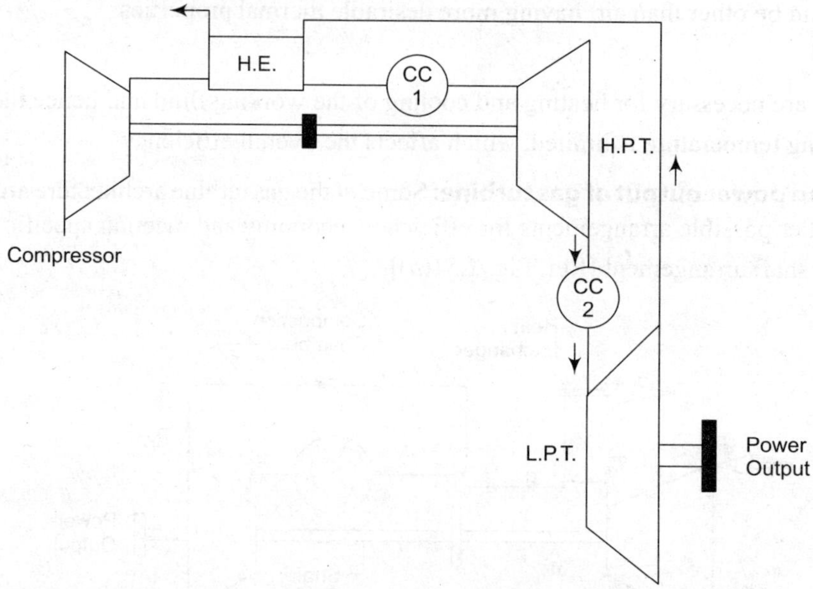

Fig. 2.51 (*c*) Gas Turbine Plant with Additional Combustion Chamber

The arrangement shown overcomes the difficulties associated with the system shown in the earlier case [Fig. 2.51(*b*)] by providing an additional combustion chamber prior to L.P.T. Therefore, the control for varying power output is now effected by adjusting fuel supply in the second combustion chamber, while maintaining constant running condition in the compressor —turbine gas generator.

The variations shown in the above configurations are all directed not only to improve the efficiency and power output, but also to meet specific requirements such as constant load/variable load conditions.

2.5 REFRIGERATION AND AIR CONDITIONING

2.5.1 Refrigeration

In the preceding section, we have studied about I.C. engines. These are heat engines, which produce mechanical work by using heat in the fuel. In refrigeration, the process may be considered reversed. Figure 2.51 shows the block diagrams of a heat engine and a refrigeration system. In the heat engine, part of the heat put into the system is **converted** to mechanical work [Fig. 2.52(*a*)]. In refrigeration, however, the mechanical work done on the system helps to **transfer** a certain amount of heat from the cold reservoir to hot reservoir [Ref. Fig. 2.52(*b*)].

Refrigerators, where refrigeration (or simply cooling) takes place, are largely used to preserve perishable food and increase its shelf life. So is the case with some of the pharmaceutical products, which may deteriorate if not refrigerated. Refrigeration is used in drinking water coolers, cold storage and low temperature applications such as in liquefaction of air. The other important application of refrigeration is in air-conditioning. Airconditioners are used in rooms to make living more comfortable. Reverse refrigeration or heating is employed to heat room during winter.

Refrigeration systems:

Following are the most widely used :

Vapour Compression Refrigeration System and

Vapour Absorption Refrigeration System.

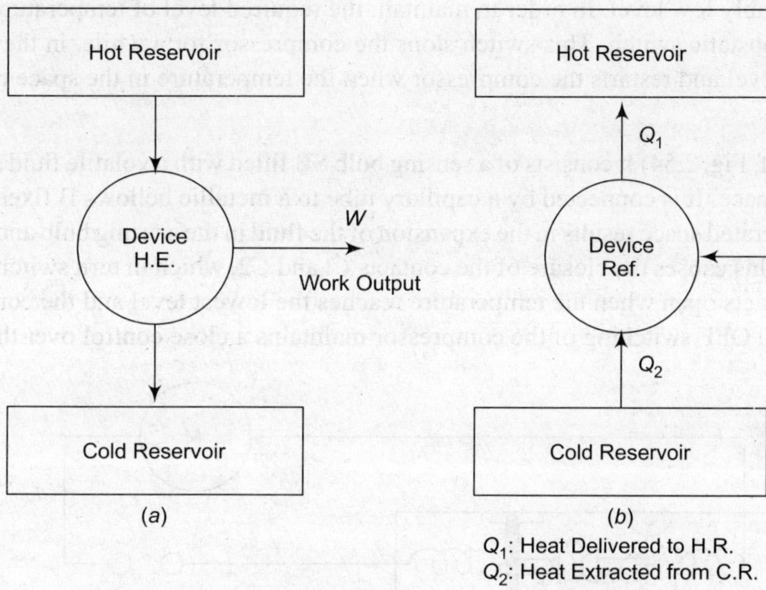

Fig. 2.52 Block Diagram of (*a*) Heat Engine and (*b*) Refrigeration Systems

2.5.2 Vapour Compression System (VCS)

In this system (Ref. Fig. 2.53) the working fluid, a refrigerant, transfers heat from the space surrounding the evaporator (REFRIGERATED SPACE) to the cooling medium circulated through the condenser.

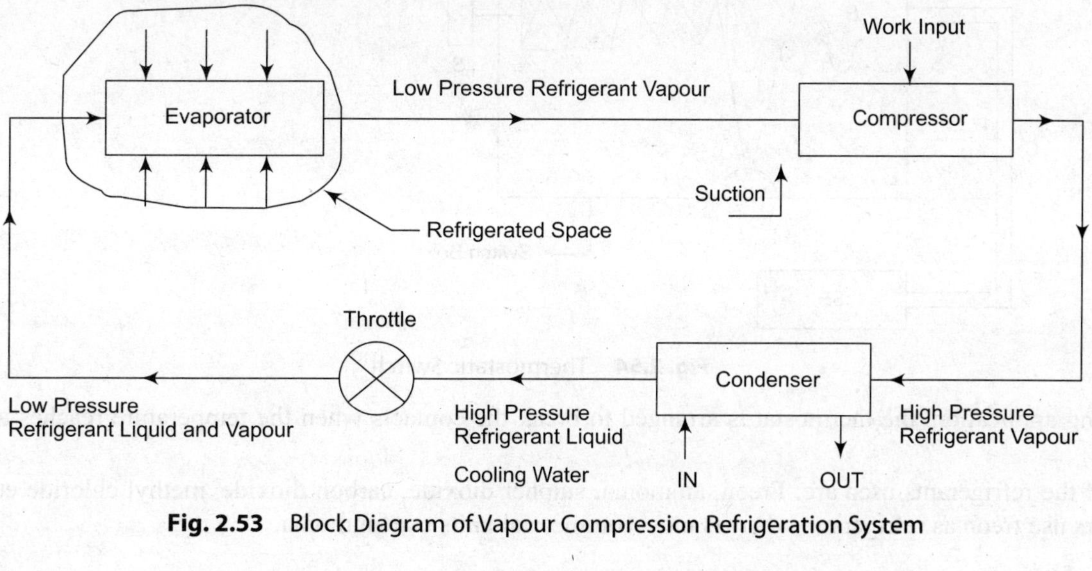

Fig. 2.53 Block Diagram of Vapour Compression Refrigeration System

Working of the vapour compression refrigerator: Referring to Fig. 2.53, the low pressure vapourised refrigerant is sucked by the compressor and is delivered as high pressure and temperature vapour. This compressed vapour enters the condenser and gets cooled to assume a liquid state. Here the latent heat of vapourization of the refrigerant is taken away by the cooling water. (The cooling may even sub-cool the refrigerant some times.) The high pressure liquid refrigerant from the condenser then passes through a throttle valve. Throttling of the liquid refrigerant brings about a large reduction in pressure and temperature of the refrigerant converting it into vapour and liquid droplets. The liquid droplets vapourize in the evaporator taking the heat from the surrounding space. Thus the refrigeration of space around the evaporator takes place. The fully vapourized low pressure refrigerant then enters the compressor to continue the refrigeration cycle.

In domestic refrigerators, air cooling of refrigerant is arranged and thus the heat is transferred from the refrigerated space to the atmosphere, by the air passing over the condenser coil.

Temperature control: In a situation where the operation of the refrigeration cycle goes on without break, the temperature of the space may reach undesirably low level. In order to maintain the required level of temperature, one of the methods employed is to provide a thermostatic switch. This switch stops the compressor for a while, in the event of temperature reaching the lowest desirable level and restarts the compressor when the temperature in the space reaches the maximum desirable temperature.

The thermostatic switch: (Ref. Fig. 2.54) It consists of a sensing bulb SB filled with a volatile fluid such as the refrigerant and is kept in the refrigerated space. It is connected by a capillary tube to a metallic bellows B fixed in the switch box. A higher temperature in the refrigerated space results in the expansion of the fluid in the sensing bulb and in turn in the bellows pushing the lever to the right. This causes the closure of the contacts Cl and C2, which in turn switches on the compressor motor. The other way, the contacts open when the temperature reaches the lowest level and the compressor motor is put off. Thus the automatic ON and OFF switching of the compressor maintains a close control over the temperature.

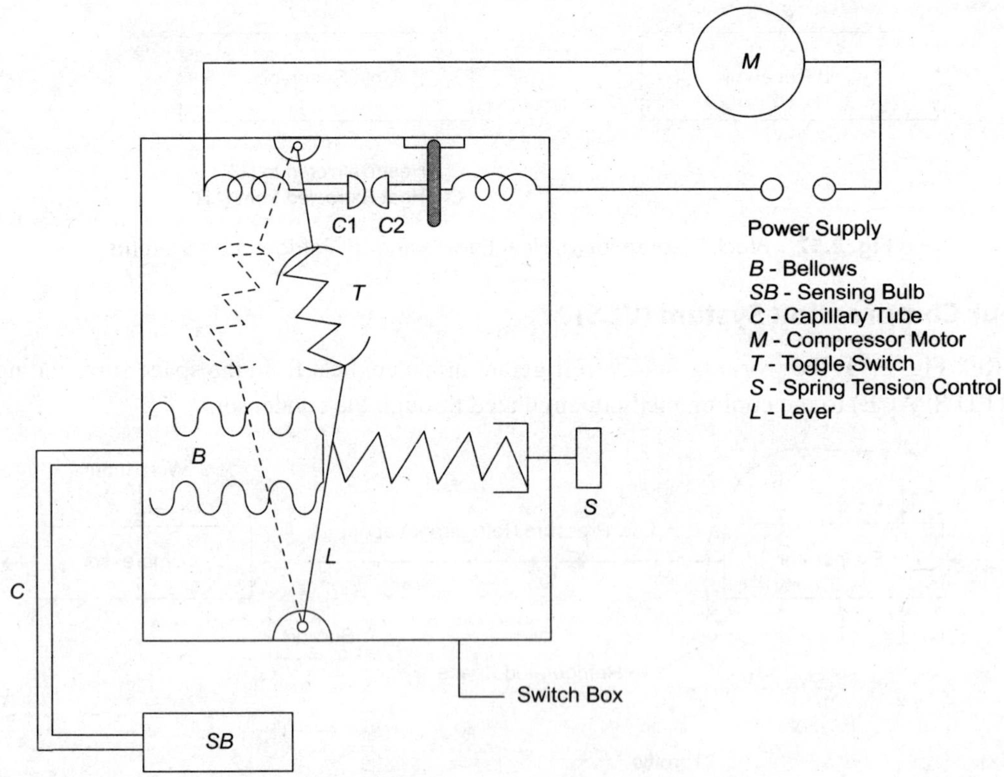

Fig. 2.54 Thermostatic Switch

In heating application, the thermostat is arranged to break the contacts when the temperature reaches a preset high value.

Some of the refrigerants used are: Freon, ammonia, sulpher dioxide, carbon dioxide, methyl chloride etc. Domestic refrigerators use freon as refrigerant, whereas most commercial units use ammonia.

Selection criteria for refrigerants include toxicity, corrosiveness, inflammability etc., besides favourable physical parameters.

Coefficient of Performance (COP): It is an index of the performance of a Refrigerator. It is defined as the ratio of

$$\frac{\text{Refrigeration provided (heat taken out) per cycle}}{\text{Work input to compressor per cycle}}$$

Problem: A refrigerator plant is of one ton capacity. It is driven by a motor of 0.5 kW. Find the COP of the plant considering the system to be 100% efficient.

Solution: Capacity of a refrigerator is given in terms of tons of refrigeration. Originally, 1 ton of refrigeration meant the production of 1 ton of ice in 24 hours. Now it has been standardized to mean a rate of removal of heat = 210 kJ/minute.

$$\text{COP} = \frac{210 \text{ kJ/min}}{0.5/0.0167} = 7.0 \text{ (1 kJ/min} = 0.0167 \text{ kW)}$$

2.5.3 Vapour Absorption System (VAS)

In this system, there is no compressor. The suction effect is created by using an absorbent which has affinity for a particular refrigerant. The solution of refrigerant in the absorbent is pumped into a generator where it is heated using steam or hot water. In the alternative a gas/kerosene burner is used to heat the solution. The refrigerant vapour separated out from the absorbent is at a comparatively higher pressure and this when cooled, is converted into high pressure liquid refrigerant. The other processes like throttling the liquid refrigerant to reduce its pressure to the level in the evaporator and also evaporating the low pressure liquid using the heat surrounding the evaporator, are identical with vapour compression system. Therefore, the schematics (Fig. 2.55) may be used to explain the vapour absorption system. In the absorber, heat is liberated as the absorption of refrigerant takes place. This heat is called heat of absorption. Also the refrigerant vapour condenses in the absorbent releasing the latent heat of vapourization. In order to remove this heat, cooling is arranged in the absorber also. In the setup, the cooling water is first passed through the absorber and then through the condenser. (*a-b-c-d*).

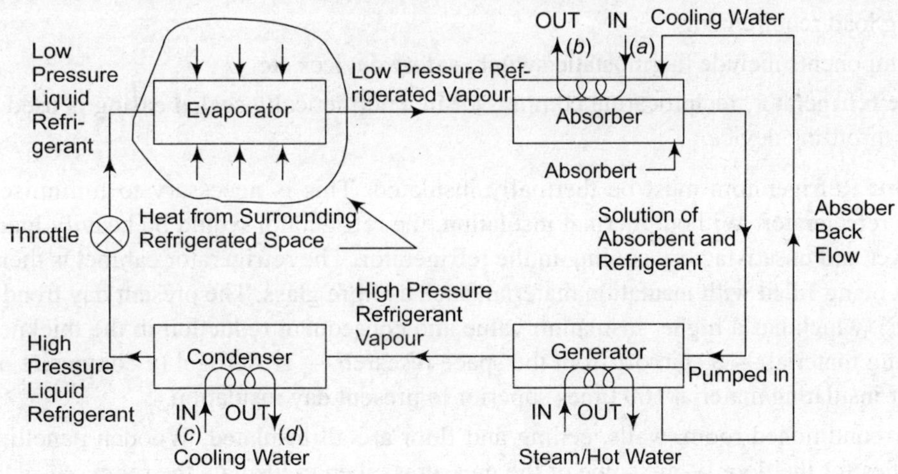

Fig. 2.55 Schematics of Vapour Absorption Refrigeration System

Types of vapour absorption system

1. Ammonia absorption system
2. Lithium bromide absorption system
3. Electrolux

Ammonia absorption system: Uses ammonia as refrigerant and water as absorbent, since water has great affinity for NH_3 and can therefore dissolve large amount of NH_3. Both air cooled and water cooled versions have been developed air-cooled system is used for domestic refrigeration.

Lithium bromide absorption system: Here lithium bromide salt solution is used as absorbent and water as refrigerant. Since water freezes at 0°C, the evaporator operating temperature has to be more than 0°C.

Electrolux: This is again an ammonia absorption system but with certain differences. It is unique because of the absence of any mechanical moving parts. It also does away with thermostat. The circulation of the refrigerant in the system is maintained by a heating source rather than using a pump to do the job. The heating is done by an electrical heater or gas/ kerosene burner. The whole system is sealed and the refrigeration process goes on uninterrupted as long as the heat is supplied to the system.

The reliability and maintainability of the system is almost 100%. However, the COP of the system is very small compared to other systems of refrigeration and this perhaps has hindered its popularity. For further information on this topic students may refer to 'Thermodynamics for Engineers' by Crawford et al., or any other book on refrigeration.

2.5.4 Components of a Refrigeration System

1. Compressor, either reciprocating or rotary
2. Condenser:
   (*a*) Air cooled has refrigerant passing through finned tubes. Air is blown over the tubes using a fan/blower to take away the heat.
   (*b*) Water cooled : Shell and tube type is used, refrigerant flowing through the tubes surrounded by cooling water.
3. Evaporator:
   (*a*) Shell and tube type, in which vapourizing refrigerant flows through the tubes surrounded by brine (salt solution). The chilled brine is circulated in a pipe kept in the refrigerated space.
   (*b*) Finned tube type over which air to be conditioned passes.
   (*c*) Zig-zag tube kept directly in a refrigerator.
4. Throttling device: A needle valve may serve the purpose. Alternatively, a capillary tube offering resistance to the flow of liquid refrigerant, acts as a throttling device. Also sometimes, automatic expansion valves are used to cater to the varying load requirements.

   The other components include thermostatic switch, safety devices etc.

   In a domestic refrigerator, reciprocating compressor in a hermetically sealed casing is used with a capillary tube acting as the throttling device.

Thermal Insulation: Refrigerators must be thermally insulated. This is necessary to minimise heat flow from the atmosphere into the refrigerator. Without thermal insulation, the refrigerator would be heavily loaded resulting in large consumption of power and unsatisfactory cooling in the refrigerator. The refrigerator cabinet is therefore double-walled, the space inbetween being filled with insulation material, such as fibre glass. The present day trend has been to use poly urathene foam (PUF), which has a higher insulation value and consequent reduction in the thickness of insulation. The more recent insulating material — a spin off from the space research — is Aerogel (a composite of silica, alumina and carbon) — a wonder insulating material 100 times superior to present day insulation.

In the case of air-conditioned room, walls, ceiling and floor are all insulated. Wooden panelling of the walls, false ceiling and plastic tiles for the floor — are some of the measures taken to insulate the room.

2.5.5 VAS vs VCS

	VAS	*VCS*
Energy	Operated by putting heat into system; uses steam/hot water or gas/kerosene flame	Operated by putting mechanical work into system. Compressor (reciprocating or rotary type used)
Reliability	More	Major overhauls required
Working Pressure	Low	High
Power requirement	Less; uses small C.F. pumps	More; uses heavy compressor.
COP	1.1 approx.	4 to 5
Heat rejection	More, hence cooling tower and pump cost more	Less

2.5.6 Air Conditioning

Necessity: Air conditioning is necessary for

1. Human comfort and 2. Controlled environment for certain activities.

1. *Human comfort:* Vagaries of nature are such that the outside temperatures are at extremes in most parts of the globe. Similarly the humidity level also varies from very dry to very humid spells. Comfort temperature is around 20°C and the humidity at about 60 R H. In our own country, during summer, some parts register very high temperatures in the order of 45°C and more. On the other hand in J&K, Delhi and some parts in North India temperatures dip to below 0°C. Either way, living in these extremes, becomes very uncomfortable and difficult. With an all round development in the economy, at least people in the upper strata can now afford A.C. units to maintain comfort temperature and humidity. Now-a-days many of the cinema theatres, conference halls, hotels etc. are all air-conditioned.
2. *Controlled environment:* We have certain areas like Operation theatres, ICUs, Clean rooms for assemblies of precision components and production of miniature items as in horology, electronics industry and now in the field of computers involving I.C. and VLSI chip manufacturing, space research etc., where the atmosphere must be kept free of dust and air conditioned. Also controlled environment is a must for precision measurement as in metrology to ensure accuracy of measurement.

From the above discussion, we see that air-conditioning involves the following tasks:

(*a*) Heating or cooling the area of activity to maintain comfort temperature
(*b*) Maintaining comfort humidity level
(*c*) Controlling dust to keep clean atmosphere and
(*d*) Keeping air movement at 5 to 7 m/min. (To provide soothing breeze and maintain uniform distribution of temperature.)

2.5.7 Types of A.C. Units

1. Window type (also called room A.C.) mostly for living rooms
2. Commercial A.C. units catering to industrial and large scale applications. Here a central A.C. plant delivers conditioned air to be circulated in the work areas through a system of ducts.

   We shall examine here a window type A.C. unit only.

2.5.8 Window Type/Room Air Conditioner

Figure 2.56 shows the schematics of a window type air conditioner. A common motor *M* drives a fan at one end and a blower at the other end. Refrigerant vapour at high pressure and temperature, coming from the compressor at *A* enters the condenser and gets cooled. Cooling changes the state of the refrigerant and the refrigerant leaves the condenser in a liquid state at *B*. Drier filter removes any trace of moisture from the liquid refrigerant. The refrigerant then passes through a capillary tube (acting as a throttling device) which reduces the pressure of the refrigerant to the level prevailing in the evaporator. Air, drawn over the evaporator, loses its heat to the refrigerant vapourising in the evaporator (cooling coil). The air filter at the front helps to clean the air passing over the evaporator and the cooled air is sent into the room. The low pressure refrigerant vapour is then sucked back at *C*. The cycle gets repeated as the low pressure vapour gets compressed to high pressure. The accumulator introduced before the compressor serves to capture any liquid refrigerant coming from the evaporator, which otherwise, may flood the compressor.

All the components – compressor, condenser, evaporator, motor running the fan and the blower, capillary tube etc., – are assembled inside a casing suitable for installation into a wall or window opening and thus the name window type.

The refrigerants commonly used are:

R-12 (CCl_2F_2 – Dichloro difluoro methane) and

R-22 ($CHClF_2$ – Monochloro difluro methane).

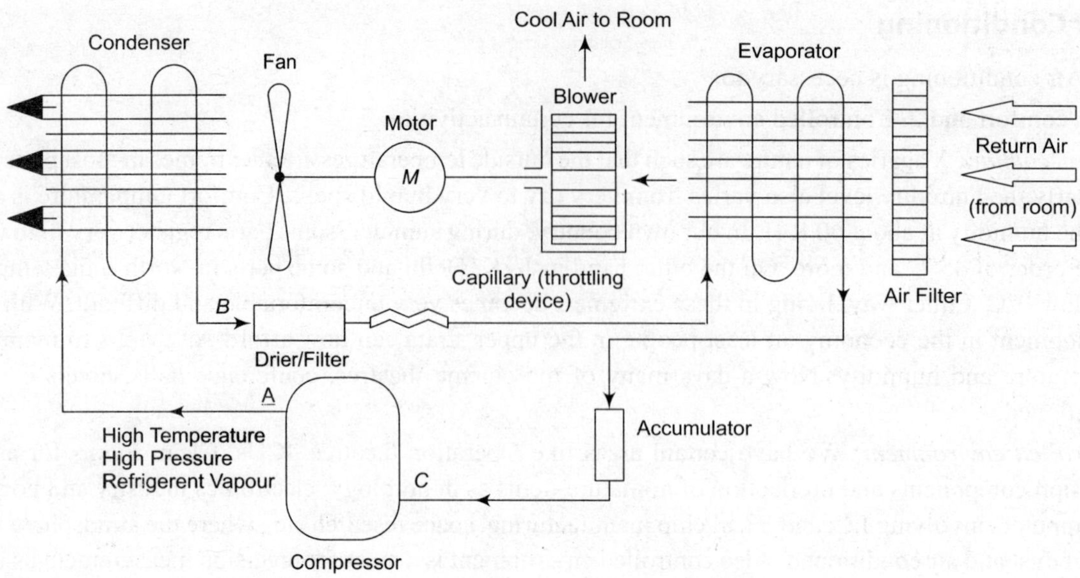

Fig. 2.56 Schematics of Window Type Air-Conditioner

The desired temperature in the room is maintained using a thermostatic switch, whose sensing bulb is kept in the return air passage. Switching on or off of the compressor is controlled by the thermostat according to the return air temperature. Figure 2.57 shows the arrangement of the various components of the window unit and its placement in the wall or window opening.

As explained earlier, the blower sucks the air from the room to be air-conditioned through the air filter and the evaporator and delivers the cool and dehumidified air back to the room. This air mixes with the room air and brings down the temperature and humidity levels in the room. Arrangement is made to admit a certain percentage of fresh air along with return air to cater to the physiological and psychological requirement of the people in the room. Damper *D* is used to regulate the fresh air. Usually there is no separate arrangement to control the humidity level in the room. Excess moisture, if present, condenses on the evaporator coil, since the return air—fresh air mixture is cooled below its dew point while passing over the evaporator coil. Thus the dehumidification takes place automatically and a certain level of R.H. is maintained in the room.

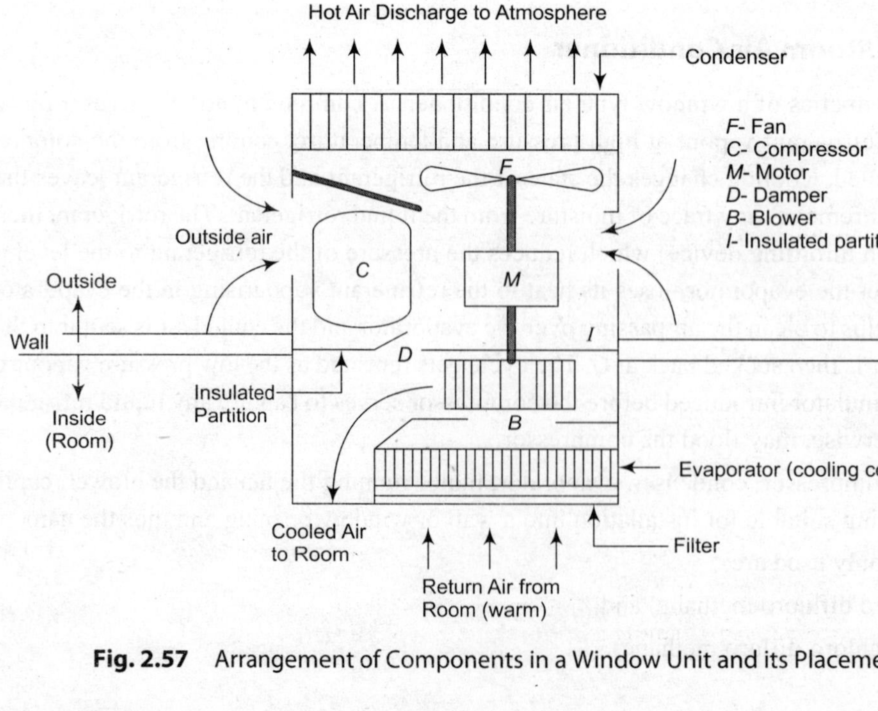

Fig. 2.57 Arrangement of Components in a Window Unit and its Placement

The window units are designed to operate on 230 V single phase supply and have ratings ranging from 0.5 to 3 ton capacity.

2.5.9 Room Heating

It is not that cooling of the room is always necessary. When the atmospheric temperature goes below the comfort temperature as in winter, we require room heating. For this, the same window unit can be used by incorporating a reversing valve (Ref. Fig. 2.58). The reversing valve has four ports and two – position valve. Figure 2.58(*a*) shows the position of the valve for cooling the room and Fig. 2.58(*b*) shows its position for heating the room. As is seen from Fig. 2.58(*b*), the discharge from the compressor is lead in to the evaporator, which now functions as a condenser. Here the refrigerant condenses losing its heat to the air in the room. This refrigerant is then throttled through the capillary and is sent to the condenser, which now acts as an evaporator. Here the heat is taken from the atmosphere and the vapourization of the refrigerant takes place. The vapourized refrigerant is then sucked back into the compressor and the cycle repeats.

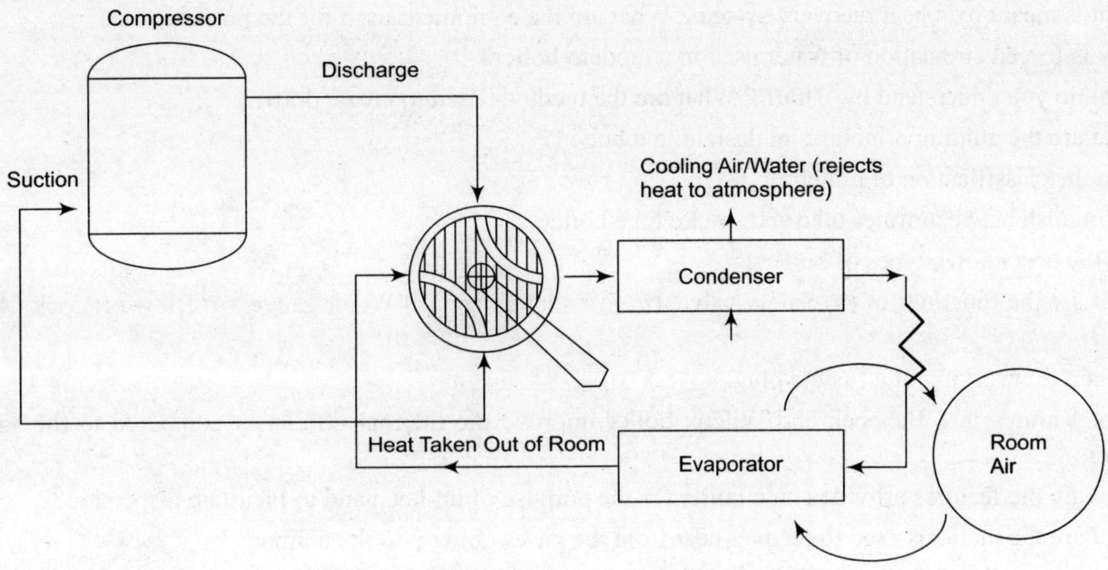

Fig. 2.58 (*a*) Position of Valve for Cooling the Room

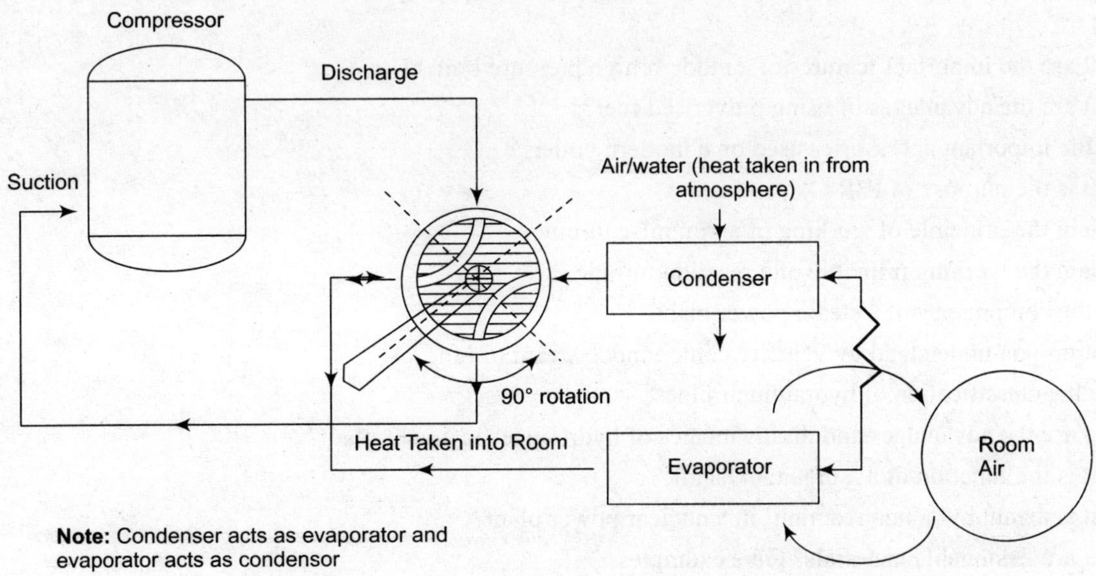

Fig. 2.58 (*b*) Position of Valve for Heating the Room

The reversing valve can be operated manually or automatically using an electric solenoid controlled by a thermostat. The advantage of using the same cooling system for heating is that the energy consumed for heating is much less compared to direct electric heating.

TOPIC: UTILIZATION OF CONVENTIONAL ENERGY

QUESTION BANK

2-Marks Questions

1. What is a power plant? Name the conventional power plants.
2. Name the sub-systems ot a steam power plant.
3. What are the components of a fuel system in a P.F. fired steam power plant?
4. What are the modes by which heat is transferred from the hot gases to the water and steam in a boiler?
5. What is meant by a heat recovery system? What are the equipment used for the purpose?
6. Why is forced circulation of water used in a modern boiler?
7. What do you understand by 'Draft'? What are the methods used to create draft?
8. What are the guiding principles in designing a boiler?
9. Give the classification of boilers.
10. Distinguish between water tube and smoke tube boilers.
11. List the best known types of boilers.
12. What are the functions of (*a*) Safety valve, (*b*) Pressure gauge, (*c*) Water gauge, (*d*) Blow off cock (Answer any Two).
13. Explain the working of a fusible plug.
14. What features in a Babcock and Wilcox boiler improve the thermal efficiency compared to the smokes tube boilers?
15. What are the features provided on a boiler for the purpose of up-keep and to facilitate inspection?
16. What are the methods used to recover heat from the gases going into the chimney?
17. What is super heated steam? What are its advantages?
18. What are the functions of (*a*) Anti-priming pipe, (*b*) Damper, (*c*) Stoker, (*d*) Feed check valve. (Answer any two).
19. What are the important features of a modern high pressure boiler?
20. What are the advantages of using pulverised fuel?
21. List the important accessories used on a modern boiler.
22. What is the purpose of IBR?
23. Explain the principle of working of an impulse turbine.
24. Explain the working principle of a reaction turbine.
25. List the components of a steam power plant.
26. What do you understand by velocity compounded steam turbine?
27. Give the classification of hydraulic turbines.
28. What are the advantages and disadvantages of hydro-electric power plant?
29. What is the function of a 'surge tank/shaft'?
30. What is meant by 'chain reaction' in a nuclear power plant?
31. What are fissionable materials? Give examples.
32. What are fertile materials? Give examples.

33. What are the major components of a nuclear reactor?
34. Explain the terms: (*a*) moderator and (*b*) reflector. Give examples.
35. What are the materials used for (*a*) Coolant and (*b*) Control rods.
36. How is the fission process regulated in a nuclear plant?
37. List the various coolants used in a nuclear reactor.
38. What is meant by 'Biological shield'? How does it work?
39. Give the classification of nuclear power plant.
40. What is meant by 'breeder reactor'?
41. Give the advantages and disadvantages of a nuclear power plant.
42. What is nuclear fusion?
43. What are the advantages of obtaining energy by nuclear fusion?
44. What are the factors to be looked into for a sustained energy production by nuclear fusion?
45. What are the methods by which nuclear waste is disposed off?
46. How do you classify internal combustion engines?
47. List the systems which constitute an I.C. engine.
48. Name the fuels used in an I.C. engine.
49. Differentiate between the fuel system in a diesel engine with that in a petrol engine.
50. Explain the need for lubrication in engines.
51. What are the parts requiring lubrication in an I.C. engine?
52. What is 'splash lubrication'?
53. What is petroil lubrication?
54. Name the methods of cooling I.C. engines. Why cooling is necessary?
55. What is meant by atomization of fuel oil? How is it accomplished?
56. What is the purpose of a radiator in an I.C engine?
57. What are the components of engine cooling system?
58. What are the methods used to control the quantity of fuel admitted into a diesel engine?
59. Bring out the differences in initiating combustion in petrol and diesel engine.
60. What is ignition delay?
61. List the various applications of an I.C. engine.
62. What is a carburettor?
63. What is meant by combined cycle power plant?
64. What are the uses of gas turbine?
65. What is meant by 'open cycle gas turbine system'?

10-Marks Questions

1. Make a neat layout of a steam power plant and briefly explain its features.
2. Explain the working of a Cochran boiler with a neat diagram.
3. Make a neat sketch of a Lancashire boiler and name its parts. Explain how the heat transfer takes place.
4. Explain the working of a locomotive boiler with the help of a neat diagram.
5. With the help of a neat sketch explain the working of Babcox and Wilcox boiler.
6. Draw the schematics of a LaMont boiler and explain its functioning.
7. Make a neat schematics of Benson boiler and explain its working.

8. Show schematically a modern boiler using pulverised fuel and explain its working with reference to heat transfer, heat recovery and combustion process.
9. Explain with neat sketches the working of (*a*) Spring loaded safety valve, (*b*) Bourdon pressure gauge.
10. Name the various boiler mountings and accessories. Explain any one in each category.
11. Draw the schematics of a velocity compounded impulse turbine and explain its working.
12. Make a schematics of a reaction turbine and explain its working.
13. Make a diagrammatic sketch of a hydro-electric power plant using pelton wheel. Explain the arrangement briefly.
14. Show with the help of a neat diagram the arrangement of a hydro-electric power plant using a Kaplan turbine. Explain how the energy in water is transformed to mechanical energy available at the turbine shaft.
15. Make a neat sketch of any one turbine and explain its working (*a*) Pelton wheel, (*b*) Kaplan turbine, (*c*) Francis turbine.
16. Make a neat sketch of a nuclear reactor and explain the function of each of the major component.
17. Distinguish between fission and fusion processes. Explain how energy is released in fission process.
18. How are nuclear reactors classified?
19. Discuss the merits and demerits of nuclear power plants.
20. How does a boiling water reactor work? Make a neat sketch and explain.
21. Explain fusion process and show how energy is released using Einstein's equation.
22. Make a neat sketch of the layout of a diesel engine power plant and explain the salient features.
23. Describe the working of a 4-stroke diesel engine with the help of neat sketches.
24. Compare 2-stroke and 4-stroke engines.
25. Describe the working of a 2-stroke diesel engine. Make neat sketches and explain.
26. Explain with the help of neat sketches the working of a 4-stroke petrol engine.
27. Explain the working of a simple carburettor with a neat diagram.
28. Bring out the differences between petrol and diesel engines.
29. Make a sketch of the ignition system for a 4-stroke petrol engine and explain.
30. Explain the working of a simple open cycle gas turbine plant.

State True or False

1. Kaplan turbines is used in a high head hydro-plant.
2. Cooling towers are used to cool the feed water.
3. Condensers are meant only for condensing exhaust steam for reuse of feed water.
4. Reaction turbines are used in large steam power plants.
5. Steam plants in India are mostly coal-based.
6. Small boilers are generally of smoke tube type.
7. U_{233} is a fertile material.
8. Breeder reactor can produce more fuel than it consumes.
9. Gas turbine is more efficient than diesel engine.
10. Heavy water is used both as moderator and coolant.
11. Complete pressure drop occurs in nozzles in the case of impulse turbine.
12. Expansion occurs in the moving blades of impulse turbine.
13. Pressure drop occurs in the moving blades in the case of reaction turbines.
14. Co-generation helps both electrical power generation and production of process steam.

15. Two stroke engines have fewer components.
16. Two stroke petrol engine is more efficient than four stroke engine.
17. For the same bore, stroke and speed, 2-stroke engine may produce almost double the power of similar 4-stroke engines.
18. Diesel engines are less efficient compared to petrol engines.
19. Diesel engines are heavier than petrol engines.
20. Diesel engine is smoother in operation and has less vibration.
21. Locomotives use petrol engines.
22. Ships use diesel engines.
23. Small air-crafts uses petrol engines.
24. Petrol engines are governed by varying the quantity of fuel-air mixture and not air-fuel ratio.
25. Diesel engines are controlled by varying the amount of fuel injected.
26. Spark should occur after the end of compression.
27. Inlet valve should close at the end of suction stroke.
28. Exhaust valve should open after the end of expansion/power stroke.
29. Inlet port should open after the exhaust port for effective scavenging of exhaust gases.
30. Air cooling is suitable for large engines.
31. Radiator is suitable for medium size engines.
32. Two stroke engine cylinders of small size are lubricated by mixing oil with fuel.

Answers

1. F	**2.** F	**3.** F	**4.** T	**5.** T	**6.** T	**7.** F	**8.** T	**9.** F	**10.** T
11. T	**12.** F	**13.** T	**14.** T	**15.** T	**16.** F	**17.** T	**18.** F	**19.** T	**20.** F
21. F	**22.** T	**23.** T	**24.** T	**25.** T	**26.** F	**27.** F	**28.** F	**29.** T	**30.** F
31. T	**32.** T.								

Fill up the Blanks

1. In steam power plants coal is burnt in .... form.
2. In high head hydro-electric plants the type of turbine used is ....
3. In medium head hydro-electric plant .... is used.
4. In low head hydro-electric plant the type of turbine used is ....
5. Condensers are used (1) to improve.............. (2) to conserve ....
6. Boiler feed water should be treated to prevent .... and ....
7. In the fire/smoke tube boilers flue gases pass .... the tubes.
8. In water tube boilers flue gases pass .... the tubes.
9. Fire tube boilers are used mainly for ....
10. Water tube boilers are used mainly for ....
11. Dust is collected in steam power plants using ....
12. Grate is used (1) to support .... (2) to provide .... air.
13. Pollutants emanating from thermal power plants are ...., ..., ..., ...
14. Sulphur dioxide is removed from gases by ....
15. Chimney is used in small plants to produce ....
16. Chimney in large plants is used to disperse .... over a wider area.

17. Slowing of neutrons is done using ....
18. The fission rate is controlled by ....
19. Nuclear fuels are ...., ...
20. Fertile materials are .... , ...
21. Materials used as moderators are ...., ..., ..., ...
22. Materials used as coolants are ...., ..., ..., ...
23. Materials used for fuel rods are ....
24. Pressure drop in the case of impulse turbine takes place in ....
25. Pressure drop in reaction turbine takes place over ...., ...
26. Gas turbines are .... efficient compared to diesel engines.
27. Co-generation is the production of ...., ...
28. Kaplan turbine is .... turbine.
29. Francis turbine is .... turbine.

I.C. Engine:

30. For two revolution of a 4-stroke engine .... power stroke is available.
31. In a 2-stroke engine one power stroke is available for .... rev.
32. Both valves should be .... during compression and expansion.
33. In a 2-stroke engine the inlet port should be uncovered .... the exhaust port is uncovered.
34. In a petrol engine spark should occur .... TDC.
35. Fuel injection should start .... TDC in a diesel engine.
36. The voltage in the secondary of .... at the time of spark is around .... volts.
37. Condenser prevents .... of breaker points.
38. Cam shaft in a 4-stroke engines rotates at .... the speed of crank.
39. The inlet valve should open .... TDC and close .... BDC.
40. The exhaust valve should open BDC and close TDC.
41. Fins are provided on the cylinder and cylinder head in the case of engines.
42. Cooling of cylinder is required for preventing damage to cylinder by very high temperature and maintaining proper ....
43. To maintain optimum temperature in the engine cylinder .... is used.
44. Atomization of fuel in the engine is done using ....
45. Pressure required for proper atomization is of the order of .... atm.
46. For cold starting of a petrol engine a .... is operated to give a .... mixture.
47. Fuel pump and injector are used on .... engines.
48. Carburettor is used on .... engines.
49. Spark ignition is used in a .... engine.
50. Compression ignition is used in a .... engine.
51. Calculate the brake power output of a single cylinder four stroke petrol engine, which runs at a speed of 400 rpm. The load on the brake drum is 24 kg and the spring balance reading is 4 kg. The diameter of the brake drum is 600 mm and the rope diameter is 30 mm.
52. A single cylinder 4-stroke I.C. engine develops 22 kW BP. IP = 26 kW. $N = 400$ rpm. Fuel consumption = 0.33 kg/BP.hr. CV = 44300 kJ/kg. Calculate (*i*) Mechanical efficiency, (*ii*) Indicated thermal efficiency and brake thermal efficiency.

Answers

1. Pulverised
2. Pelton wheel type impulse
3. Francis type-reaction
4. Kaplan type axial flow reaction turbine
5. (*a*) thermal efficiency (*b*) feed water
6. scaling and foaming
7. Inside
8. outside
9. Process
10. power generation
11. Electrostatic precipitators
12. (*a*) burning fuel, (*b*) to provide combustion air through its openings
13. dust, ash, sulpur dioxide and carbon dioxide
14. water spray
15. draft
16. pollutants
17. moderators
18. control rods
19. U_{235}, Pu_{239}, U_{233}
20. thorium and U_{238}
21. graphite, heavy water and beryllium
22. ordinary and heavy water, liquid metals like sodium and potassium, CO_2, helium, etc.
23. boron or cadmium
24. nozzles
25. moving and fixed blades
26. less
27. electric power and process steam
28. axial flow reaction
29. mixed flow reaction.

I.C. Engines:

30. one
31. each
33. closed
33. after
34. before
35. before
36. induction coil, 14000
37. pitting
38. half
39. before, after
40. before, after
41. air cooled
42. lubrication
43. thermostat
44. injector/nozzle
45. 200
46. choke, rich
47. diesel
48. petrol
49. petrol
50. diesel.
51. BP = 2.58 kW.
52. = 84%, = 29.2%, = 24.7%

TOPIC: REFRIGERATION AND AIR-CONDITIONING

QUESTION BANK

2-Marks Questions

1. What is a refrigerant?
2. Name any two commonly used refrigerants.
3. What is the principle of refrigeration?
4. What is the difference between heat engine cycle and a refrigeration cycle?
5. Name two important applications of refrigeration.
6. What is meant by air-conditioning?

7. Name the two important refrigeration systems.
8. What are the factors to be considered in selecting a refrigerant?
9. How is the temperature controlled in a refrigerator?
10. What is the function of a throttling device?
11. What are the two types of condensers generally used in a refrigeration system?
12. What is the function of an evaporator?
13. Define COP.
14. How is the capacity of a refrigeration plant specified?
15. What is the purpose of an absorbent in a VAS?
16. What factors are controlled in A.C.?
17. Name two industrial applications of A.C.
18. What is the function of an accumulator?
19. Why is a damper used in A.C.?
20. Why is a reversing valve introduced in an A.C. system?

10-Marks Questions

1. Explain with a neat sketch the working of a vapour compression refrigeration system.
2. Describe the vapour absorption refrigeration system with a neat sketch.
3. Compare and contrast vapour compression and vapour absorption refrigeration systems.
4. Make a neat sketch of a window type A.C. unit and explain how it functions.
5. Describe with a neat sketch the working of a thermostatic switch.

Fill in the blanks/ Say Yes/No/Tick the right answer

1. Vapour absorption system has a compressor. (Yes/No)
2. Thermostatic switch is a safety device. (Yes/No)
3. Capillary tube in a refrigeration system acts as ________ device.
4. Accumulator in a suction line is used to prevent flood back to compressor. (Yes/No)
5. Thermostatic expansion valve is used to control/throttle the flow of liquid refrigerant.
6. Fins on an evaporator help faster dissipation/extraction of heat from the surrounding.
7. CFCs affect earth's ozone layer. (Yes/No)
8. A human being is considered equivalent to a 100W bulb with reference to heat loss from the body. (Yes/No)
9. Pressure developed by a fan is expressed in mm of water/(kg/cm^2).
10. Cooling of absorber is done to take away ________.
11. In a lithium bromide absorption system ________ acts as absorbent and ________ as refrigerant.
12. Capacity of a refrigerator is expressed in terms of power of compressor in kW. (Yes/No)

Answers

1. No	2. No
3. Throttling	4. Yes
5. Control and Throttle	6. extraction
7. Yes	8. Yes
9. mm of water	10. heat of absorption
11. Lithium bromide water	12. No.

GLOSSARY (PART I)

Chapters I and II (Power Plants)

Aerofoil section: Section similar to that of the wings of an aeroplane.

Air preheater: A heat exchanger used to heat combustion air getting into the furnace/fire place. This takes the heat from the exhaust gases going into the chimney.

Anti-priming pipe: A piece of perforated pipe fixed at the exit of steam from the boiler, which while collecting steam sheds the water particles to some extent and thus reduces the moisture carry over.

Ash softening temperature: It is the temperature at which the ash becomes plastic. Coal with low ash softening temperature necessitates discharge of ash in a molten condition. For the design of ash handling equipment this is an important information.

Bio-gas: It is the gas obtained by digesting animal and vegetable wastes by anaerobic action. In India it is called Gobar gas. The gobar gas plants use largely the dung of animals.

Boiler efficiency: Also called the thermal efficiency of boiler, it indicates the percentage of heat transferred from the fuel to water/steam in the boiler.

Calorific value/Heating value of a fuel: Defined as the quantity of heat produced by complete combustion of unit quantity of fuel. (*i*) Higher heating value is defined as the gross heat released (say in kJ/kg) from a fuel. This includes the latent heat of vapourisation of moisture in the burnt gases when the fuel is burnt with dry air. (*ii*) Lower heating value is the net heat released when the heat used in vapourising the moisture is not considered.

Chimney: Chimney is a device to let the noxious gases and dust up into the atmosphere from where they get distributed over a wide area and thus the intensity of pollution is reduced. In a smaller boiler it also creates necessary draft (called natural draft). This draft is created by the pressure difference between the column of hot gases inside the chimney and a column of air outside the chimney.

Cogeneration: When a boiler is used to produce steam both for the purpose of producing power and also providing process steam the system is called cogeneration.

Combined cycle power plant: When a power plant uses both gas turbine as well as steam turbine for production of power it is called combined cycle power plant.

Compression ratio: It is the ratio of the volume at the beginning of compression stroke to the volume at the end of compression stroke. In other words, it is the ratio of the sum of the clearance volume and the stroke volume (swept volume) to clearance volume.

Condenser: It is a heat exchanger, which is used to condense the exhaust steam from the turbine/engine for reuse as feed water. It also helps to maintain a low pressure, by which a higher expansion ratio is obtained. This means higher work output and therefore higher efficiency.

Convection: It is a mode of heat transfer in fluids by the actual movement of heated molecules.

Cyclone: Spiral movement of wind towards a low pressure centre.

Dead centres (Top and bottom dead centres): Refer to positions of the crank pin when the crank aligns with the connecting rod. In either positions (TDC/BDC) any amount of force on the piston will not be able to rotate the crank.

Dew point: It is the temperature of air below which the water vapour contained in air begins to condense.

Draft: A stream of air admitted into the furnace is called Draft. When a chimney creates this stream, it is called 'Natural Draft'. If a fan (induced draft fan/forced draft fan or both) is used for the purpose it is called 'Artificial Draft'.

Dry steam/Dry saturated steam: When the steam at saturation temperature is totally free from moisture, it is called dry saturated steam or dry steam. Any addition of heat to the dry steam will increase the temperature of steam above the saturation temperature.

Dryness fraction (x): It is the amount of dry steam contained in a unit quantity of wet steam.

$x = W/(W + w)$ where W = Weight of dry steam and w = weight of moisture contained in a unit quantity of wet steam.

Economiser: This is a heat exchanger used to heat feed water using the exhaust gases coming out from a boiler. This enables recovery of the heat otherwise going waste and therefore the thermal efficiency of the boiler is improved.

Electrostatic precipitator: It is a pollution control device used to arrest fine dust particles going out into the atmosphere.

Estuary: The wide lower tidal part of a river.

Fast breeder reactor: It is a reactor in which fuel is produced at a greater rate than it is consumed. Fast neutrons are captured by fertile materials like thorium and U-238 to become fissile materials, which can be used as fuels.

Feed water: Water used in a boiler is called feed water. Feed water should conform to certain standards and any raw water before being introduced into the boiler is treated to remove salts etc.

Fertile material: Thorium, U-238, etc., are called fertile materials because they can be converted in a fast breeder into fissile materials.

Fire point: It is the temperature at which the heated oil continues to burn after the flammable vapour-air mixture catches fire.

Fissile material: Material that can be fissioned or split in a nuclear reactor. Example: U-235.

Flash point: It is the temperature at which the vapours released from heated oil forms a combustible mixture with air. This is determined by passing a lighted wick over the surface of the heated oil and the temperature is recorded when the vapours burn off with a flash.

Flue: It is a tube or passage in which hot air, smoke or gases flow.

Forced circulation: Circulation of a fluid maintained by a mechanical pump. In a modern boiler, forced circulation is used, which makes better heat transfer possible.

Forced draft fan: It is a blower kept at the entrance of the furnace, which forces atmospheric air into the furnace creating an artificial draft.

Grate/Grating: It is a support for the burning fuel, which allows combustion air through the slots made in it.

Grindability Index: It is an index of the power required to grind coal. Higher the index, lower is the power required.

Heating surface: It is the surface which has hot gases/fire on one side and water on the other side.

Hurricane: Wind blowing at very high speed greater than 120 km/h.

Impulse turbine: A turbine in which high pressure fluid (steam/gas or water) is expanded in a set of nozzles to a high velocity jet and the jet is made to impinge on blades fixed to a rotor. The kinetic energy of the jet is absorbed by the blades and is made available at the turbine shaft.

Induced draft fan: It is a suction fan kept at the entrance to the chimney. This fan creating a partial vacuum in the furnace makes the atmospheric air rush into it. Thus an artificial draft is created.

Isotopes: Isotopes of an element have different mass numbers. They exhibit the same properties as the element, but their nuclear behaviour is different.

Kinetic energy: It is the energy possessed by a body by virtue of its motion.

Latent heat of steam: It is the amount of heat to be added to unit quantity of water at boiling point to convert it into dry saturated steam.

Maintainability: It is the ease and effectiveness with which a system can be kept in good operating condition.

Moving chain grate stoker: Stoker is a coal feeding device. A chain grate stoker continuously moves at a slow speed and the coal fed at one end burns almost completely as it comes to the other end of the chain grate.

Muffler: Also called silencer, it reduces the noise emanating from the running engine. It is fitted to the exhaust of the engine.

Natural draft: See 'draft'.

Potential energy: It is the energy possessed by a body by virtue of its position.

Pour point: It is the temperature below which the oil loses its fluidity. It is an important information in the selection of lubricating oils.

Primary air: In a P.F. boiler, primary air is used essentially for carrying pulverised fuel. Secondary air is used at the burner to help complete combustion.

Radiator: It is a heat exchanger in which the jacket water from the engine is cooled by the forced flow of atmospheric air around the tubes of radiator carrying the water.

Reaction turbine: In this type of turbine, the conventional nozzles are not used. On the other hand the blade passages are so made that they act as nozzles. As the pressurized fluid passes through the passages, velocity of the fluid increases. This high velocity fluid leaving the passages on blades mounted on the rotor exerts a reactive force on the blades making the wheel/rotor to rotate. It is for this reason the turbine is called a reaction turbine.

Refrigerant: A cooling or freezing agent.

Relative Humidity: It is the ratio of the actual amount of moisture contained in unit volume of air at a given temperature to the maximum amount of moisture air could hold at that temperature. This ratio is expressed in %.

Reliability: Refers to the probability that a device will perform as planned for an intended length of time.

Saturated steam: As long as the steam is in contact with the boiling water its (steam) temperature remains the same as that of boiling water. The steam is said to be saturated and its temperature is called saturation temperature.

Sensible heat: It is the heat/thermal energy that causes a change of temperature of a solid or fluid.

Silencer: See 'muffler'.

Smoke tube: A flue tube surrounded by water being heated.

Sub-cooling: If the temperature of the refrigerant liquid is less than its saturation temperature the liquid is said to be in a sub-cooled condition.

Superheated steam: When heat is added to dry saturated steam, its temperature rises above the saturation temperature. The steam is then said to be superheated.

Superheater: It is a heat exchanger which superheats the steam. The increase in temperature of steam improves the overall efficiency and also reduces moisture in the last stages of expansion.

Surge tank/Surge shaft: Sudden stoppage of water in a penstock results in build up of very high pressure (surge) due to deceleration of the large flow of water. This may cause damage to pen stock. Provision of a surge tank, however, allows the water to rise up in it, thus reducing pressure build up.

Thermo-syphon: Circulation of water in a boiler brought about by purely differential temperature.

Thermodynamic efficiency: It is the percentage of thermal energy converted to mechanical energy in an engine or a turbine.

Thermostat: It is a device to automatically maintain a constant temperature. In an I.C. engine an optimum temperature of about 60EC is maintained in the cylinder jacket water with the help of a thermostat. Thermostat also helps a quick warm up of the engine during starting by keeping thermostatic valve closed.

Tornado: Very violent whirling wind and thunder storm.

Water jacket: Passages between double walls of cylinder and cylinder heads are called water jackets.

Water tube: Tube in which water passes and which is surrounded by hot gases.

Wet steam: It is the steam which is in contact with boiling water or contains water particles at saturation temperature. Any addition of heat goes to vaporize water particles without increasing the temperature.

Wind farm: A vast stretch of open land where a number of wind mills are erected to derive wind power and generate electricity.

Part-II

Production Processes

INTRODUCTION

Much more than the participation in the activity of large scale production of electrical power, mechanical engineers have greater role to play in the production of components and equipment required for various industries, power sector, transport, defence, communication systems, etc. In all this, the guiding factors are:

1. Economics of production or simply cost of production.
2. Functional aspects, i.e., how good the component/product serves the purpose. This involves the design of components based on strength, type of fit and other requirements.
3. The quality, in keeping with the type of function for which the component is designed.
4. Rate of production, which in turn determines the cost of production.
5. Safety, reliability and maintainability aspects.

Economics of production being one of the prime considerations, material and production process alternatives must be examined to begin with.

In the chapters to follow we shall learn a few things about the production alternatives only, so that we may select the right process for a given job.

The main production alternatives are:

- Forming Process
  - Casting
  - Forging
    - Hot forging
    - Cold forging
  - Cold forming
- Machining process and
- Metal joining by welding, Brazing and Soldering.

Chapter-3

Casting

Casting is a manufacturing process, which may be defined as pouring of molten metal into a mould and taking it out after it has solidified. The resulting solid will have the shape of the cavity in the mould. The product is called a casting.

The place where castings are made is called Foundry.

The mould material is usually sand but metal is also used to cast low-temperature-melting alloys.

The sand generally used for moulding called Foundry sand, is a mixture of river sand, clay and moisture in proper proportions. The mixing is done normally in a machine called muller, something similar to the arrangement in a tilting type wet grinder. The river sand, which is mostly silica, provides a good refractory material to withstand the high temperatures of many molten metals. Also it provides a porous structure necessary for the escape of gases and vapours produced during casting process.

Clay and moisture provide cohesion among the sand particles and also adhesion of the sand to the moulding box. Clay gives also the foundry sand necessary plasticity to shape the mould.

3.1 MOULD MAKING

Moulds are made by hand or machine. Hand moulding is slow and is used for job works, where one or a few castings are made. Machine moulding is used for large scale or mass production.

First let us take up hand moulding and see some of the tools and equipment used for the purpose.

(*a*) Moulding boxes: (Fig. 3.1) The boxes, usually in pair, are square or rectangular in shape, both ends being open. Wood or metal is used in making the boxes. The top box is called COPE and the bottom box is called the DRAG.

(*b*) Shovel (Fig. 3.2): This is used for transferring the moulding sand into the box.

(*c*) Moulding board (Fig. 3.3): This provides a level and firm surface on which a mould is made.

(*d*) Riddle (Fig. 3.4): Used for removing foreign material from the sand.

(*e*) Sand rammer (Fig. 3.5): This is used for ramming the sand in the moulding box.

(*f*) Strike-off bar (Fig. 3.6): After filling of the moulding box with sand and ramming, the moulding box is levelled off using this bar.

(*g*) Vent wire (Fig. 3.7): It is a thick stiff wire sharpened at one end and bent on the other end to form a handle. The mould is vented using this wire. Vents provide passage for the gases and vapours to escape from the mould and help complete filling of the mould with molten metal.

(*h*) Sprue pin (Fig. 3.8): It is a wooden pin used to provide passage for the molten metal to get into the mould.

(*i*) Riser pin (Fig. 3.8): This is yet another wooden pin made of suitable size used to provide passage for the molten metal to come up after filling the mould. Riser opening acts as a reservoir of molten metal and supplies metal in to the mould as the metal shrinks by solidification.

(*j*) Draw spike (Fig. 3.9): This is a sharp steel pin used for taking out wooden pattern from the mould. Sometimes a screw type is used.

(*k*) Trowel (Fig. 3.10): Trowels are used for finishing the mould. There are several types of trowels.

(*l*) Slick (Fig. 3.11): Slick is a double ended tool used for finishing and repairing the mould.

(*m*) Lifter (Fig. 3.12): This is made of a thin blade of steel bent to shape shown and is used to clean up and repair narrow openings in the mould.

(*n*) Gate cutter (Fig. 3.13): This is made of steel and serves to cut-open gates in the mould.

(*o*) Swab (Fig. 3.14), Mallet (Fig. 3.15), etc. are other tools used in mould making.

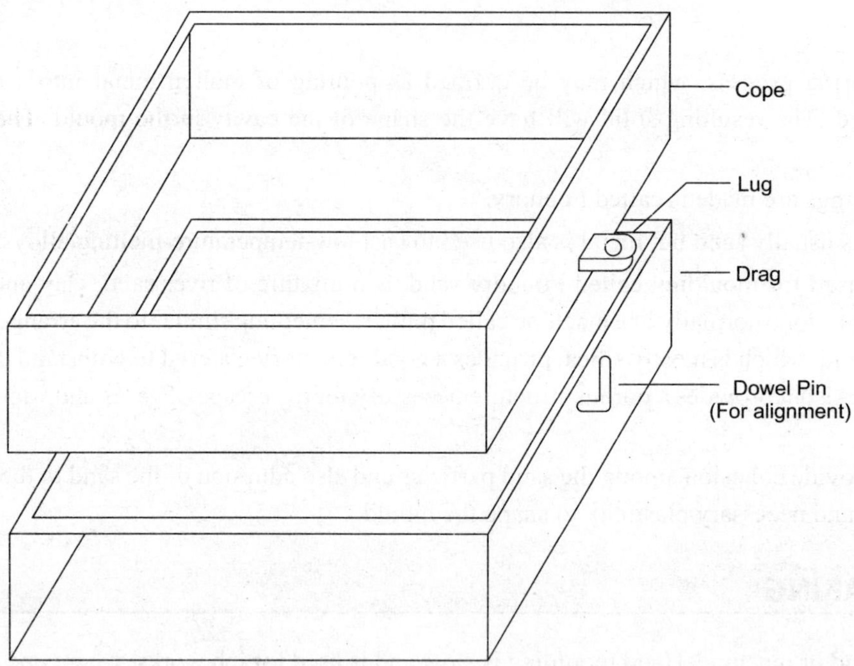

Fig. 3.1 Moulding Boxes

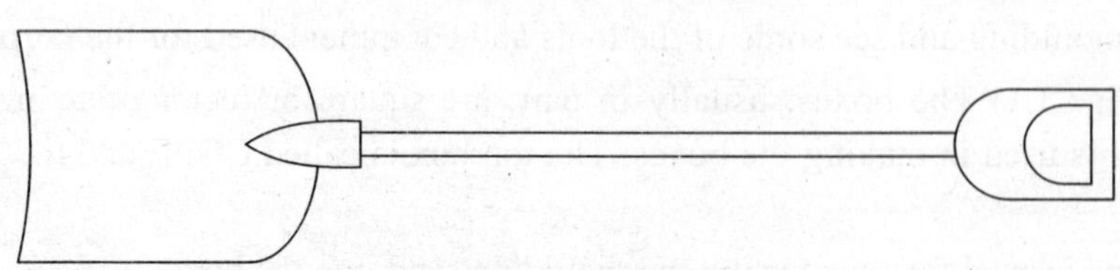

Fig. 3.2 Shovel

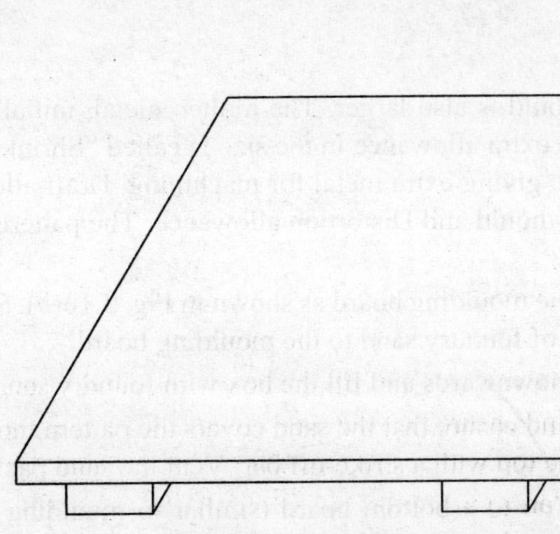

Fig. 3.3 Moulding Board/Bottom Board

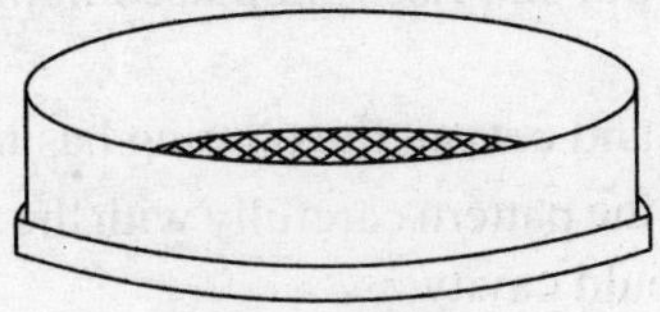

Fig. 3.4 Riddle

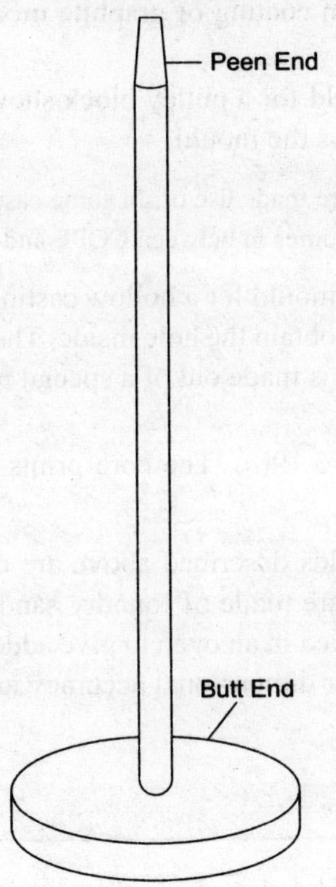

Fig. 3.5 Sand Rammer

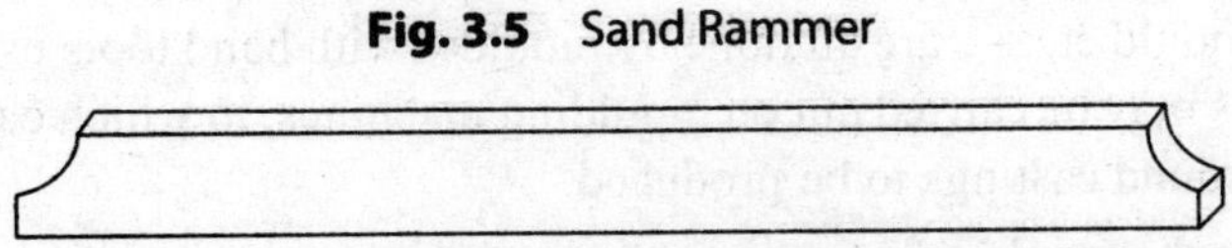

Fig. 3.6 Strike-off Bar

The process of mould making

Example 1: Consider making a mould for a component, whose pattern is shown in Fig. 3.16(*a*). The pattern is a replica of the component (final product), but with certain differences. One of the important differences is that the size of the pattern is slightly more, so that the cavity left in the mould is also larger. The molten metal, initially occupying this volume, shrinks after it solidifies to the desired size. This extra allowance in the size is called "Shrinkage allowance". The other allowances considered are: Machining allowance giving extra metal for machining, Draft allowance giving slight taper on the sides of pattern for its easy removal from mould and Distortion allowance. The pattern considered here is in one single piece and is called a solid pattern.

I. Preparation of drag: Place the pattern on the moulding board as shown in Fig. 3.16(*b*). Sprinkle some parting sand (clean, dry river sand) to prevent sticking of foundry sand to the moulding board.

   Place the drag with its dowel pins facing downwards and fill the box with foundry sand.

   Ram the sand initially with the peen end and ensure that the sand covers the pattern thoroughly. Then fill the box completely with sand, ram and level up the top with a strike-off bar. Vent the sand packing using vent wire.

II. Preparing Cope: Now roll over the drag on to a bottom board (similar to moulding board) [Fig. 3.16(*c*)] and assemble the cope with the lugs engaging the dowels on the drag. Sprinkle parting sand inside to prevent cope and drag clinging to each other. With sprue pin and riser pin placed in position, fill the cope with foundry sand, ram and vent.

   Slowly take out the sprue and riser pins and cut out the pouring basin at the top of the sprue opening.

   Separate the cope from the drag, remove the pattern carefully with the help of Draw spike, cut out a gate connecting the bottom of sprue passage with the mould cavity.

III. Reassemble the cope and drag [Fig. 3.16(*d*)]. The mould is now ready for casting.

   Usually the mould surface is given a thin coating of graphite mixture using a brush. This improves the surface finish of the casting.

Example 2: Now let us consider making a mould for a pulley block shown in Fig. 3.17. A split pattern (pattern in two pieces) is required in this case. Figure 3.18 shows the mould.

Note: In the above examples, two moulding boxes are made use of. In some cases, where removal of pattern from the mould poses problem, a third box, called CHEEK is also used. It comes in between COPE and DRAG.

Example 3: Let us now see how we can make a mould for a hollow casting. Taking the example of a pipe casting shown in Fig. 3.19(*a*), a CORE is required to be used to obtain the hole inside. The mould with the core is shown in Fig. 3.19 (*b*). The core in this case is a cylindrical piece and it is made out of a special mixture of sand called CORE SAND. The core is invariably baked to give it strength.

The pattern for the casting is shown in Fig. 3.19(*c*). The core prints shown leave hollow spaces in the mould for supporting the core later.

Green sand and Dry sand moulds: The moulds described above are called Green sand moulds for the reason they contain moisture. For quality castings, moulds are made of foundry sand containing special binders and after they are made just as the Green sand moulds, they are baked in an oven to give added strength. These are called Dry sand moulds. Castings made using drysand moulds have better dimensional accuracy and surface finish. Also thy are free from blow holes and other defects.

3.2 MACHINE MOULDING

The moulds in the examples discussed above are hand made, in the sense, the packing of the moulding boxes with sand, removing of the pattern from the mould etc. — are all done by moulder with hand tools using his skill and muscle power. On the other hand, these operations may be carried out on moulding machines, in which case the quality of the mould will be consistent and good, enabling sound castings to be produced.

The two important operations in the moulding process are (*a*) ramming and (*b*) careful removal of the pattern from the mould. Moulding machines differ in the techniques adopted for ramming and removal of the pattern from the mould.

Machines for ramming/compacting the sand: There are three types: (*i*) Squeeze type machine, (*ii*) Jolt machine and (*iii*) sand slinger.

Squeeze machine uses a platen to squeeze the sand in the mould as shown in Fig. 3.20.

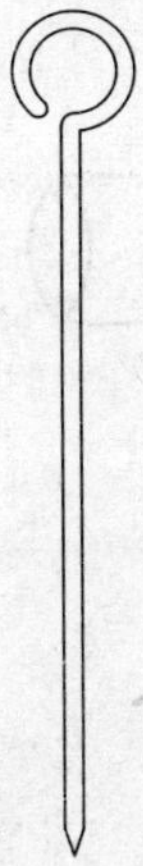

Fig. 3.7 Vent Wire

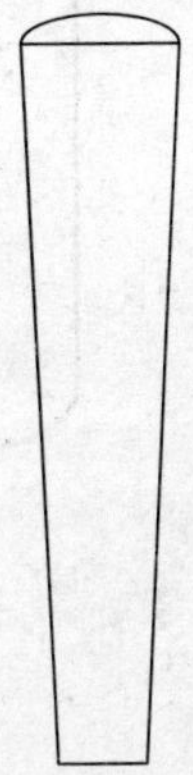

Fig. 3.8 Sprue/Riser Pin

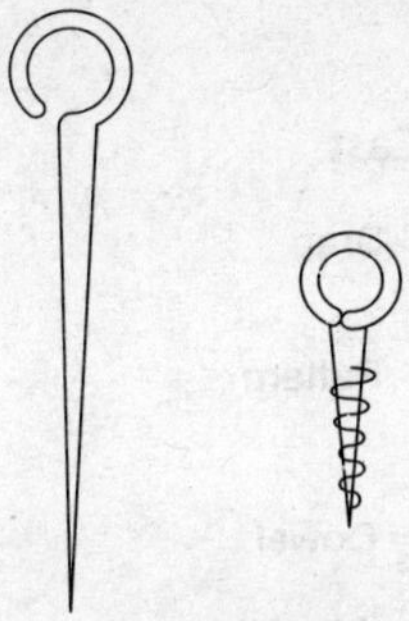

Fig. 3.9 Draw Spike

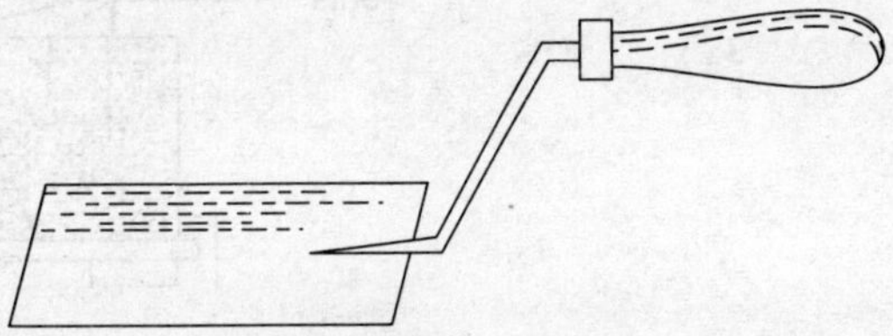

Fig. 3.10 Trowel

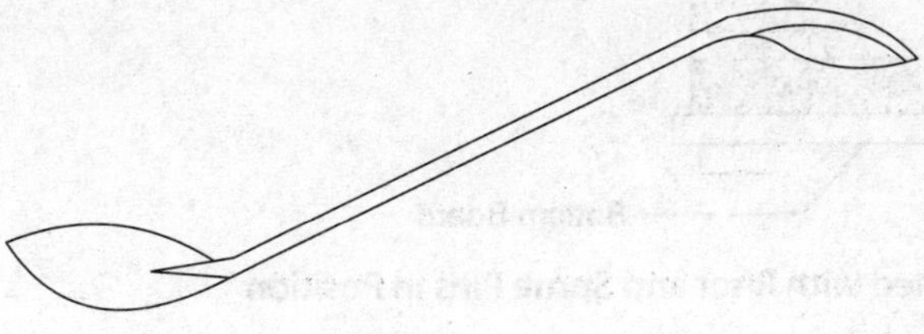

Fig. 3.11 Slick

Fig. 3.12 Lifter

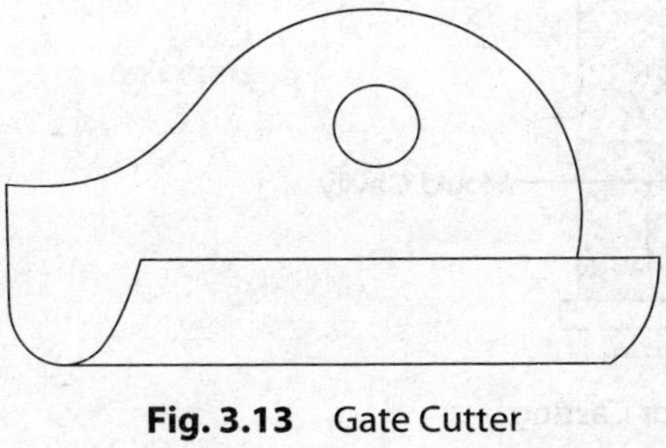

Fig. 3.13 Gate Cutter

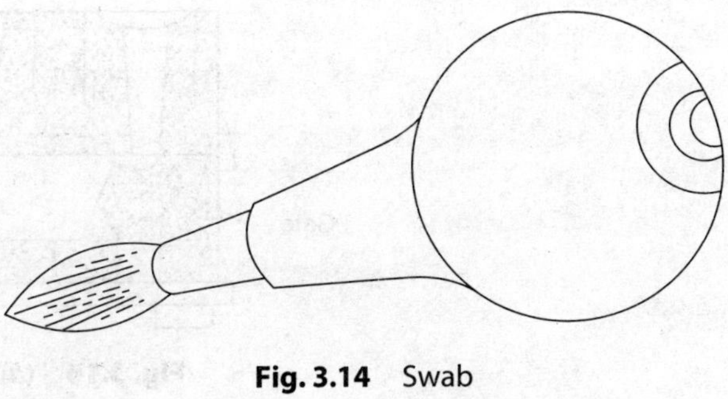

Fig. 3.14 Swab

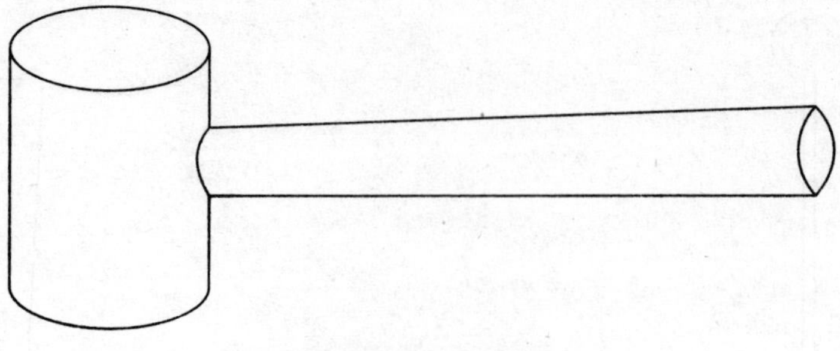

Fig. 3.15 Mallet

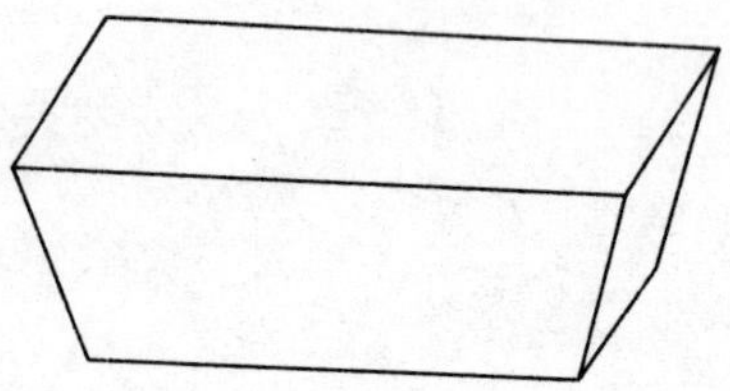

Fig. 3.16 (*a*) Pattern of a Component to be Cast

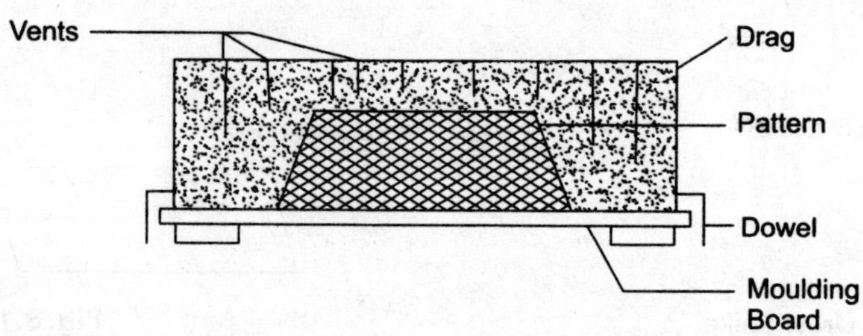

Fig. 3.16 (*b*) Preparation of Drag

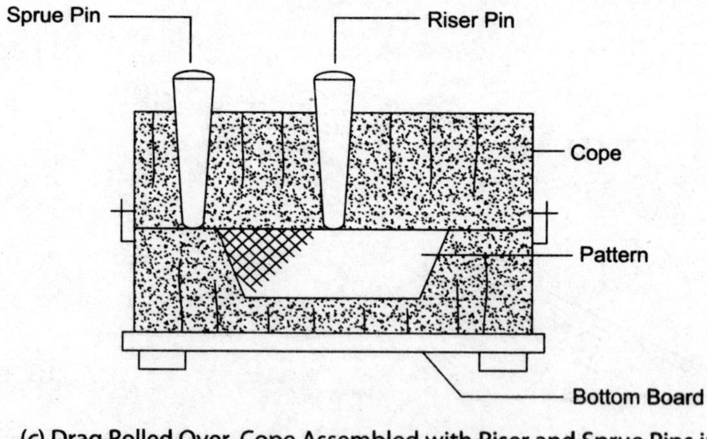

Fig. 3.16 (*c*) Drag Rolled Over, Cope Assembled with Riser and Sprue Pins in Position

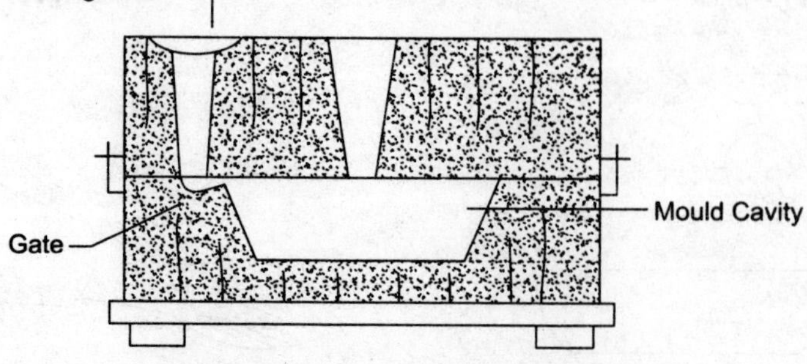

Fig. 3.16 (*d*) Mould Ready for Casting

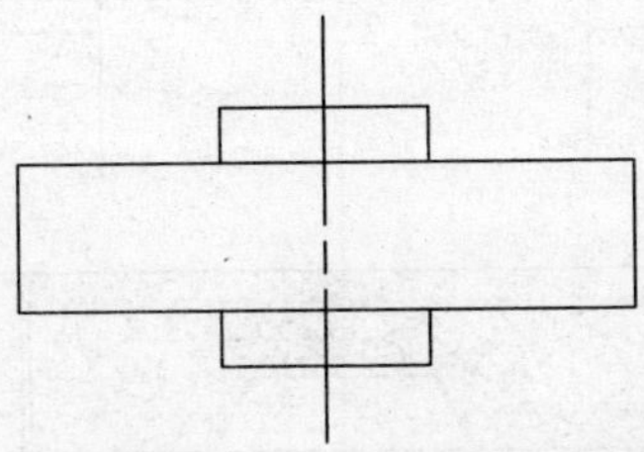

Fig. 3.17 (*a*) Pulley Block

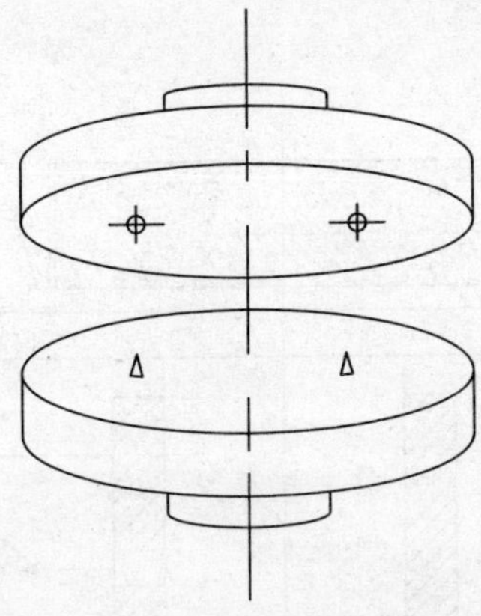

Fig. 3.17 (*b*) Split Pattern for Pulley Block

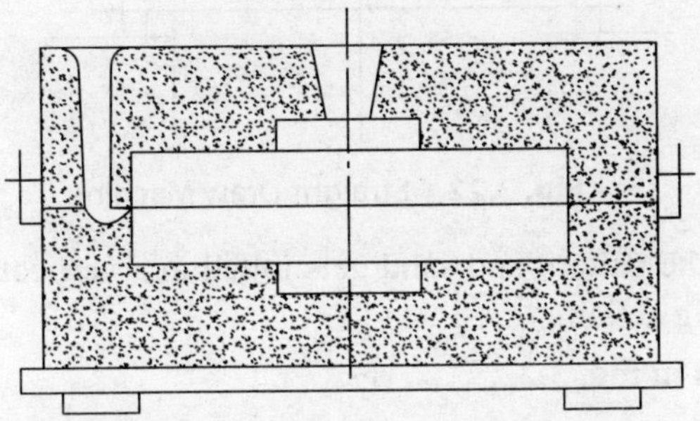

Fig. 3.18 Mould for Pulley Block

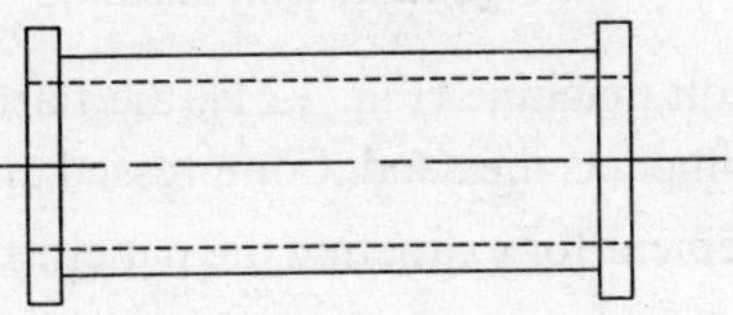

Fig. 3.19 (*a*) Pipe Casting

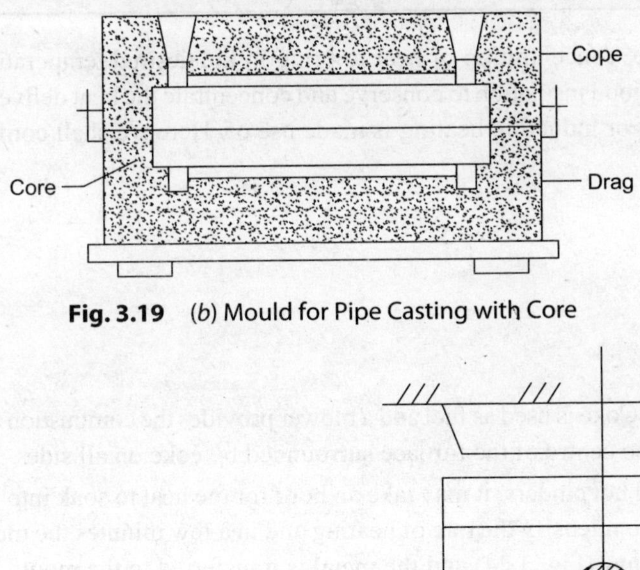

Fig. 3.19 (*b*) Mould for Pipe Casting with Core

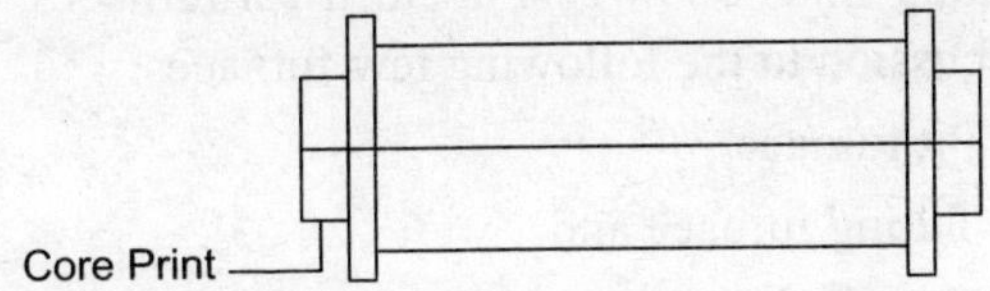

Fig. 3.19 (*c*) Pattern for Pipe Casting

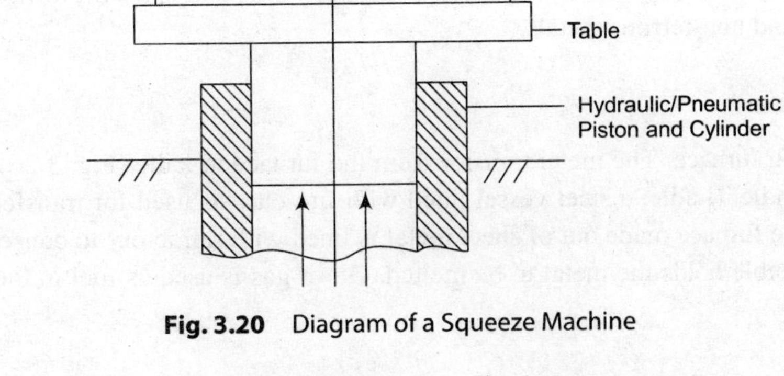

Fig. 3.20 Diagram of a Squeeze Machine

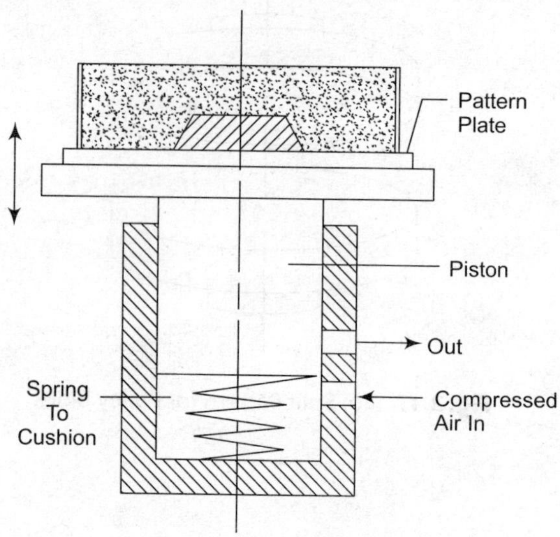

Fig. 3.21 Jolt Machine

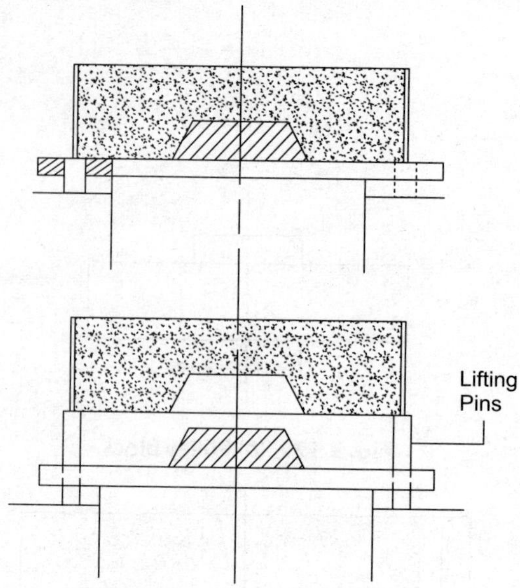

Fig. 3.22 Straight Draw Machine

In the Jolt machine (Fig. 3.21), the flask (moulding box) with the pattern and sand gets jolted in a vertical direction and this compacts the sand. Compressed air is used to get the jolting action.

Arrangement for extracting the pattern from the mould is shown in Fig. 3.22.

3.3 MELTING OF METALS

Most metals melt at high temperatures and require therefore, concentration of heat to bring in the desired temperature. Furnaces are used for this purpose and they are provided with good insulation to conserve and concentrate the heat delivered by burning coke, oil or gas. In electrical furnaces, resistance or induction heating is made use of. Here we shall confine our discussion to the following few furnaces:

(*i*) Pit furnace

(*ii*) Tilting furnace and

(*iii*) Cupola.

3.3.1 Pit Furnace: (Fig. 3.23)

As the name implies, it is constructed in a pit dug in the ground. Coke is used as fuel and a blower provides the combustion air. The metal to be melted is put in a graphite crucible kept in the centre of the furnace surrounded by coke on all sides.

Once the fire is made and the whole pit is glowing with red hot cinders, it may take an hour for the heat to soak into the metal. At the end of this soaking period, the blower is put on to intensify the rate of heating and in a few minutes the metal will be in molten condition. The crucible is lifted out with tongs (Fig. 3.24) and the metal is transferred to the mould.

Pit furnaces provide only a few kg of metals (30 to 100 kg) at a time and is useful for job works only. They are used for melting both ferrous and non-ferrous metals.

3.3.2 Tilting Furnace

Figure 3.25 shows a tilting furnace. The metal transfer from the furnace to ladle (Fig. 3.26) is easily done by tilting the furnace with a geared handle. (Ladle, a steel vessel lined with fire-clay, is used for transferring molten metal from the furnace to the mould). The furnace made out of sheet metal is lined with refractory to conserve and direct the heat on to the metal. A graphite crucible holds the metal to be melted. Oil or gas is used as fuel in the furnace. Combustion air is supplied by a blower.

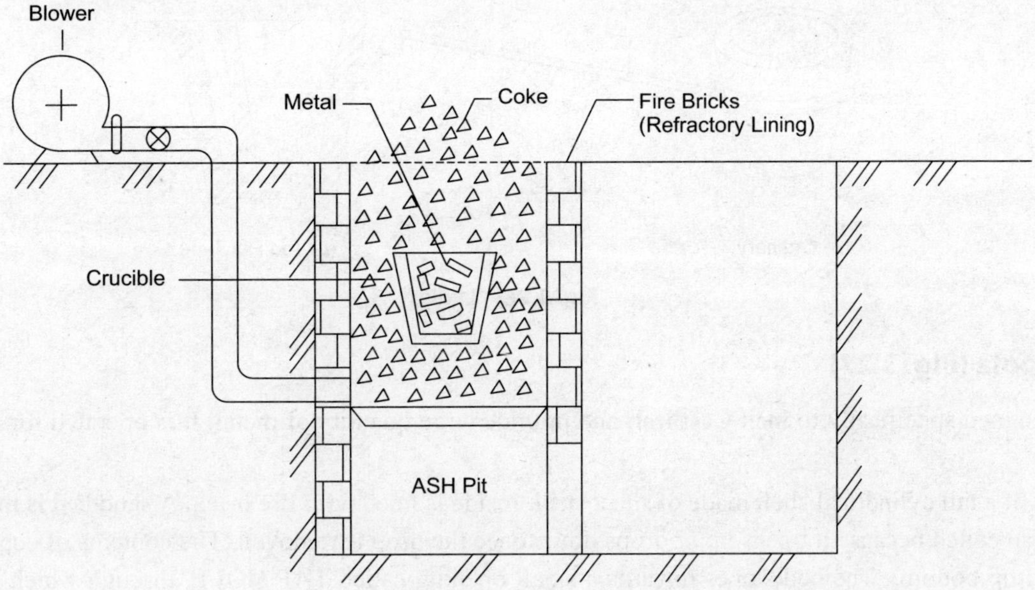

Fig. 3.23 Pit Furnace

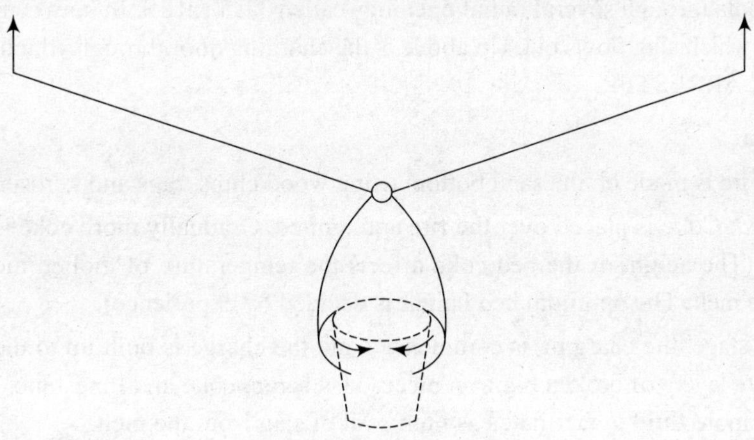

Fig. 3.24 Lifting Tongs

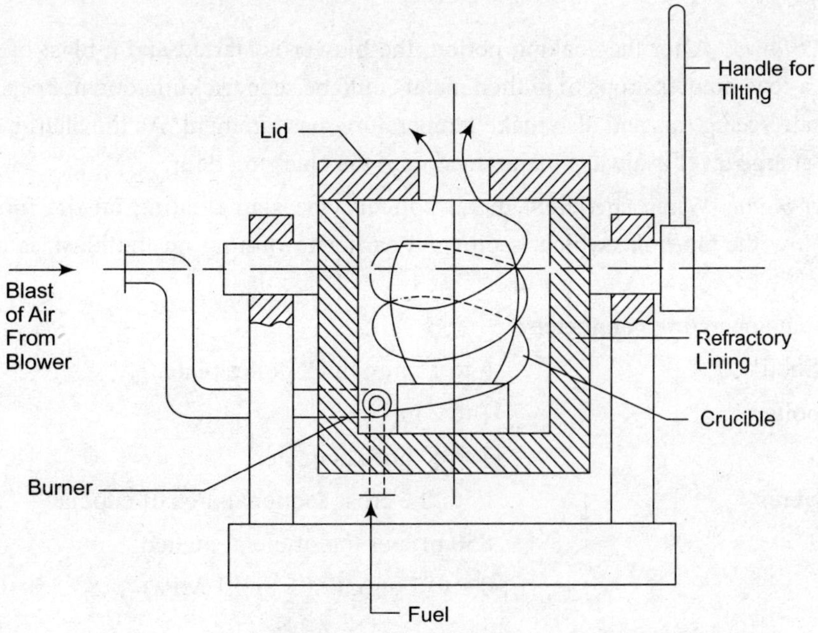

Fig. 3.25 Tilting Furnace

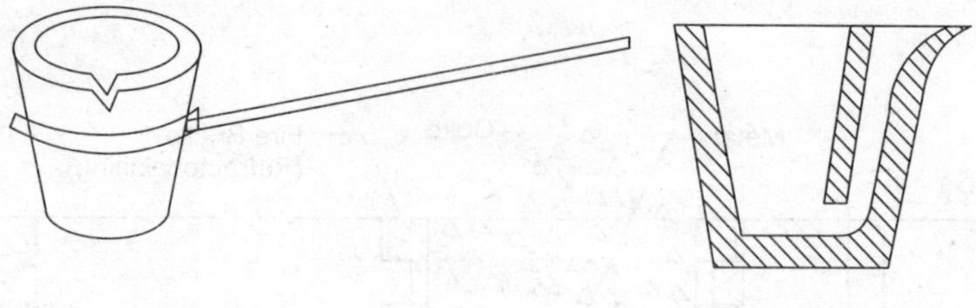

(*a*) Ordinary (*b*) Tea Pot Type

Fig. 3.26 Ladles

3.3.3 Cupola (Fig. 3.27)

Cupola is designed specifically to melt Cast Iron and provide large quantity of metal. It is operated for several hours at a stretch.

It consists of a tall cylindrical shell made of sheet steel. Inside is lined with fire brick. A sand bed is made on the drop bottom (It is so called because it opens up or drops down once the prop is removed. The contents of cupola are emptied through the drop bottom). The bed slopes towards a small opening called TAP HOLE, through which molten metal is tapped out. About 500 to 600 mm above the tap hole is the WIND BELT. Air from a blower gets into the wind belt and finds its way into the Cupola through several radial openings called TUYERES. In between the tap hole and the tuyeres is the SLAG HOLE from which slag flows out. Up above is the charging door through which charging is done. At the top of the shell is the SPARK ARRESTER.

Operation of the cupola:

1. *Starting the fire:* Fire is made on the sand bottom using wood chips, rags and kerosene.
2. *Preparing coke bed:* Coke is placed over the fire and ignited. Gradually more coke is added and a bed of red hot coke is prepared. (The height of the bed coke affects the temperature of molten metal, rate of melting and the composition of the melt. The optimum bed height is decided by experience).
3. *Charging:* At this stage, the charging is commenced and the charge is built up to the charging door. The charge consists of alternate layers of broken pig iron pieces iron scrap, coke and limestone. The limestone acts as a flux rendering the slag more fluid to facilitate easy removal of slag from the melt.
4. *Soaking in the heat:* The whole content of the cupola is now allowed to soak in using natural draft. This may take an hour or longer.
5. *Turning on the Blower:* After the soaking period, the blower is started and a blast of air rushes in through the tuyeres. Within a few minutes drops of molten metal could be seen trickling down. Peep holes are provided on the cupola to facilitate seeing this and also make temperature measurement. As the charge sinks additional charging is done and the charge level is always maintained upto the charging door.
6. *Tapping molten metal:* When sufficient metal collects, the slag floating on the molten metal issues out of the slag hole. Now the tap hole, which is plugged earlier to turning on the blast, is opened to get the molten metal.

Some design details and operating parameters:

- Sheet metal of Shell : 6 to 12 mm thick boiler plate.
- Diameter of cupola : 1 to 2 m
- Height : 3 to 5 times diameter.
- Total area of tuyeres : 1/5th the cross sectional area of cupola
- Air requirement : 850 m^3 per ton of metal melted.
- Blast pressure : 0.5 to 1 atm. (0.05 to 0.1 MPa)

- Temperature attained : 1600 to 1700°C.
- Iron to coke ratio : 10 : 1
- Proportion of flux added : 2 to 3% of metal charge.

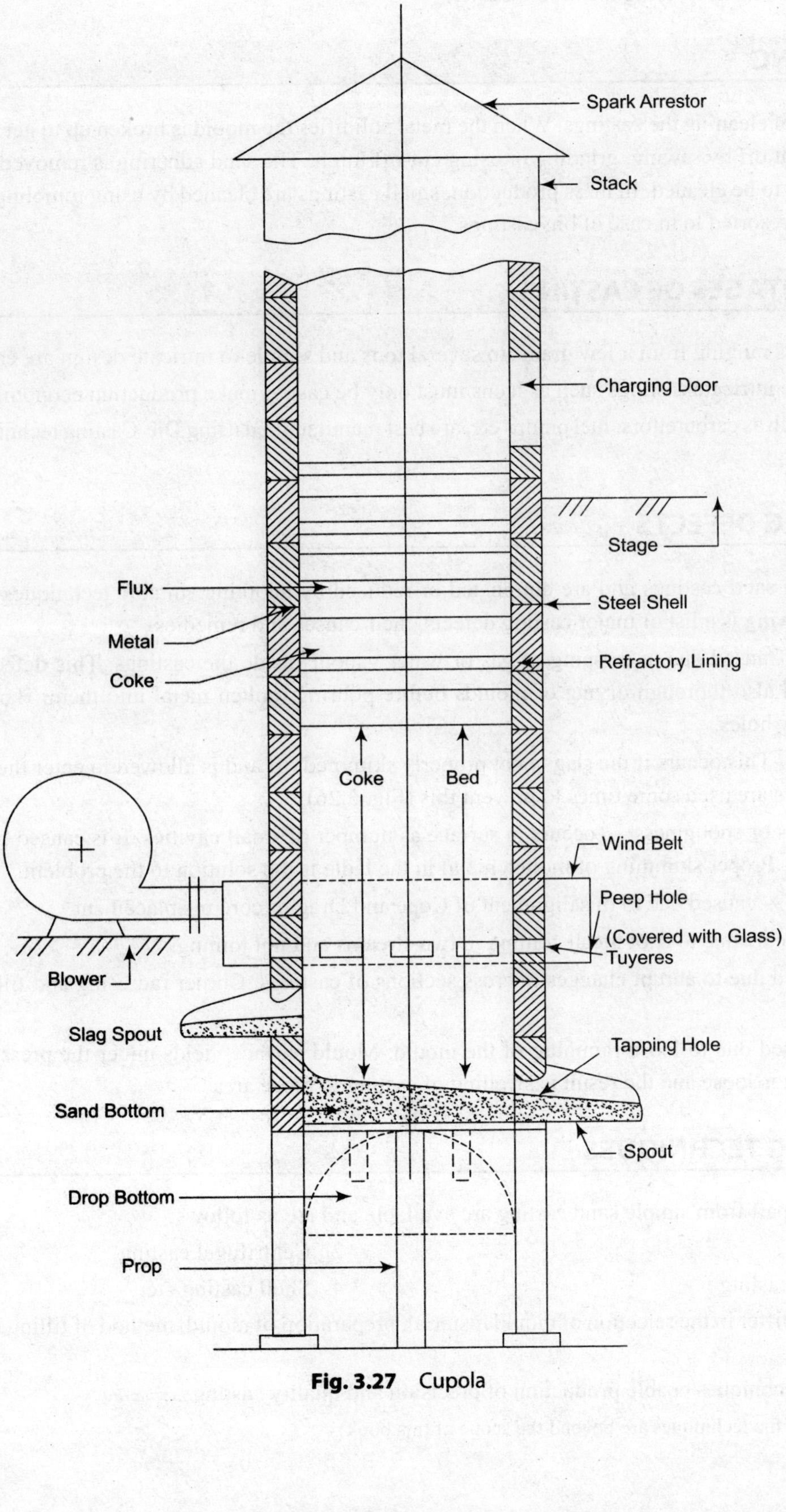

Fig. 3.27 Cupola

3.4 POURING MOLTEN METAL INTO THE MOULD

This should be done steadily and without break until the metal fills the riser completely. Care must be taken to see that the slag is prevented from entering the mould cavity.

3.5 FETTLING

It is the operation of cleaning the castings. When the metal solidifies the mould is broken up to get the casting. The runner and the riser are cut off by sawing, grinding or using cut-off torch. The sand adhering is removed by a wire brush if one or two castings are to be cleaned. In mass production, small castings are cleaned by using tumbling barrels. Sand blasting or shot blasting is resorted to in case of big castings.

3.6 ADVANTAGES OF CASTING

1. Components ranging from a few grams to several tons and simple to intricate design are easily cast.
2. Some of the intricate castings such as icons must only be cast to make production economical.
3. Castings such as carburettors, fuel pumps etc. are best manufactured using Die-Casting technique out of low-melting alloys.

3.7 CASTING DEFECTS

Defects do arise in sand castings and are eliminated or reduced by adopting suitable techniques and observing certain precautions. Following is a list of major casting defects, their causes and remedies.

Blow Holes — Caused due to trapping of gas or water vapour inside the castings. This defect can be corrected by proper venting and also thorough drying of moulds before pouring molten metal into them. Hard ramming is another factor causing blow holes.

Slag Inclusion —This occurs if the slag is not properly skimmed off and is allowed to enter the mould along with the metal. Teapot ladles are used some times to prevent this (Fig. 3.26).

Honey combings or sponginess — occurs in surface as number of small cavities. It is caused by dirt in molten metal entering the mould. Proper skimming of molten metal in the ladle is the solution to the problem.

Lifts and Shifts — caused due to misalignment of Cope and Drag or core misplacement.

Cold Shut — occurs due to cool metal coming in two streams and not joining.

Cracks — caused due to abrupt changes in cross sections of castings. Corner radiusing and filleting help reduce this happening.

Swelling —caused due to loose ramming of the mould. Mould surface yields under the pressure of metal at places where the ramming is loose and the result is swelling of casting in those areas.

3.8 CASTING TECHNIQUES

Many techniques apart from simple sand casting are available and are as follows;

1. Die casting
2. Centrifugal casting
3. Investment casting
4. Shell casting etc.

The techniques differ in the selection of mould material, preparation of mould, method of filling the cavity with molten metal etc..

Some of these techniques enable production of precision and quality castings.

(**Note:** The details of the techniques are beyond the scope of this book)

QUESTION BANK

2-Marks Questions

1. What is casting?
2. Name some of the casting techniques.
3. What is moulding?
4. What is a sprue/runner?
5. What is a riser?
6. Name the important tools used in hand moulding.
7. What is a pattern?
8. What are the various allowances considered in making pattern?
9. What is a core?
10. What are the chief constituents of foundry sand?
11. What are the important characteristics of foundry sand?
12. What is parting sand?
13. What is a core print?
14. Name some of the moulding machines.
15. What is the flux used in cupola? How does it function?
16. In what ratios pig iron, coke and flux are introduced into cupola?
17. Name the chief defects in casting.
18. What are the causes of blow holes?
19. What is Honey-combing and what are its causes?
20. What is Cold shut? How is it caused?
21. What is shift?
22. What is swelling and how is this caused?
23. Name the methods used for cleaning castings.
24. What are the advantages of casting as a manufacturing process?

10-Marks Questions

1. Explain the preparation of a mould for a pulley block with neat sketches.
2. Explain the procedure for making a mould for a pipe casting. Add neat sketches.
3. Briefly describe the construction and working of a cupola with a neat sketch.

State True or False

1. Pattern should be made to the exact size of the component.
2. Very intricate castings can be made by lost wax process.
3. Shrinkage allowance is highest for aluminium.
4. Moulds should be well vented to avoid blow holes.
5. Wood is commonly used as the material for pattern making.
6. Flux is used to raise the temperature of molten metal.
7. Coke is used to supply carbon to cast iron.
8. Fettling is the process of finishing the castings.

9. Cope is the bottom half of the mould box.
10. Clay in foundry sand acts as a binder.

Answers

1. F **2.** T **3.** T **4.** T **5.** T **6.** F **7.** F **8.** F **9.** F **10.** T.

Fill in the Blanks

1. The lower box in moulding is called ....
2. The upper box in moulding is called ....
3. In three box moulding, the central box is called ....
4. The property of sand to withstand high temperature is called ....
5. The property of sand to allow gas/vapour to escape easily is called ....
6. A hole in the casting can be obtained using a ....
7. The opening in the mould for pouring molten metal is called ....
8. The opening for the excess molten metal to come up is called ....
9. Casting in permanent mould (mould made in metal) is called ....
10. In jolting machines, sand is rammed by ....
11. A .... rams sand by throwing it at great velocity.
12. Cast Iron is generally produced in ....
13. The charge of cupola consists of ....
14. The impurities floating on the molten metal is called ....
15. The air supply passages into cupola are called ....
16. The charge metal to coke is ....
17. The charge ratio metal to flux is ....
18. The carbon content of C.I. is ....
19. Blow holes are caused by .... being trapped.
20. When the metal does not fuse where two stream of molten metal meet, the defect is called ....
21. When the two halves of the cast part are displaced, the defect is called ....

Answers

1. Drag
2. Cope
3. Cheek
4. Refractoriness
5. Permeability porosity
6. Core
7. Sprue/runner
8. Riser
9. Die casting
10. Vibration
11. Sand slinger
12. Cupola
13. Pig iron, coke and flux
14. Slag
15. Tuyeres
16. 10 : 1
17. 30 : 1
18. 2.5 to 4%
19. Gas/vapour
20. Cold shut
21. Shift

Chapter-4
Forging

Forging is the process of shaping metals by impact or very high pressure. The metal in this process is squeezed to attain the desired shape and this requires the metal to be in plastic condition. This plastic state or plasticity is brought about either by heating or increasing the stress level beyond elastic limit.

When the metal is heated to plastic temperature and shaped, it is called Hot Forging or simply Forging.

When the metal is deformed and shaped at room temperature raising the stress level beyond elastic limit, the process is called Cold Forging.

Forging as a manufacturing process is intended to produce components of high strength out of wrought iron, steel, aluminium alloys, brasses, etc. It must be remembered, however, that not all metals can be forged. For example, brittle metals like Cast Iron (C.I.) cannot be forged.

Components such as connecting rod, crankshaft, leaf springs etc. are all forged. Examples of heavy forgings include rotors for turbines and generators, propeller shaft for big ships, columns for big presses etc.

4.1 HAND FORGING AND MACHINE FORGING

Hand forging is done by blacksmiths in a section of a workshop called smithy. Heating of metal is done in an open fire and hand tools are employed for forging. Job works of smaller size are handled in the smithy section.

Machine forging is done in a section called forge, where dosed furnaces are used for heating the jobs and power hammers or presses are used for shaping them. When large sections or large number of components are to be shaped, they are sent to the forge section.

4.2 TOOLS AND EQUIPMENT USED IN SMITHY

Hearth: (Figure 4.1) shows the sketch of a hearth. Coal, coke or charcoal is used as fuel to heat the job. A blower provides blast of air to produce rapid combustion.

Anvil: (Fig. 4.2) Anvil provides a flat and rigid surface on which the heated job is kept and hammered. The pritchel hole helps in punching and bending operations. The horn (or beak) is used in bending jobs to different curvatures. The Hardie hole (square in shape) is used to hold the bottom half of various swages and other forging tools.

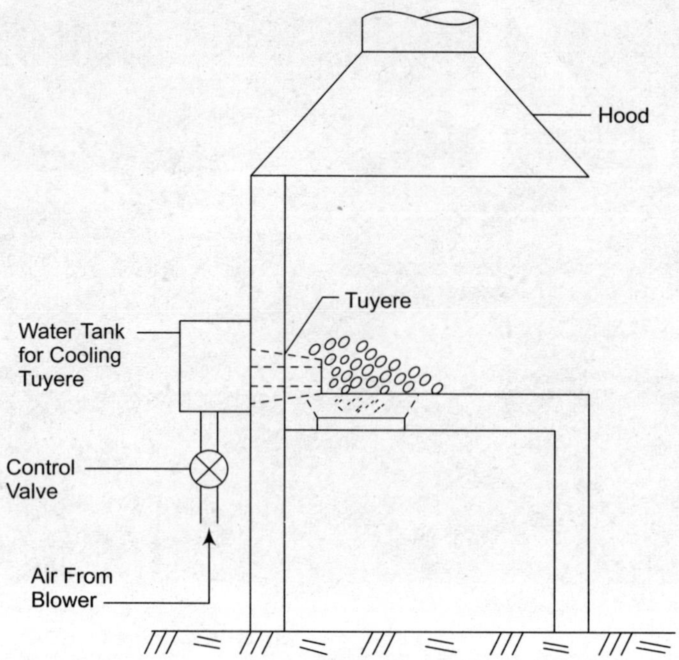

Fig. 4.1 Hearth

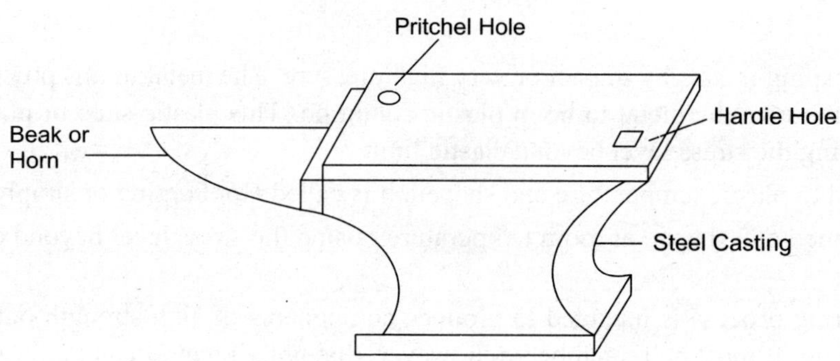

Fig. 4.2 Anvil

Hammers: (Fig. 4.3) Smaller ones weighing 0.5 to 1.5 kg are handled by the blacksmith himself. These are classified as Ball peen, Cross peen and Straight peen hammers depending on the shape of the peen. Bigger ones called Sledge hammers, weighing 5 to 8 kg, are handled by a hammerman and he delivers the blows on the job to the instructions of the blacksmith.

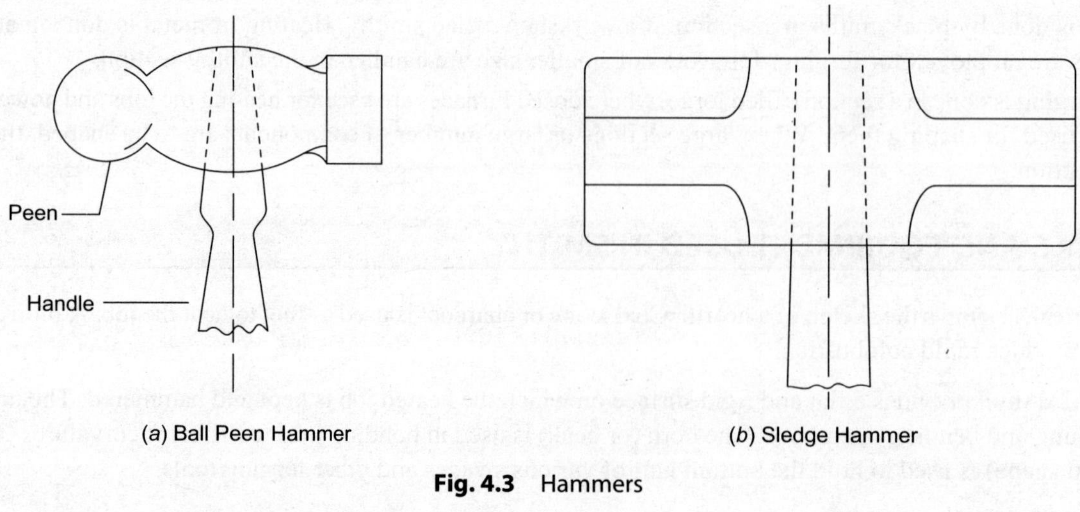

Fig. 4.3 Hammers

Tongs: (Fig. 4.4) Tongs are used to handle heated jobs and hold them during the forging operations. Tongs are of general purpose and special types.

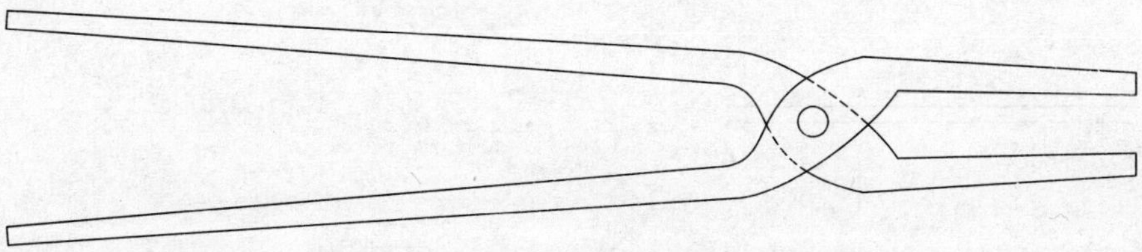

Fig. 4.4 Tongs

Swages: (Fig. 4.5) These are used to get different sectional shapes like round, square and hexagon. In the figure is shown one set of swages for giving round shape. The bottom part having square shank fits into the hardie hole on the anvil. The top part having a handle, is held by the smith.

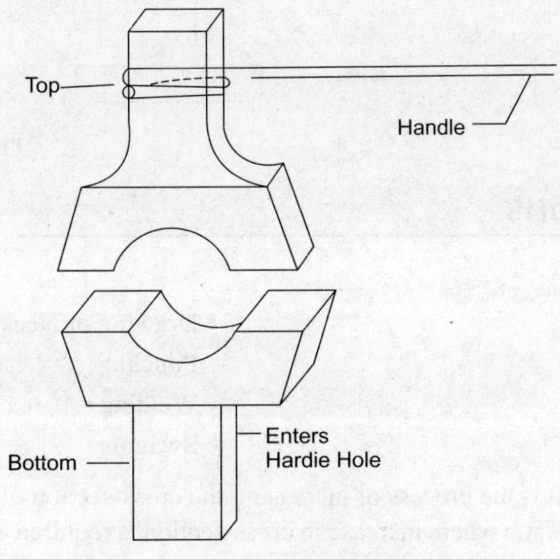

Fig. 4.5 Swages

Swage block: (Fig. 4.6) Measuring about 500 mm × 500 mm × 150 mm, the steel swage block has various shaped grooves and holes to shape, size and bend jobs.

Fig. 4.6 Swage Block

Fullers: (Fig. 4.7) These are used for reducing the cross section of a job. They are used in pairs just as swages.

Flatter: (Fig. 4.8) This is used for flattening and smoothing the surface of a job, for example, after fullering operation.

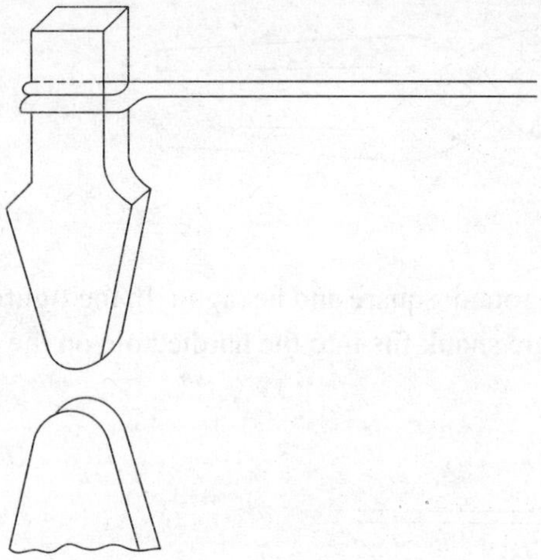

Fig. 4.7 Fullers

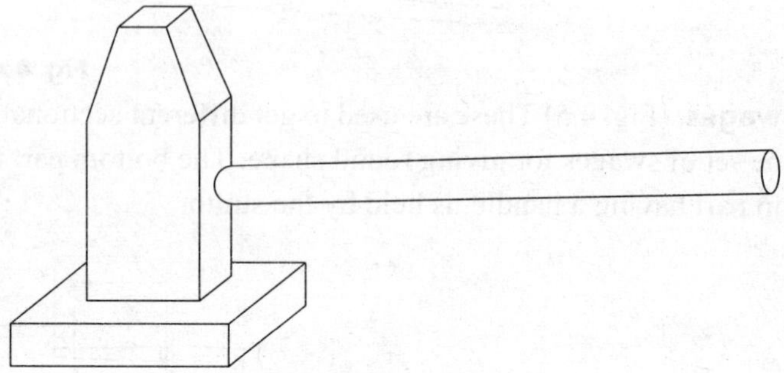

Fig. 4.8 Flatter

4.3 FORGING OPERATIONS

The main forging operations are:

- Upsetting or Jumping
- Setting down
- Bending
- Cutting
- Drawing or Necking
- Punching
- Welding
- Swaging

Upsetting: (Also called Jumping) It is the process of increasing the cross-section of a job at the expense of its length. It consists of heating that portion of the job where increase in cross-section is required and hammering in an axial direction. Figure 4.9 shows the trends of deformation in upsetting.

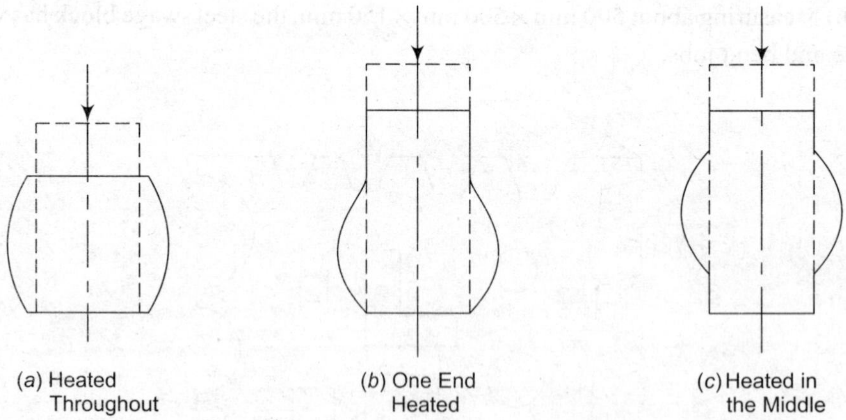

Fig. 4.9 Upsetting—Trends of Deformation

Drawing: (Also called Necking down) The process of drawing reduces the cross-section of a job. Figure 4.10 shows the drawing operation using fullers. Note that the job gets elongated.

Setting down: Localised thinning of a job is called setting down. The job is first fullered or thinned down using fullers and then finished using a flatter. Figure 4.11 shows the stages in setting down operation.

Punching: Holes are easily produced using punch when the job is hot. Steps are shown in Fig. 4.12.

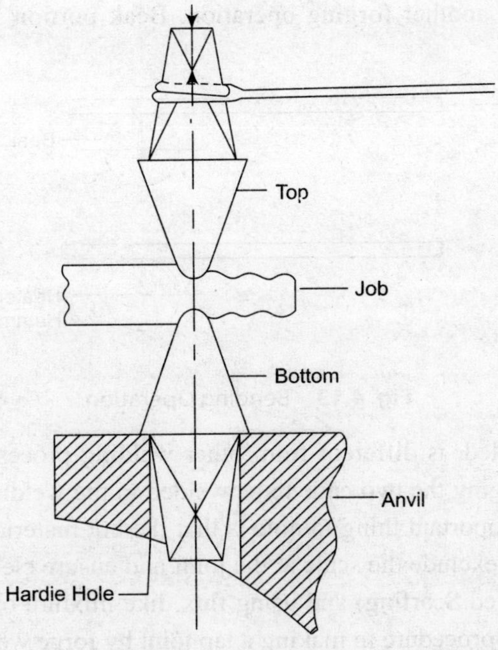

Fig. 4.10 Drawing Operation

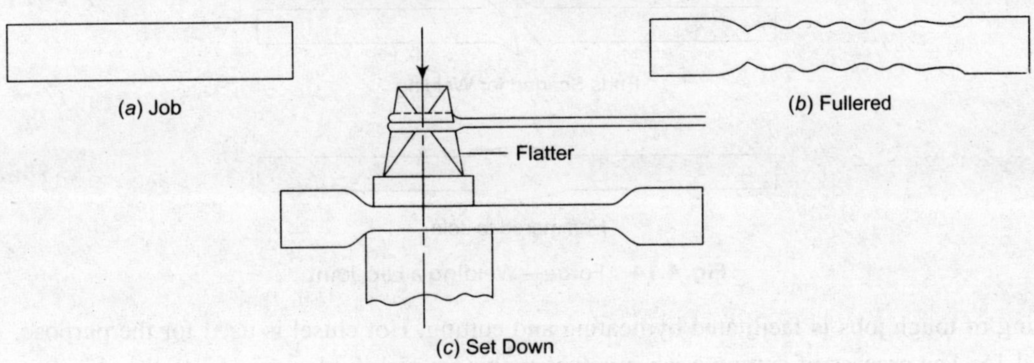

Fig. 4.11 Setting Down

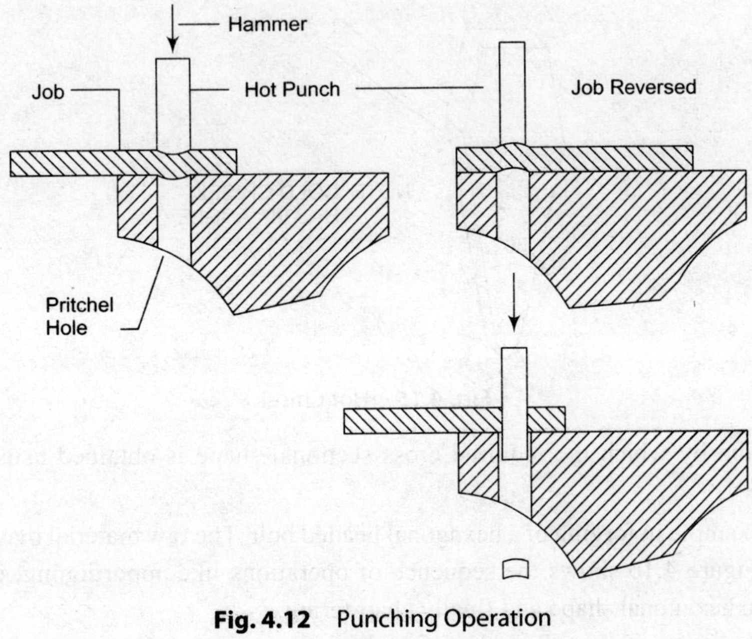

Fig. 4.12 Punching Operation

Bending: Bending rods and flats is another forging operation. Beak portion is used for this purpose as shown in Fig. 4.13.

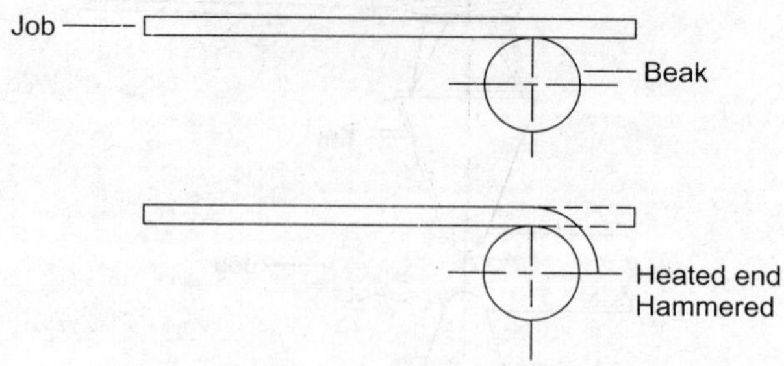

Fig. 4.13 Bending Operation

Welding: Forge-Welding as it is called, is different from other welding processes in that no filler material is used. The welding is brought about by bringing the two ends to be welded to the welding temperature (white hot) in a hearth and hammering them together. The important thing to note is that the hot material oxidizes fast and scales are formed. Therefore, the joining process should exclude the scale at the joint and ensure clean metal to metal contact. This can be done by proper edge preparation (called Scarfing) and using flux, like mixture of four parts of Borax with one part of Sal-ammoniac. Figure 4.14 shows the procedure in making a lap joint by forge welding.

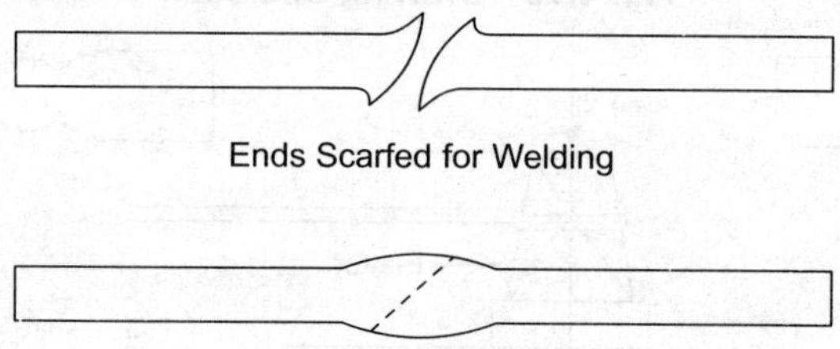

Fig. 4.14 Forge—Welding a Lap Joint

Cutting: Cutting of tough jobs is facilitated by heating and cutting. Hot chisel is used for the purpose. A hot chisel is shown in Fig. 4.15. The process of cutting a job involves making a nick (indentation with the chisel) on the heated job and breaking it.

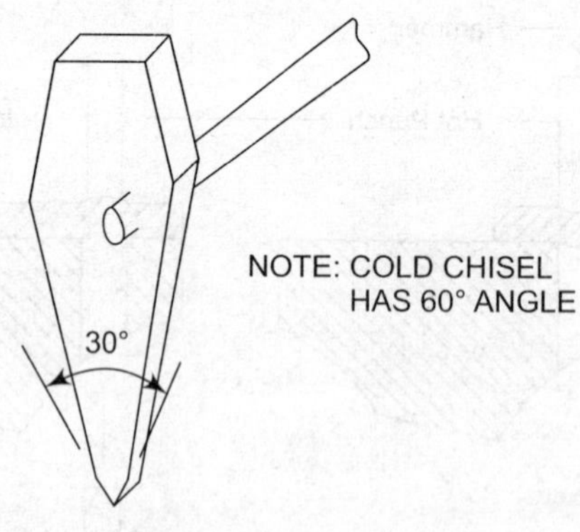

Fig. 4.15 Hot Chisel

Swaging: It is the operation by which the required cross sectional shape is obtained using appropriate swages. See Fig. 4.16.

Now let us consider an example in forging of a hexagonal headed bolt. The raw material may be a round rod of diameter equal to that of the bolt. Figure 4.16 shows the sequence of operations like apportioning, upsetting the head portion, shouldering and swaging to hexagonal shape and finally chamfering.

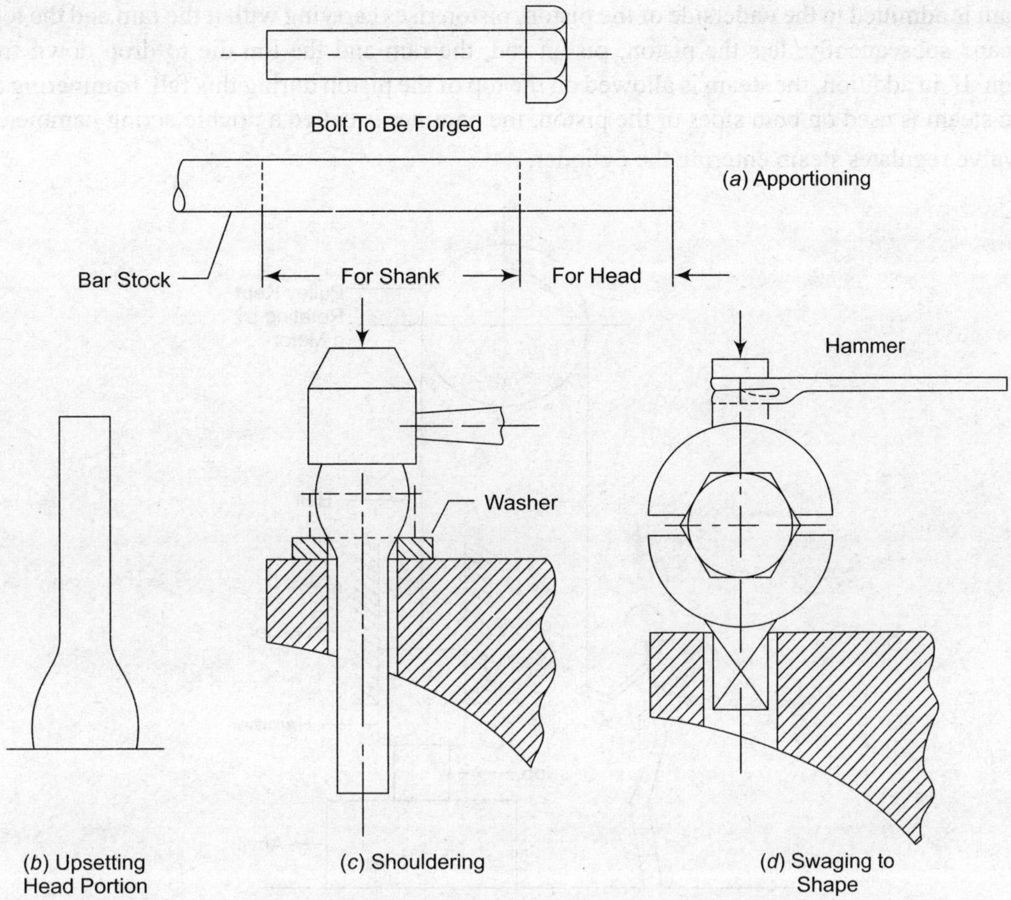

Fig. 4.16 Operations in Forging Hexagonal Bolt

4.4 EQUIPMENT IN A FORGE SECTION

The main equipment here are (*i*) Furnaces for heating and (*ii*) Power hammers/presses for shaping.

4.4.1 Heating Furnaces

Heating furnaces have closed enclosures enabling conserving and directing the heat on the job. This arrangement helps in

(*i*) Rapid and uniform heating of job

(*ii*) Protecting the job from atmospheric contamination.

Furnace usually have temperature control, which helps maintain the job at the desired temperature.

These furnaces are classified as Box or Batch type, continuous type with mechanical pushers, rotary hearth type etc.

With the availability of cheap electric power, high frequency induction furnaces are used increasingly. The advantages are fast and uniform heating and closer control of temperature. More importantly, the heating process is almost totally free from oxidation.

4.4.2 Power Hammers

Smaller capacity hammers are of mechanical type. Some are shown in Fig. 4.17. They use electric motors to power the machines. Larger capacity hammers use steam/air power.

Figure 4.18 shows a steam powered forging hammer. It has a pair of flat surfaced dies. The bottom die is fitted to the anvil block. The top die is fixed to the ram attached to the piston rod.

The **Anvil block** is mounted on an independent foundation to prevent shock and vibrations getting transmitted to the frame of the machine and surroundings.

When the steam is admitted to the underside of the piston, piston rises carrying with it the ram and the top die. Allowing the steam to escape subsequently, lets the piston, piston rod, the ram and the top die to drop down freely giving the hammering action. If, in addition, the steam is allowed on the top of the piston during this fall, hammering action is further enhanced. When steam is used on both sides of the piston, the hammer is called a double acting hammer.

The control valve regulates steam entering the cylinder.

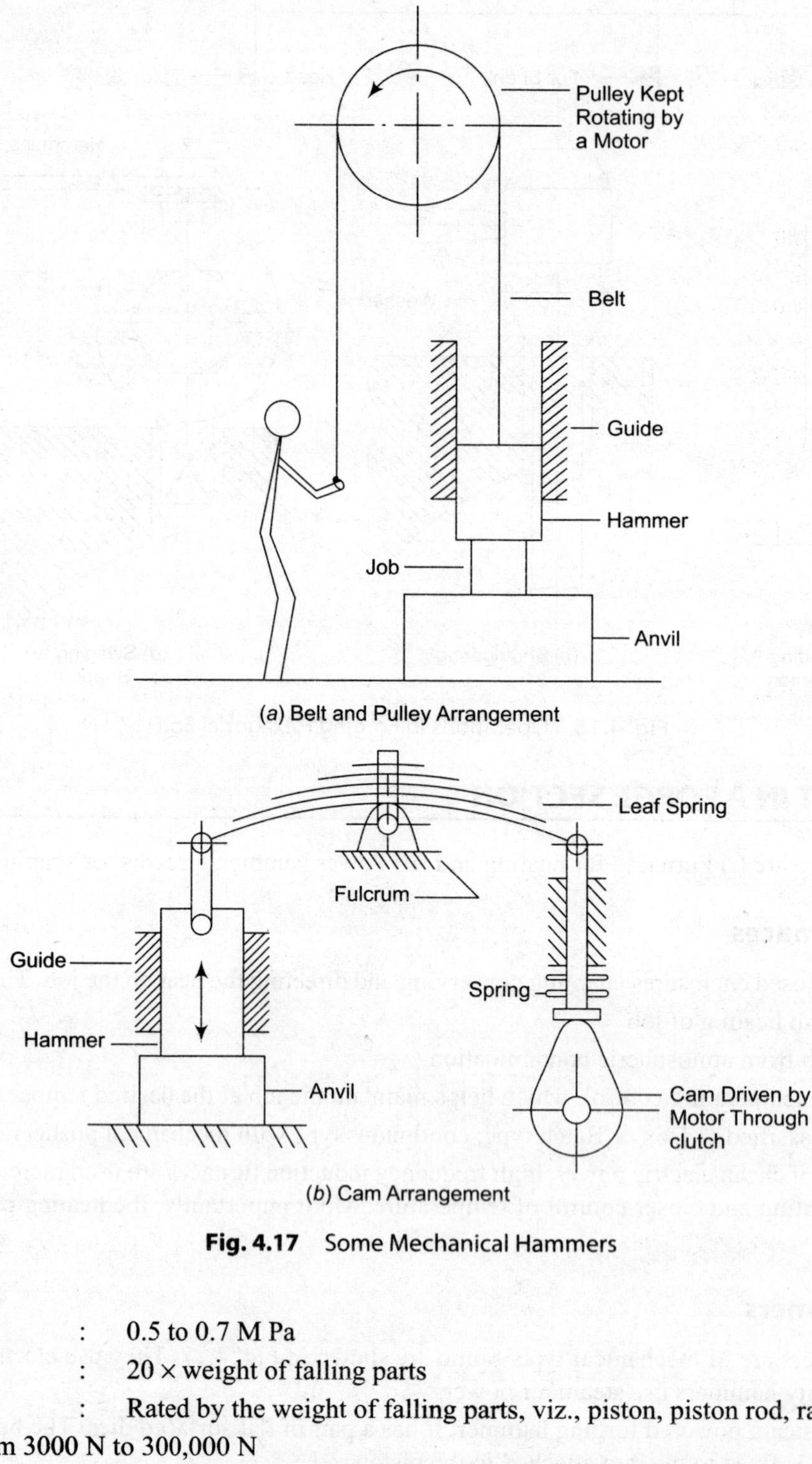

Fig. 4.17 Some Mechanical Hammers

Some design details:

Operating Pressure : 0.5 to 0.7 M Pa
Weight of anvil : 20 × weight of falling parts
Capacity : Rated by the weight of falling parts, viz., piston, piston rod, ram and the top die.
Capacity ranges from 3000 N to 300,000 N

Types of hammers: The two types are (*a*) Single frame type for smaller jobs and (*b*) double frame type for bigger jobs. Power hammer shown in Fig. 4.18 is a double frame type.

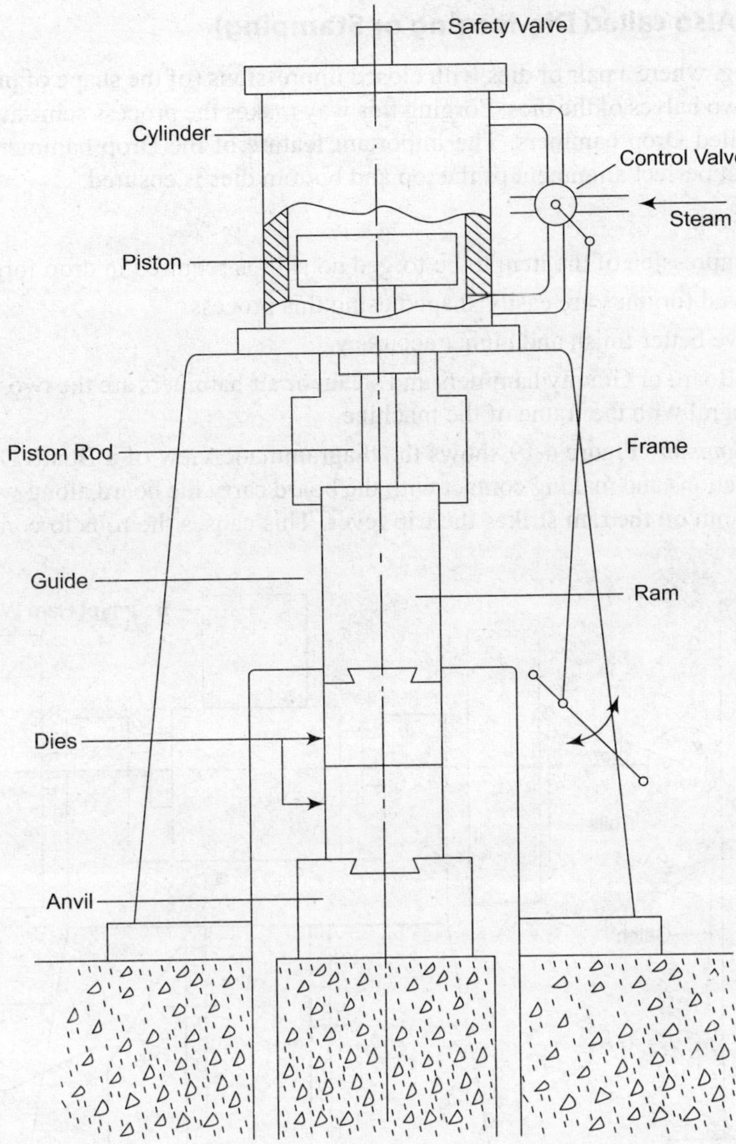

Fig. 4.18 Steam Forging Hammer

4.5 FORGING TECHNIQUES

Many forging techniques have been developed with a view to speed up production and improve quality of forgings. Also available are techniques specific to certain job requirements.

Following are some of the most commonly used forging techniques.

1. Hammer Forging
2. Drop Forging
3. Press Forging
4. Upset Forging
5. Rolling
6. Roll Forging and
7. Extrusion

4.5.1 Hammer Forging

This method employs power hammers described above. The forgings are made using flat faced dies. Good skill is required on the part of the operator to get the desired shape and finish.

Big jobs are easily forged using this method.

4.5.2 Drop Forging: (Also called Die-forging or Stamping)

It is a type of machine forging, where a pair of dies with closed impressions (of the shape of product) is made use of. The impression is shared by the two halves of the dies. Forging this way makes the process semi-automatic. Forging hammers used for this purpose are called Drop hammers. The important feature of the Drop hammers is that the anvil is made integral with the frame so that perfect alignment of the top and bottom dies is ensured.

Advantages:

- Since the dies carry impression of the item to be forged no skill is required in drop forging.
- Small and medium sized forgings are easily shaped using this process.
- Drop forged items have better finish and higher accuracy.

Types of Drop Hammers: Board or Gravity hammers and steam or air hammers are the two types used in drop forging. In both cases the anvil is integral with the frame of the machine.

Board or Gravity Drop Hammer: Figure 4.19 shows the diagrammatic view of a Board Drop hammer, in which the rolls rotating in opposite directions and making contact with the board carry the board along with the ram upwards. After reaching a certain height the pin on the ram strikes the trip lever. This causes the rolls to come down leaving the board and the ram free to drop.

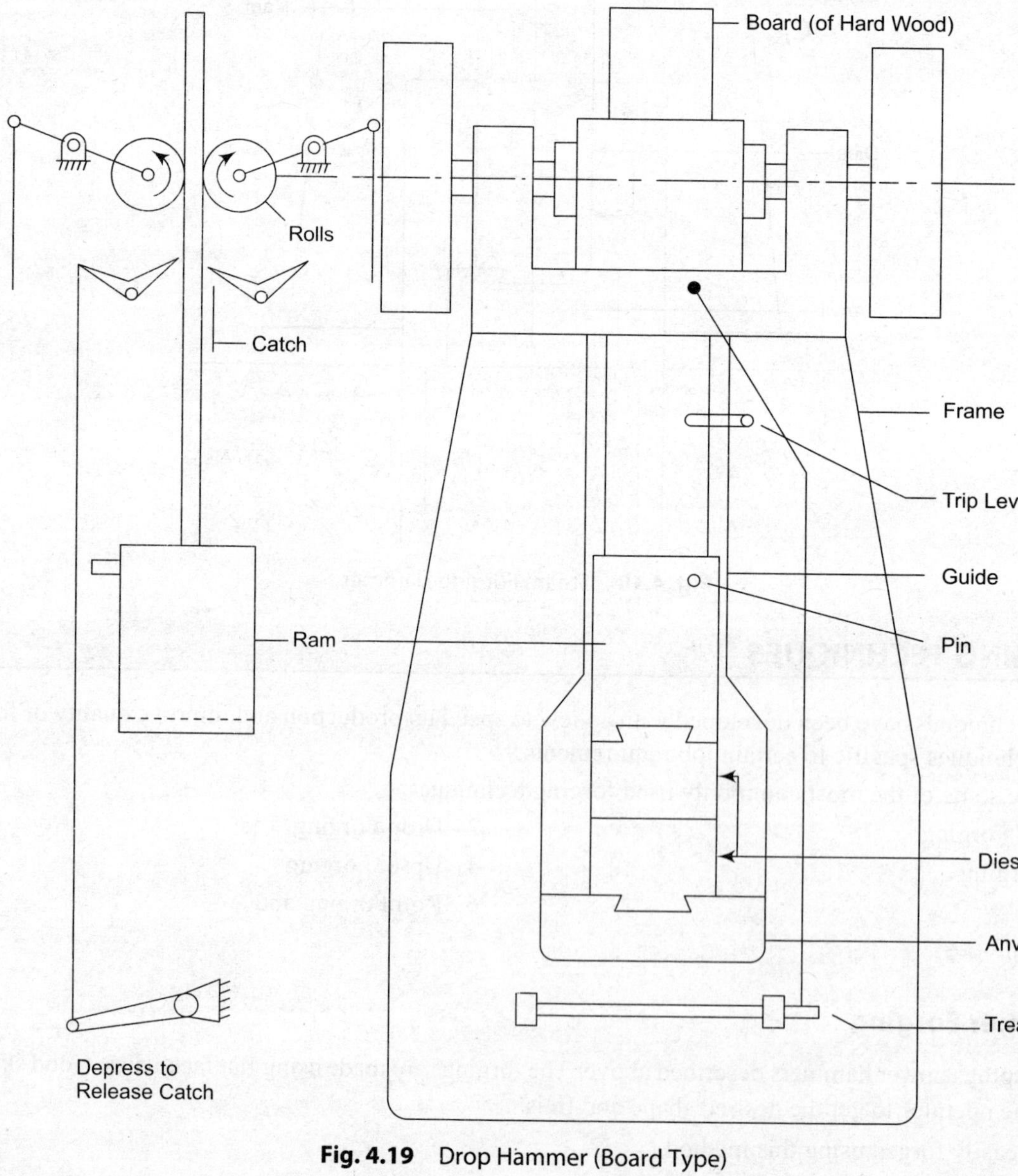

Fig. 4.19 Drop Hammer (Board Type)

As the ram drops, rolls get back into contact with the board and the ram rises again. This cycle of rising and falling of the ram repeats as long as the treadle is kept depressed by the operator. Once the treadle is released and upward motion of the ram ceases, downward motion commences. At this stage, the catches bite into the board and prevent the ram from falling. Only when the treadle is depressed again, catches release the board and hammering continues.

Rating: Drop hammers are also rated in the same way as forging hammers. The capacity is given in terms of the total weight of the falling parts. Capacity ranges from 1000 to 40,000 N, with rise in height of 1 to 2 m. No. of blows: 300 per minute.

Examples of drop forgings: Spanner, Scissors, Hammers, etc.

Steam or Air operated drop hammers: They use steam or compressed air to operate the ram. Capacity, however, goes up to 200,000 N.

4.5.3 Press Forging

Forgings made using presses are called press forgings. Hydraulic and Mechanical Presses are used for this purpose. Presses exert gradual pressure and the metal heated to plastic temperature is squeezed to flow into the cavity of the dies.

Advantages of Press forging:

- Press forgings have better accuracy.
- Each forging is completed in one single stroke. In contrast, in hammer forging or drop forging several blows are required to complete a forging.

Figure 4.20 shows the schematics of a hydraulic press. The capacity ranges from 50 to 50,000 tons. (Pressure requirement may be as high as 190 M Pa).

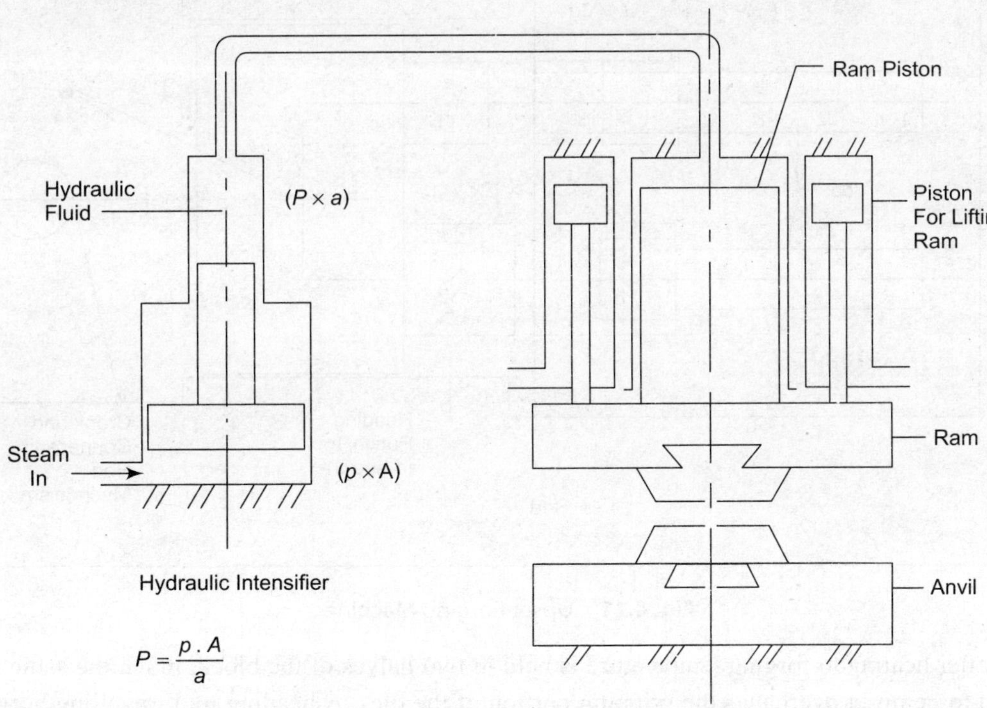

Fig. 4.20 Schematics of Hydraulic Press

Press forging vs. Hammer forging

Factors Considered	Press Forging	Hammer Forging
1. Working Principle	In one single stroke the shaping is done by a slow squeezing action	Shaping requires a series of successive blows
2. Type of Dies	Closed impression	Flat faced

3. Quality of forging		
(*i*) Accuracy	High	Low
(*ii*) Structure	Dense and more uniform	Comparatively Less
(*iii*) Finish	Smooth	Rough
4. Maneuverability	More convenient to vary the stroke length and amount of pressure	Difficult
5. Possibility to make intricate forgings	Less	More
6. Capacity of Machine	Very high capacity possible	Limited to 40 tons because of vibrations.
7. Energy utilization	Better	Much of energy is wasted
8. Economy	For symmetrical and less intricate jobs it is economical	For irregular complicated shapes more economical

4.5.4 Upset Forging

This type of machine forging is intended for making components out of bar stock, which require upsetting operation. An upset forging machine designed for this purpose is shown diagrammatically in Fig. 4.21.

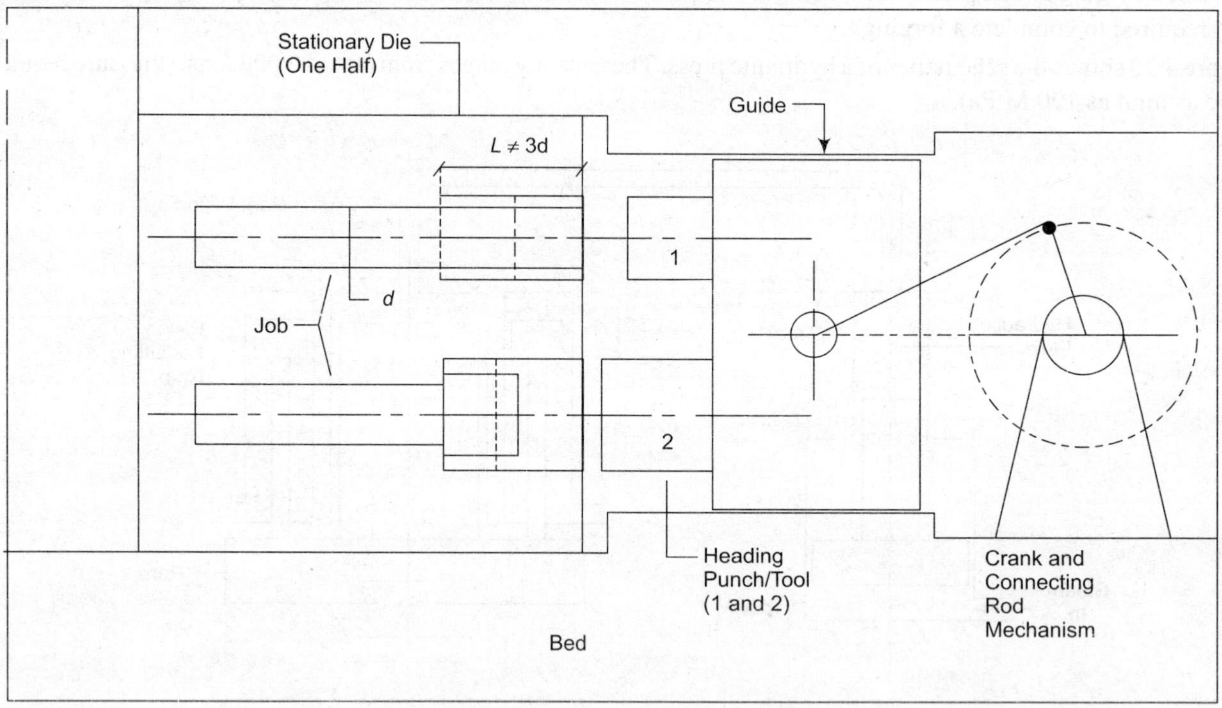

Fig. 4.21 Upset Forging Machine

The bar stock after heating to forging temperature is held in two halves of die block, much the same way as holding in a vice. The end to be upset overhangs the gripping portion of the dies. A heading tool travelling horizontally moves against the heated end of the job and upsets it.

For the sake of simplification a 2-stage upsetting and forming of a bolt head is shown in the figure. The job successively moves from top to bottom grooves after each stroke of the heading tool, and is completed as it leaves the last set of grooves.

Examples of upset forgings: Bolts, Mushroom type valves Rear axle of automobiles etc.

4.5.5 Rolling

Rolling is a forging process used for making rods, sheets and various other structural items like I-Section, T-section, Channels, Angles etc., Starting from Blooms, Billets and Slabs.

Figure 4.22. shows the production sequence in getting a rolled product. Figure 4.23 shows the stages of rolling in bringing down the billet to small section and shape it into a round rod. Figure 4.24 shows the rolling operation. We can see how the grains elongate and orient themselves in the direction of rolling.

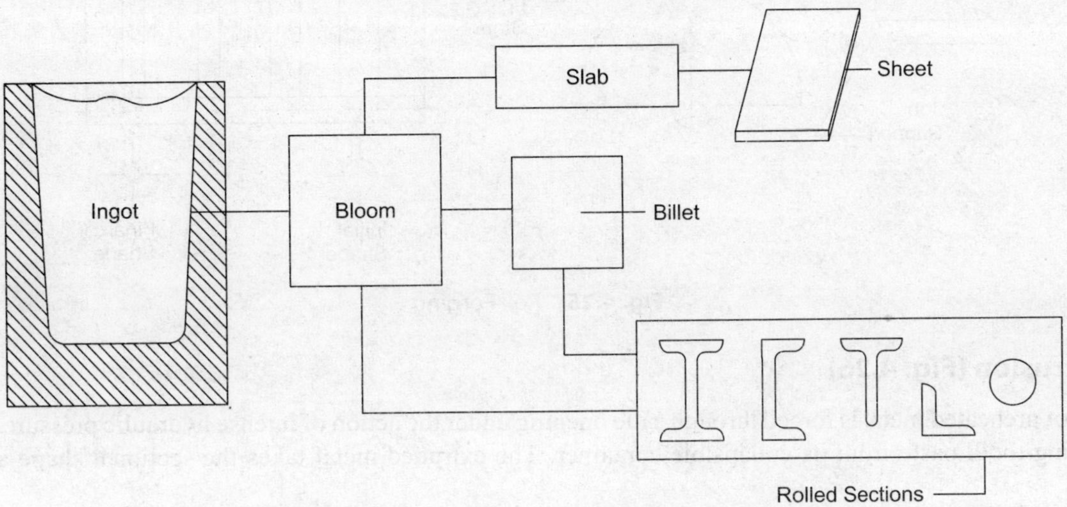

Fig. 4.22 Production Sequence in Getting Rolled Products

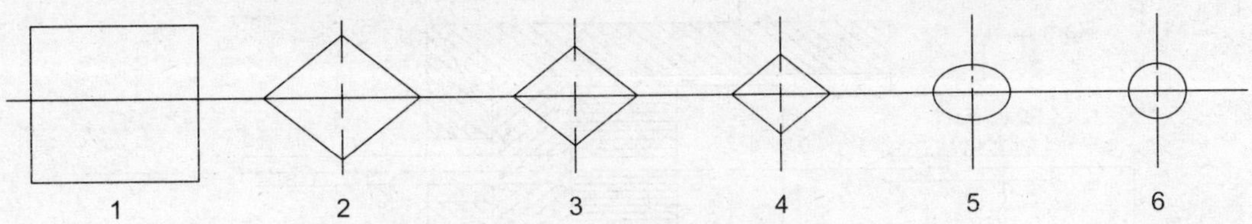

Fig. 4.23 Stages in Rolling a Billet into a Round Rod

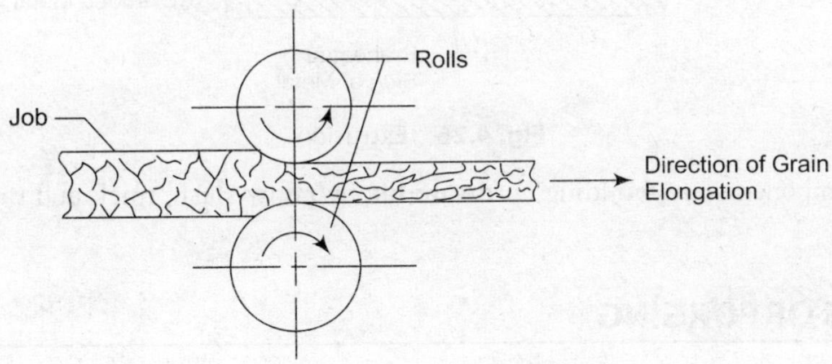

Fig. 4.24 Rolling Operation

4.5.6 Roll Forging

This process helps to reduce the section and bring about the desired shape in smaller jobs.

Roll forging machine consists of two semicircular rolls as shown in Fig. 4.25 with a series of grooves cut on them for specific operations.

As the rolls rotate, the operator places the heated rod between the rolls in the first set of grooves. In the 180° rotation, a rod undergoes certain change. In the next 180° rotation of the rolls, the job (rod) is withdrawn and is then shifted to the next set of grooves and so on sequentially and the required shape is obtained as the rod comes out from the last set of grooves.

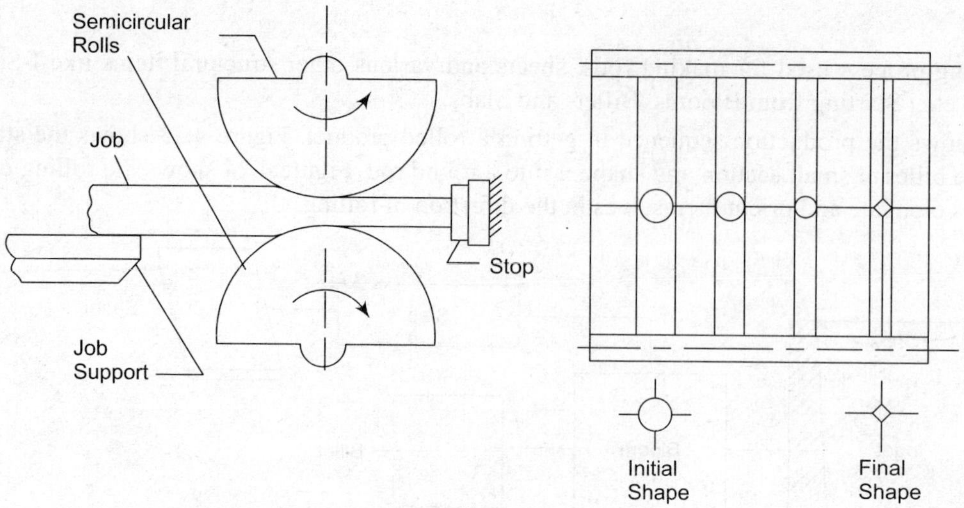

Fig. 4.25 Roll Forging

4.5.7 Extrusion (Fig. 4.26)

In this a slug of preheated metal is forced through a die opening under the action of intense hydraulic pressure. The process is like squirting tooth paste from its collapsible container. The extruded metal takes the sectional shape as that of the die.

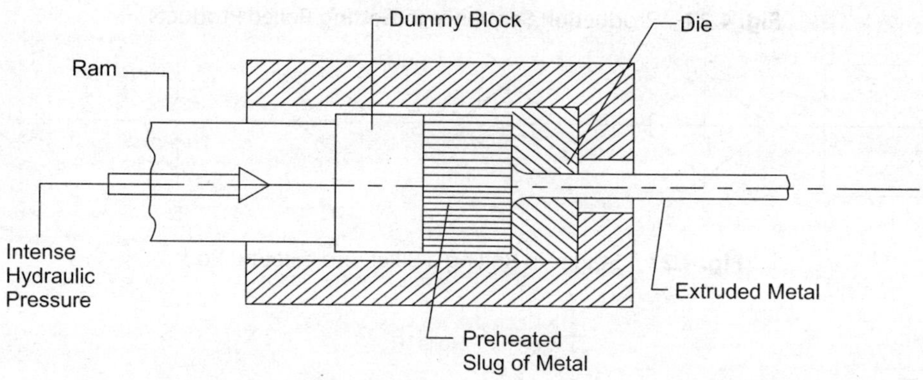

Fig. 4.26 Extrusion

Example of extruded components are: building hardware like window slash, track and rails, structural elements like angles, channels etc.

4.6 ADVANTAGES OF FORGING

The most important aspect of forging is that the grain flow lines could be so oriented that the component develops maximum strength.

A rolled component such as a steel rod has grain elongation in the direction of rolling (Fig. 4.24) and hence we can visualize flow lines in that direction.

Consider the following examples:

I. A bolt machined out of a solid bar [Fig. 4.27(*a*)] may fail at the corner under tension as shown.

On the otherhand a forged bolt [Fig. 4.27(*b*)] shows the flow lines oriented in such a way that the bolt is stronger than the one made by machining.

II. In the case of a forged crank shaft shown in Fig. 4.28, the flow lines are so oriented that the directional property of forging strengthens the web connections to the crank pin and the shaft.

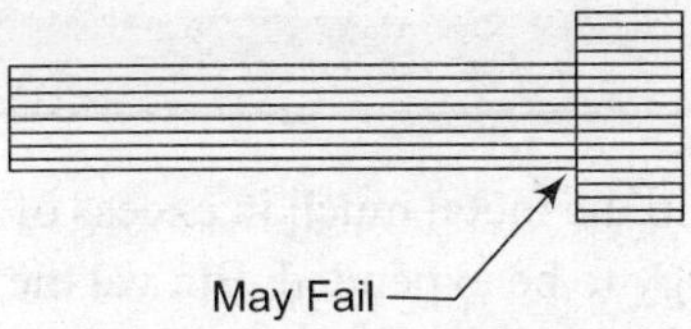

Fig. 4.27 (*a*) Machined Bolt

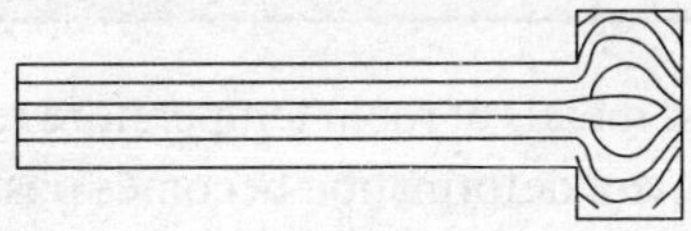

Fig. 4.27 (*b*) Forged Bolt

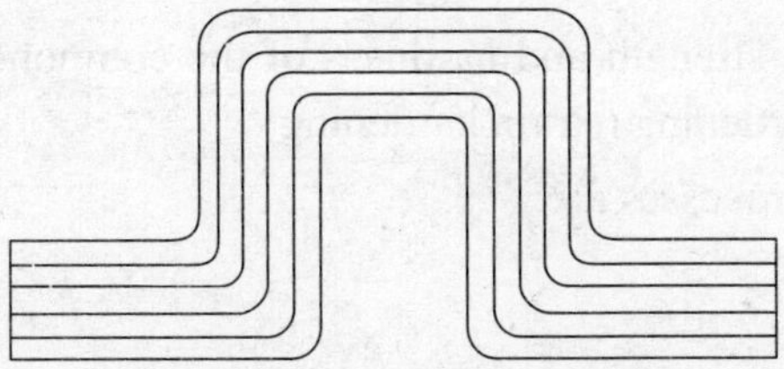

Fig. 4.28 Forged Crank Shaft

The other advantages are lighter weight and better finish when compared with casting.

Forging Vs. Casting

Forging:

- Forging refines structure of metal by breaking up large grains and closing up cavities that may be present.
- Forged components exhibit directional properties, *i.e.*, they are stronger in the direction of flow lines.
- The impurities are broken up and distributed.
- Resistance to shock and vibration is improved.

Casting:

- Castings are always hard and brittle with some exceptions.
- Castings are easily broken up by shock and vibration.
- Poor in tensile strength.
- No bending is possible.
- Cracks and blow holes make castings weaker and so unsuitable for many applications.

4.7 DEFECTS IN FORGINGS

Defects in forging are attributable to the following factors.

1. Raw Material
2. Forging Temperature and
3. Hammering Action

Raw Material: Defects arise because of the inclusion of dirt and slag in the raw material. Ingot defects such as pipes, cracks etc., are carried into the forgings.

Forging Temperature: Too high a forging temperature causes continued growth of grain size even after forging operation is stopped. This will result in coarse grained structure, with poor mechanical property. Finishing of forgings at too low a temperature (below the recrystallization temperature; in the case of steel it is about 700°C.) may result in hair cracks. Over heating (beyond 900°C in the case of steel) makes the grains very coarse and hammering results in tears and cracks. Also in this condition oxidation is very rapid and excessive scales are formed. Consequently loss of material occurs. High temperature also causes decarburization to occur affecting the mechanical property.

Hammering Action: Improper hammering action leads to cracks and laps.

4.8 COLD FORGING

Forging of most metals at room temperature requires stresses to be induced in the metal much in excess of the elastic limit so that a permanent deformation becomes possible. This requires lot of energy to be expended. But yet the process of cold forging is used where accuracy and high finish are called for.

Many hot rolled products are cold rolled in the end to get better dimensional accuracy and surface finish.

Also cold forging causes increase in strength and hardness of the component. This method of increasing hardness without heat treatment is called work hardening (strain hardening).

Some of the important cold forging processes are:

(*i*) Cold rolling
(*ii*) Cold extrusion
(*iii*) Wire drawing
(*iv*) Coining
(*v*) Heading

4.8.1 Cold Rolling

This process is similar to hot rolling and employs rolling mills. But during the process, work hardening may take place and further rolling becomes difficult. When this happens the job is subjected to annealing process —a heat treatment process, which reduces the hardness. The annealed job is subsequently rolled to size.

4.8.2 Cold Extrusion

The process is similar to Hot extrusion, but it is applicable only to soft metals notably lead, tin and aluminium alloys.

Examples of cold extruded products are: rods, tubes, structural shapes, lead covered cables, etc.

Impact extrusion is a type of cold extrusion brought about by the impact of a punch on soft metal blank placed in a blind die.

Figure 4.29 shows the action of impact extrusion.

Toothpaste tubes, soft drink cans etc., are produced by this method.

4.8.3 Wire Drawing

Wires are cold-drawn starting with rod of diameter 6.0 mm or less. Figure 4.30 shows the process.

4.8.4 Coining

This process is used for making coins, medallions etc. The blank is squeezed between dies and the increase and decrease of cross section takes place. Figure 4.31 shows the scheme of coining process.

4.8.5 Heading

Nails, rivets, small bolts etc., are made by this process. Figure 4.32 shows the schematics of making rivets.

4.9 COLD FORMING

Cold forming methods are essentially used for manufacturing products out of sheet metal. Following are some of the important processes.

(*a*) Punching and Blanking (*b*) Drawing/Cupping

(*c*) Spinning and (*d*) Embossing

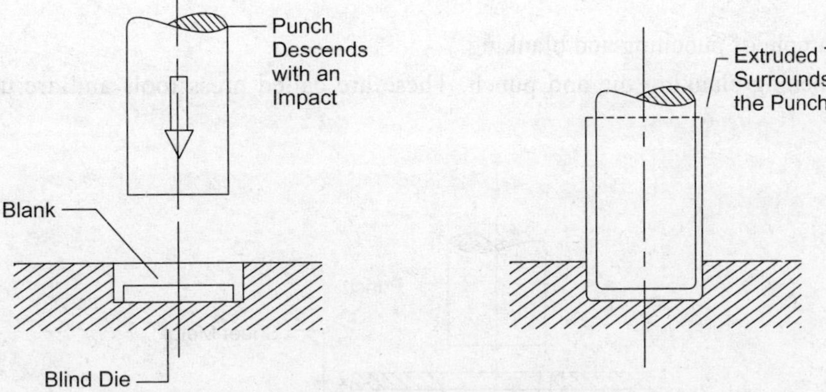

Fig. 4.29 Impact Extrusion

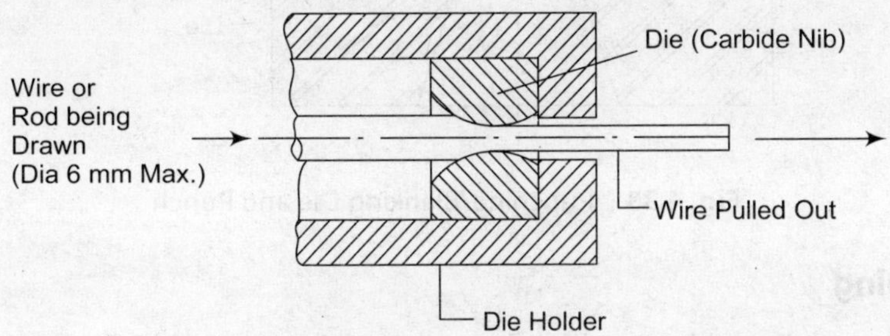

Fig. 4.30 Wire Drawing

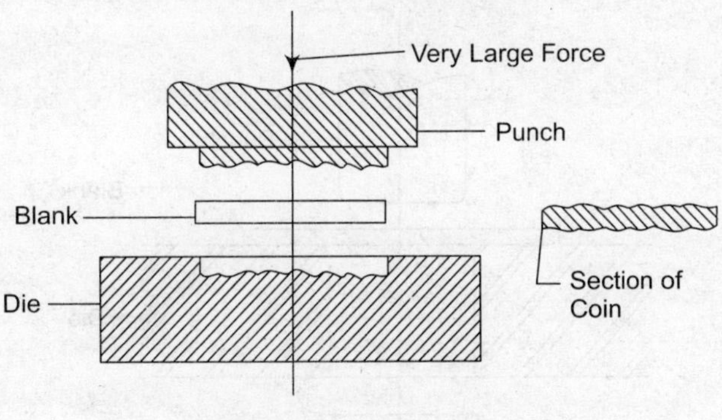

Fig. 4.31 Coining Process

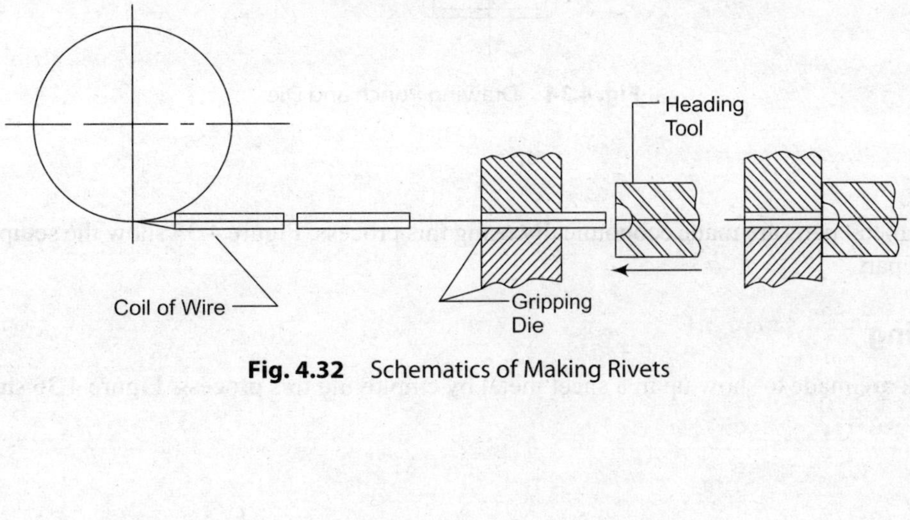

Fig. 4.32 Schematics of Making Rivets

4.9.1 Punching and Blanking

Punching is an operation to produce a desired shape and size of hole in a sheet metal, whereas Blanking is cutting out a piece of required shape and size from out of a sheet metal. The blank obtained is used for making sheet metal products in subsequent operations.

A washer is a good example of punching and blanking.

Figure 4.33 shows punching/blanking die and punch. These are called press tools and are used in mechanical or hydraulic presses.

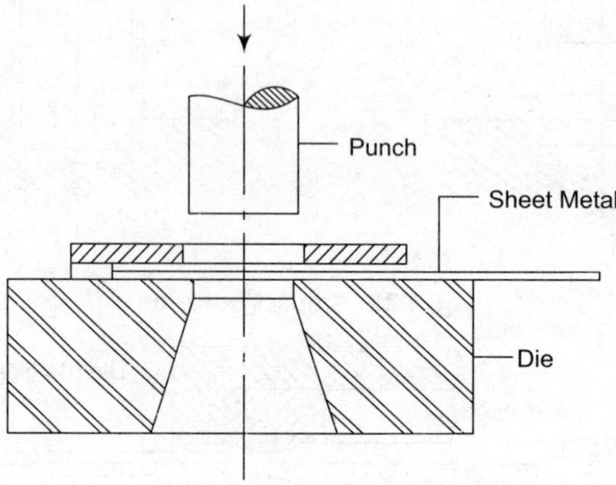

Fig. 4.33 Punching/Blanking Die and Punch

4.9.2 Drawing/Cupping

This sheet metal operation is used for making cup like objects out of sheet metal blanks. Figure 4.34 shows the press tools used for drawing with the blank placed for operation.

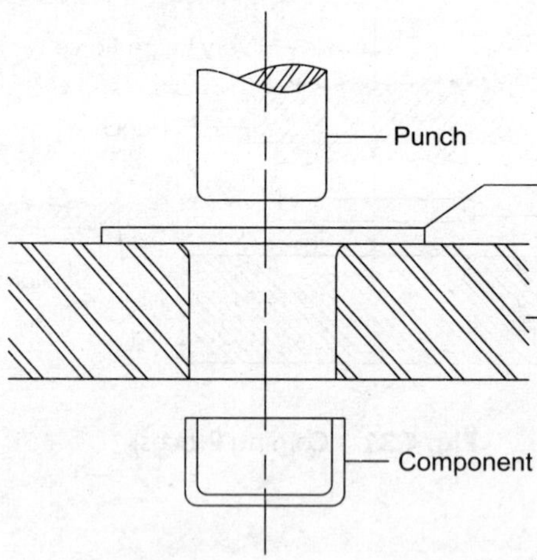

Fig. 4.34 Drawing Punch and Die

4.9.3 Spinning

Large hollow sheet metal parts are made economically using this process. Figure 4.35 show the setup for spinning a sheet metal into a hollow part.

4.9.4 Embossing

Letters and art work are made to show up in a sheet metal by employing this process. Figure 4.36 shows a set of dies for embossing 'MCE'.

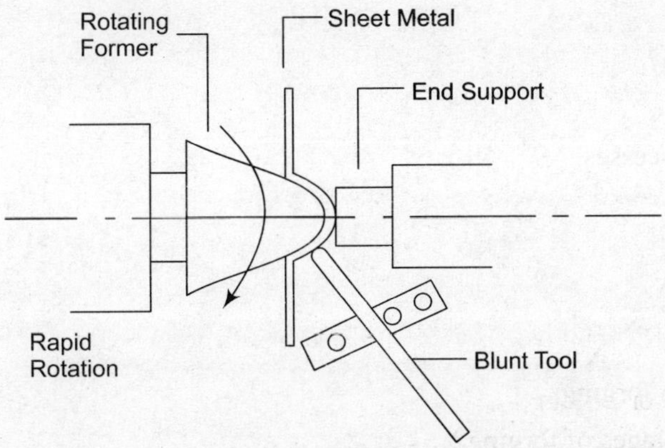

Fig. 4.35 Spinning Setup

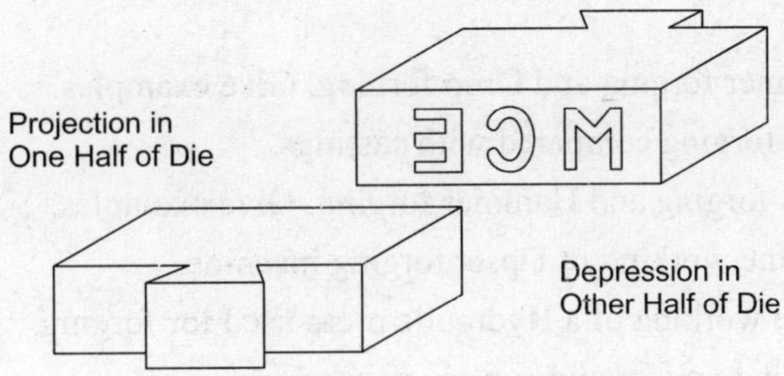

Fig. 4.36 Embossing Dies

QUESTION BANK

2-Marks Questions

1. What is forging?
2. What is cold forging?
3. Give a few examples of forging.
4. What is hand forging? Where is it done?
5. What is machine forging? What is the work place called?
6. What is a hearth?
7. What is an anvil?
8. What are swages?
9. What are fullers?
10. List the main forging operations.
11. What is 'upsetting'?
12. What is 'drawing'?
13. What is hammer forging?
14. What is drop forging?
15. Give a few examples of drop-forged components.
16. What is press forging?
17. What is upset forging?
18. What is roll forging?
19. What is rolling?
20. What is extrusion?

21. What is cold rolling?
22. What is impact extrusion?
23. What is wire drawing?
24. List the cold forming processes.
25. What is spinning?
26. What is embossing?
27. What is punching?
28. What is blanking?
29. List the defects arising in forging.
30. What are the main advantages of forging?

10-Marks Questions

1. Differentiate between Hammer forging and Drop forging. Give examples.
2. What are the advantages of forging compared with castings.
3. Differentiate between Press forging and Hammer forging. Give examples.
4. With a neat sketch explain the working of Upset forging machine.
5. Explain with schematics the working of a Hydraulic press used for forging.
6. Make a neat sketch of a Drop hammer and explain its working.
7. Explain the working of a Steam forging hammer with a neat sketch.
8. Give a step-by-step procedure for making a hexagonal bolt by forging.

State True or False

1. Forging damages the material and reduces the strength of the component.
2. Hot forging produces poorer surface finish.
3. Cold forging requires less power compared with hot forging.
4. Cup-like component can be obtained by deep drawing.
5. Hot chisel has an angle of 60 degrees.
6. Localised thinning of job by forging is called Setting down.
7. Drop forged items have better finish and higher accuracy.
8. Forging hammers use flat faced dies.
9. Spanner is an example of hammer forging.
10. In drop forging each forging is completed in one single stroke.
11. Forged items are less resistant to shock and vibration.
12. Cold forging causes increase in strength and hardness of the component.
13. Toothpaste tubes are produced by cold extrusion.
14. Embossing is a cold forming process.
15. Blanking is an operation to produce a hole of desired shape and size.

Answers

1. F **2.** T **3.** F **4.** T **5.** F **6.** T **7.** T **8.** T **9.** F **10.** F
11. F **12.** T **13.** T **14.** T **15.** F.

Fill in the Blanks

1. The processes using compressive forces to shape components are ....
2. Wire drawing uses .... force.
3. Tubes can be produced by .... or ....
4. In a forge section the two important equipment used are .... and ....
5. Structural items like I-section etc. are produced by ....
6. In drop hammer the anvil is made .... with the frame.
7. The capacity of a drop hammer is given in terms of the total weight of ....
8. Dies used in drop forging have .... impression of the product.

Answers

1. Forging, Rolling, Extrusion
2. Tensile
3. Rolling, Extrusion
4. Furnaces, Forge hammers
5. Rolling
6. Integral
7. Falling parts
8. Closed.

Chapter-5

Metal Joining

Metal joining is one of the important production activities contributing to the success of the latest engineering achievements. Although the processes —welding, brazing and soldering together with riveting, fastening with nuts and bolts, etc., have been used over the past hundreds of years, it is only in the recent past the one process —Welding has been developed to such an extent that some of the marvels in engineering like fabrication of high pressure boilers, nuclear reactors and many such modern advancements have become possible.

Welding —What is it? Welding may be defined as joining of similar metals by bringing them to fusion temperature and making the joint with or without filler metal. Also, pressure may or may not be applied in the process.

5.1 WELDING-TYPES

Some of the very common types of welding are:

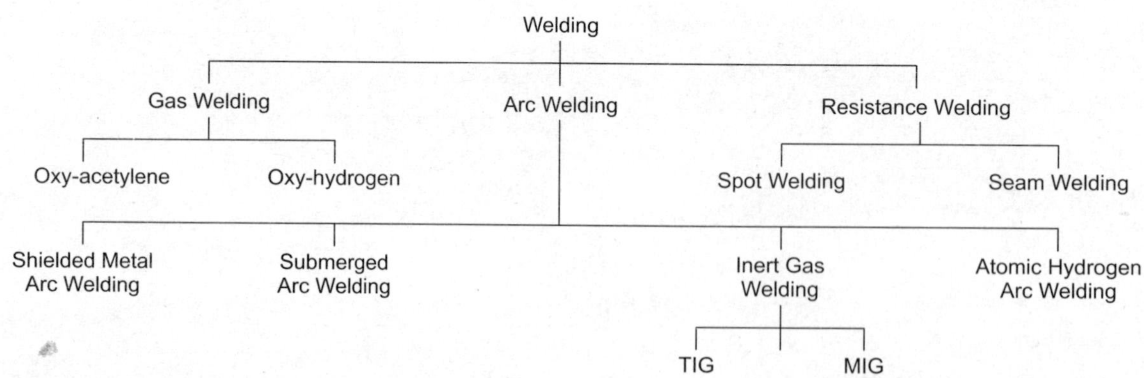

More recent developments in welding are:

(*i*) Electron beam welding

(*ii*) Laser beam welding

(*iii*) Friction welding

(*iv*) Ultrasonic welding

(*v*) Electro-slag welding

In addition, a special type of welding called 'Thermit welding' is very widely made use of in rail road work and also joining heavy sections.

5.2 GAS WELDING

In this process the fusion temperature is brought about by using oxyhydrogen or oxyacetylene flame. The oxyhydrogen flame gives about 1980°C and therefore has limited use. Whereas oxyacetylene flame producing about 3500°C, has wide applications.

Oxygen is obtained either by electrolysis or by liquefaction of air. Acetylene is generated sometimes by adding calcium carbide to water. The reaction is as follows:

$CaC_2 + 2H_2O \rightarrow Ca(OH)_2 + C_2H_2$

Calcium carbide + Water → Slacked lime + Acetylene

For industrial applications, the acetylene is bought in cylinders, in which case, the gas will be under pressure. So also oxygen is bought in cylinders.

5.2.1 Gas Welding Equipment

The following are the equipment used in gas welding.

1. Acetylene cylinder
2. Oxygen cylinder
3. Welding torch
4. Pressure regulators and connecting hoses
5. Gas lighter
6. Safety goggles, gloves, apron and safety shoes

Figure 5.1 shows the oxygen and acetylene cylinders with regulators and the welding torch.

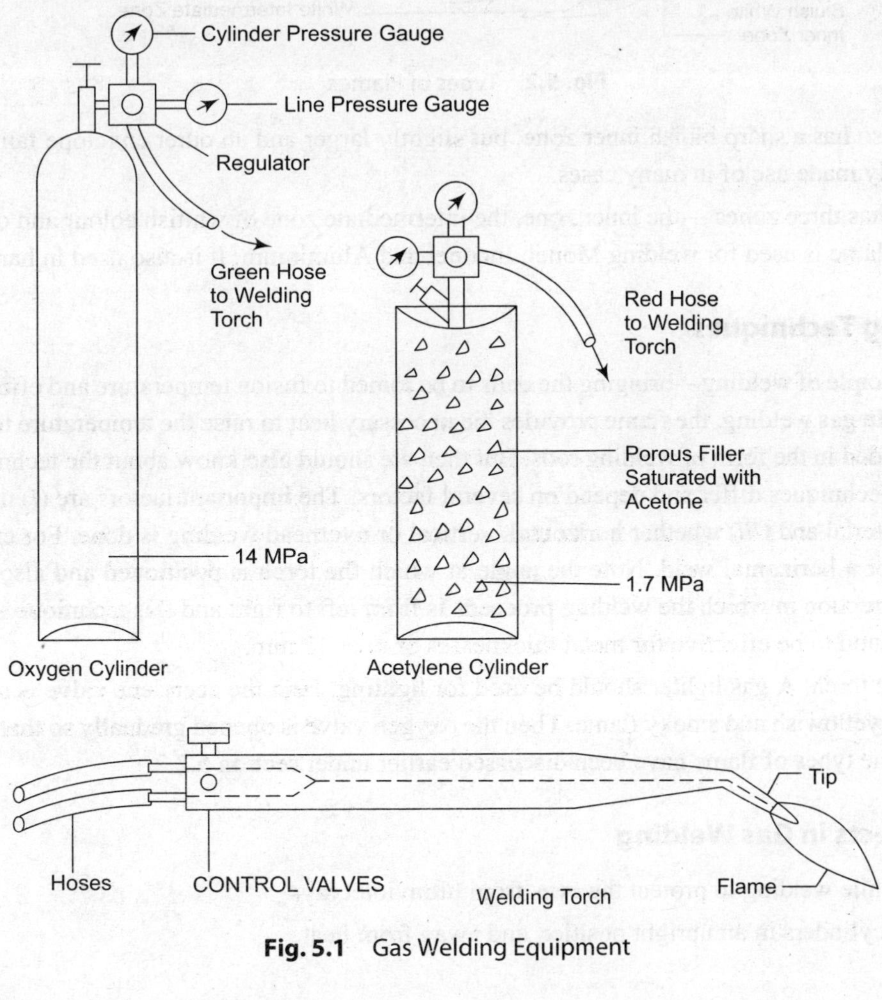

Fig. 5.1 Gas Welding Equipment

5.2.2 Types of Flames

Three types of flames are used in gas welding:

(*a*) Oxidizing flame (oxygen drawn is more than in neutral flame)

(*b*) Neutral flame

(*c*) Reducing flame (oxygen drawn is less than in neutral flame)

Figure 5.2 shows the three types of flames. The oxidising flame has a short inner zone of bluish tinge and an outer envelope. This is used in welding Brasses, Bronzes and Cast Iron.

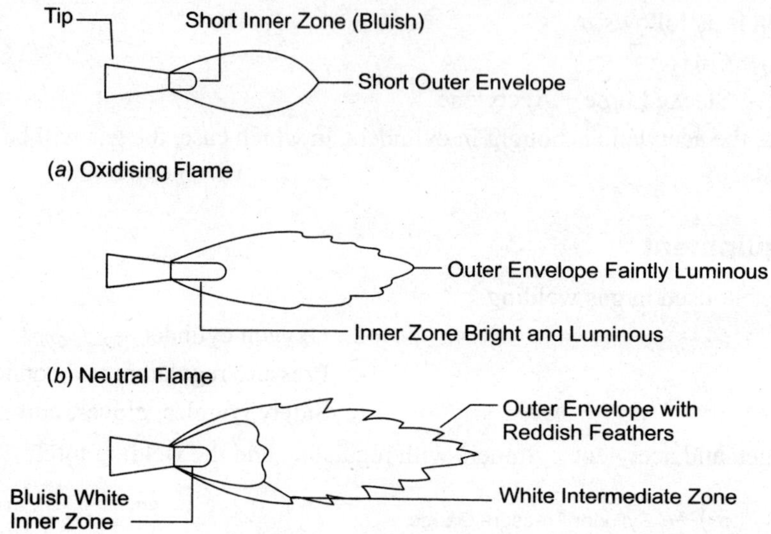

Fig. 5.2 Types of Flames

The neutral flame also has a sharp bluish inner zone, but slightly larger and an outer envelope faintly luminous. This flame is most commonly made use of in many cases.

The reducing flame has three zones —the inner zone, the intermediate zone of whitish colour and outer envelope with reddish feathers. This flame is used for welding Monel, Inconel and Aluminium. It is also used in hard facing.

5.2.3 Gas Welding Techniques

We know the basic principle of welding—bringing the ends to be joined to fusion temperature and effecting the joint with or without filler metal. In gas welding, the flame provides the necessary heat to raise the temperature to the required level and filler metal is provided in the form of welding rods. But then we should also know about the techniques by which we get satisfactory weld. Techniques differ and depend on several factors. The important factors are (*i*) the type of material, (*ii*) the thickness of material and (*iii*) whether horizontal, vertical or overhead welding is done. For example, in Fig. 5.3 is shown the method for a horizontal weld. Note the angle at which the torch is positioned and also the positioning of the welding rod. The direction in which the welding proceeds is from left to right and this technique is called Rightward Welding. It has been found to be effective for metal thicknesses of over 12 mm.

Lighting the welding torch: A gas lighter should be used for lighting. First the acetylene valve is opened and torch is lighted up. This gives a yellowish and smoky flame. Then the oxygen valve is opened gradually so that the required flame condition is reached. The types of flame have been discussed earlier under section 5.2.2.

5.2.4 Safety Aspects in Gas Welding

1. Wear goggles while welding to protect the eyes from ultraviolet rays.
2. Store acetylene cylinders in an upright position and away from heat.

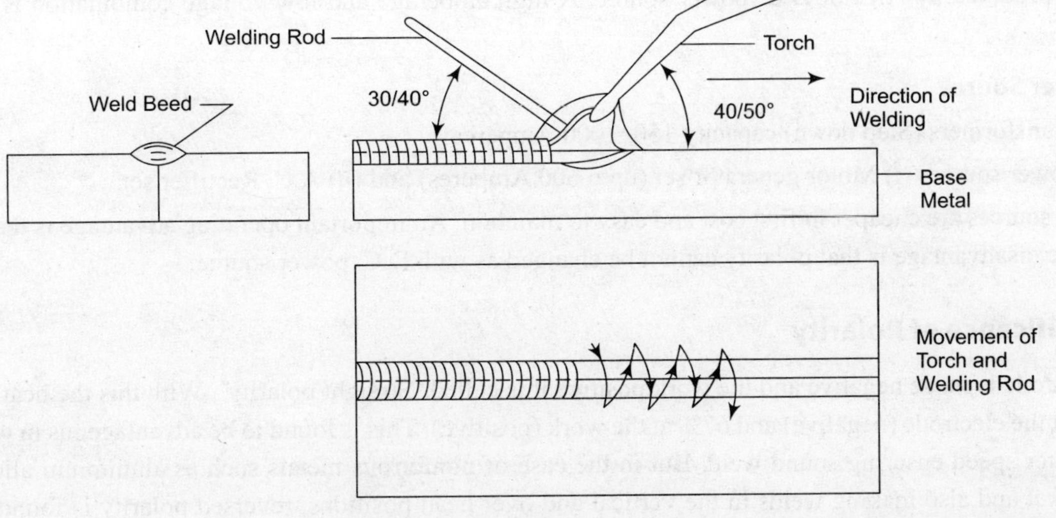

Fig. 5.3 Gas Welding Technique

3. See that oil or grease does not come into contact with high pressure oxygen, as it can become explosive.
4. Keep the welding area well ventilated.

5.2.5 Advantages of Gas Welding

1. The equipment is simple and less costly.
2. It is easily transported to the place of welding, indoor or outdoor.
3. Most metals can be welded using gas welding.
4. The heating of base metal and the filler rod can be preferentially done and so greater control over the process is possible.
5. Oxy-acetylene flame may be used in cutting of steel and also preheating of jobs for certain operations.

Limitation

It is difficult and time consuming to weld plates of thickness greater than 6 mm.

5.3 ARC WELDING

In this process the heat to bring about the required weld temperature is given by an electric arc, established between an electrode and the metal being welded. A.C. or D.C. current is made use of. The temperature attained is about 4000°C.

The electrode used in most cases is the filler metal rod itself and is consumed. The arc welding set-up is shown in Fig. 5.4.

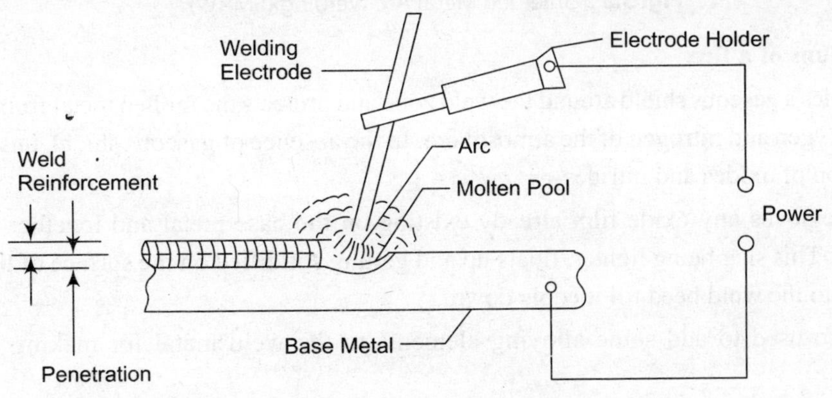

Fig. 5.4 Arc Welding Setup

5.3.1 Arc Welding Power Source

Electric arc is provided by A.C. or D.C. power source. A high amperage and low voltage combination is the requirement.

Types of Power Source

1. A.C. Transformers (Step down) capacity 150–1000 Amperes:
2. D.C. Power sources (*i*) Motor generator set (upto 600 Amperes) and (*ii*) A.C. Rectifier set.

A.C. Power sources are cheaper in first cost and easy to maintain. An important operating advantage is that there is no 'arc blow'. The disadvantage is that polarity cannot be changed as with D.C. power source.

5.3.2 Significance of Polarity

When the electrode is made negative and the work positive it is called 'Straight polarity'. With this the heat distribution is about 33% at the electrode (negative) and 67% at the work (positive). This is found to be advantageous in welding mild steel with greater speed ensuring sound weld. But in the case of nonferrous metals such as aluminium alloys, bronze, monel and nickel and also making welds in the vertical and over head positions, reversed polarity is found to be more advantageous. This is possible only with D.C. power sources, but only at a higher cost.

5.3.3 Arc Welding Techniques

(*i*) **Shielded Metal Arc Welding (SMAW).** Figure 5.5 shows the details of shielded metal arc welding. In this, the electrode providing the arc is consumed. That is, it melts and provides the filler metal. In order to protect the weld during the process, the electrode is covered with a flux, which burns and produces a gaseous protective shield around the weld zone. For this reason the process is called Shielded Metal Arc Welding.

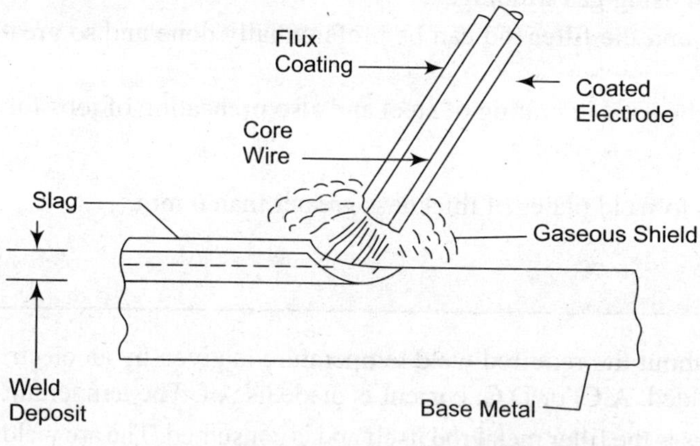

Fig. 5.5 Shielded Metal Arc Welding (SMAW)

Important functions of a flux

(*a*) Flux provides a gaseous shield around the weld zone and protects the molten metal from getting contaminated with the oxygen and nitrogen of the atmosphere. In the absence of gaseous shield, unsound weld results due to formation of oxides and nitrides.

(*b*) The flux removes any oxide film already existing on the base metal and together with other impurities forms slag. This slag being lighter, floats up and getting solidified on the surface of the weld formed, gives protection to the weld beed till it cools down.

(*c*) Flux is also used to add some alloying elements to the weld metal for making the weld ductile and strong.

Flux Materials and their use.

1. Ferro-alloys – used as deoxident.
2. Dextrin, starch, cellulose, wood powder, graphite – produce protective gas.
3. Powdered quartz, chalk kaolin, dolomite powder – used for formation of slag.

(*ii*) **Submerged-arc welding:** The process is so named because the arcing takes place under a blanket of flux. The schematics of the process is shown in Fig. 5.6. Flux is deposited continuously on the seam ahead of the electrode and the flux feeder and the electrode feeder together move as the welding proceeds.

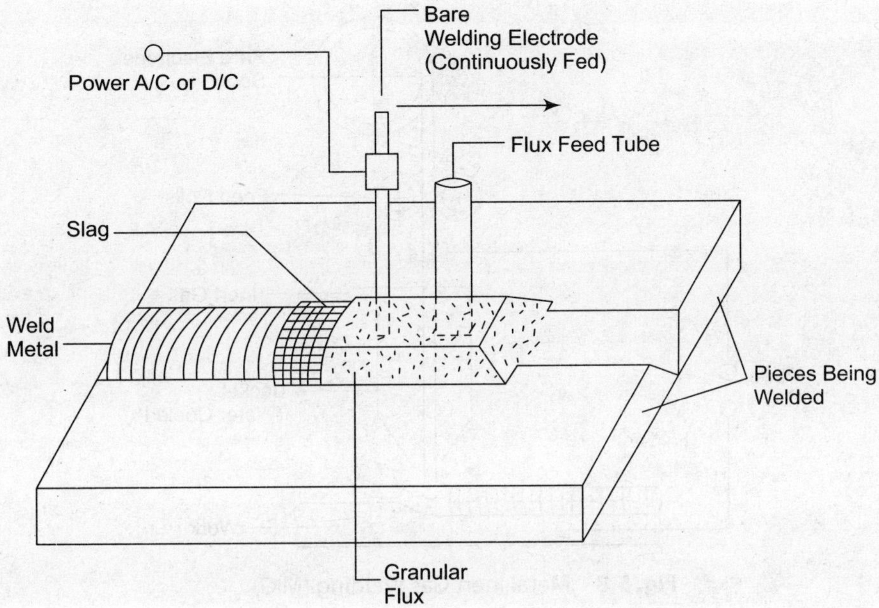

Fig. 5.6 Submerged-Arc Welding

Advantages

(*a*) Submerged arc welding is faster when compared with SMAW. (*b*) The quality of the weld obtained is excellent and uniform. (*c*) This technique enables deeper penetration. (*d*) The whole process is automatic. (*e*) The process is well-suited for welding thick plates.

(*iii*) **Inert gas shielded-arc welding:**

(*a*) *Tungsten Inert gas welding* (TIG) Fig. 5.7. In this a nonconsumable electrode made of tungsten is used. The arc is established between the tungsten electrode and the work. Generally A.C. is used for the arc. Filler metal is supplied in the form of welding rod. The weld zone is protected using an inert gas like argon, helium or carbondioxide.

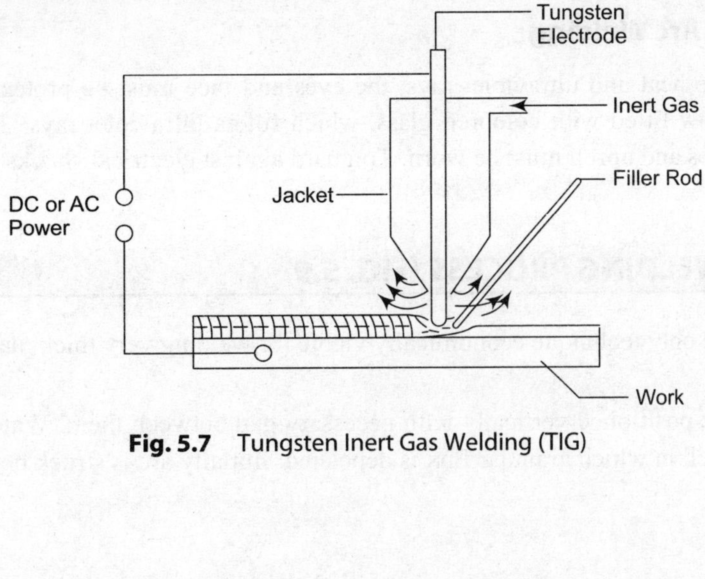

Fig. 5.7 Tungsten Inert Gas Welding (TIG)

Application: Although this is a costly process, it is warranted in quality welding of aluminium and magnesium alloys, since they have a tendency to burn off. The process is also used in high quality welding of stainless steel, cast iron and copper alloys.

(*b*) *Metal Inert Gas welding* (MIG): This is a welding process in which consumable electrode is made use of. Inert gas like argon or helium provides the shielding to prevent oxidation of the weld. Fig. 5.8 shows the schematics of MIG welding. An automatic feed mechanism continuously feeds the electrode as the weld progresses.

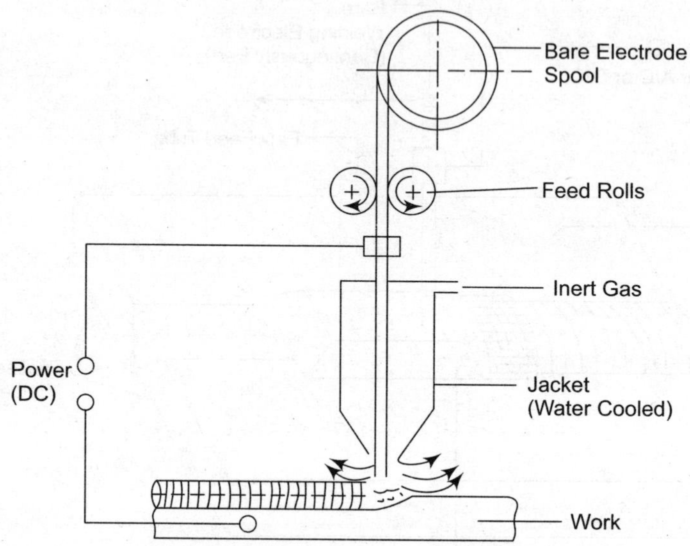

Fig. 5.8 Metal Inert Gas Welding (MIG)

Use of inert gas makes the process costlier, but then for high quality work there may not be any other option.

(*iv*) **Atomic hydrogen-arc welding:** When it is required to weld heat resistant metals like high chrome-nickel steel, the welding temperature must be higher than what is available in gas or ordinary arc welding. This technique is capable of giving temperature in the order of 4500°C. Here, the hydrogen gas is passed through an electric arc maintained between two tungsten electrodes. Hydrogen, subjected to high temperature of the arc, dissociates into atomic hydrogen and when it recombines outside the arc, a large amount of heat is released, which combining with the already intense heat of the arc provides a very high temperature.

Hydrogen provides a good protective shield and also removes oxide film from the surface of the work. This enables very fine and uniform weld to be produced.

5.3.4 Safety Aspects in Arc Welding

Since the arc gives out intense heat and ultraviolet rays, the eyes and face must be protected with a shield to avoid injury. The shield has a window fitted with coloured glass, which filters ultraviolet rays. To protect hands and body against spark and spatter, gloves and apron must be worn. To guard against electrical shocks good safety shoes must be worn.

5.4 ELECTROSLAG WELDING PROCESS (FIG. 5.9)

Developed in Russia, this is the only technique economically viable for welding very thick plates of the order of 100 mm and more.

The pieces to be welded are positioned vertically with necessary gap between them. Water-cooled copper slides on either side of the gap form a well in which granular flux is deposited. Initially arc is struck between electrode/electrodes and the bottom plate. The arc melts the electrode and the flux and a pool of molten metal with slag floating atop results. When enough slag accumulates, the arc is put off and further requirement of heat is provided by the resistance offered by the slag to the current flowing through it. The heat so generated would be sufficient to melt the electrode and also fuse the base metal.

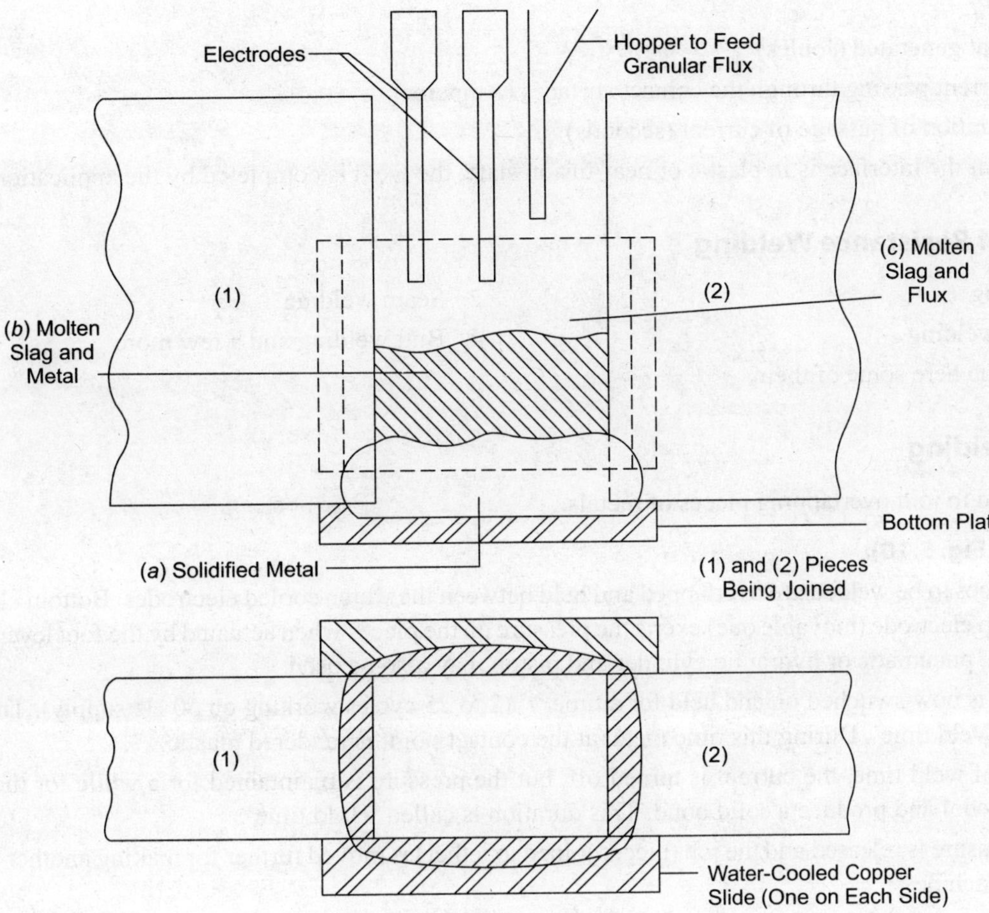

Fig. 5.9 Electroslag Welding Process

Once the process is on, we may visualize three zones as in Fig. 5.9: (*a*) solidified metal, (*b*) molten slag and metal and (*c*) molten slag and flux. As the weld progresses, the welding head supplying electrode and flux moves up and so also the water cooled copper slides confining the molten metal, the slag and the flux. The flux used is magnesium silicate and a refining process takes place in the zone (*b*).

Advantages

1. More than one electrode may be used in this process and therefore very high welding speed is possible.
2. The weld metal is totally out of contact with atmosphere and hence the best quality of weld results.
3. Very minimum joint preparation is required.

Some design details

The working voltage: 45–50 V

Current: 500–550 Amperes per 10 SWG electrode

Typical capability: Welding of 100 mm thick plates (with two 10 SWG electrodes taking 550 Amperes) takes 2 hours per metre length of weld.

5.5 RESISTANCE WELDING

In this process the heat required for welding is obtained by the contact resistance R (between the surfaces being welded) offered to the flow of high amperage current as per the equation.

$$H = I^2 Rt$$

Where H = Heat generated (Joules)

I = Current passing through the contact surfaces (Amperes)

t = Duration of passage of current (seconds)

Once the metal at the interface is in plastic or near fusion state, the weld is completed by the application of pressure.

5.5.1 Types of Resistance Welding

1. Spot welding
2. Seam welding
3. Projection welding
4. Butt welding and a few more

We shall examine here some of them.

5.5.2 Spot Welding

This process is used to join overlapping pieces of metals.

Procedure (Refer Fig. 5.10):

1. The two pieces to be welded are overlapped and held between the water-cooled electrodes. Bottom electrode being fixed, the top electrode (movable one) exerts the pressure on the pieces when actuated by the foot lever. For exerting the pressure, pneumatic or hydraulic cylinder and piston may also be used.
2. The current is now switched on and held for a time 't' (2 to 25 cycles working on 50 Hz supply). This duration is called the 'Weld time'. During this time metal at the contact point is rendered plastic.
3. At the end of weld time, the current is turned off, but the pressure is maintained for a while for the metal at the interface to cool and produce a solid bond. This duration is called 'Hold time'.
4. Now the pressure is released and the job (pieces joined together) is moved further for making another weld or taken out of the machine.

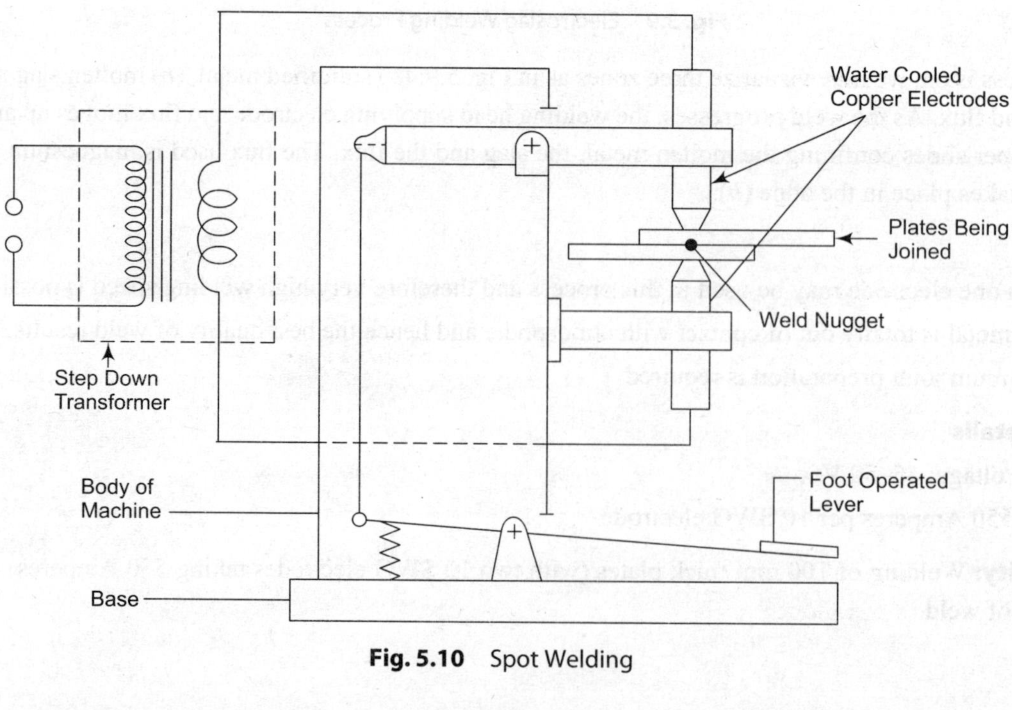

Fig. 5.10 Spot Welding

Application: This technique is used mostly in thin sheet metal work like for example in making sheet metal boxes, automobile body work, refrigerator and other cabinets. However, thicker metals upto 12 mm have been successfully spot-welded. All combinations of ductile metals and alloys can be welded using this process:

5.5.3 Seam Welding (Fig. 5.11)

In this process, a pair of circular electrodes is used. In operation, the electrodes rotate and draw in the overlapping pieces of metal to be welded. They exert a certain amount of pressure on the pieces and at the same time pass electric current intermittently. This amounts to obtaining a series of spotwelds at such close intervals that the weld nuggets overlap to give a continuous weld.

Seam welding is used in sheet metal work where a leak proof joint is desired.

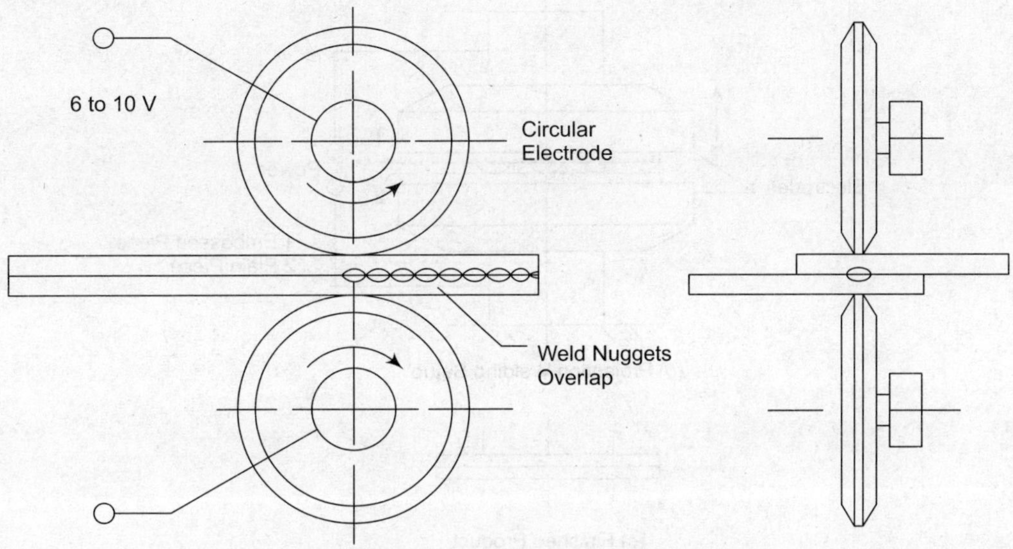

Fig. 5.11 Seam Welding

5.5.4 Projection Welding

This technique makes possible simultaneous spot welding at several locations/spots.

The process consists of 1. Embossing in one of the sheet metal pieces (to be projection welded) at locations where welding is required. This leaves projections as shown in Fig. 5.12(*a*). 2. Putting the two pieces between flat electrodes and effecting weld in much the same way as spot welding [Fig. 5.12(*b*)].

In the final product the gap between the sheet metal closes as shown in Fig. 5.12(*c*) with welding occuring at locations where projections were there.

The advantages are (*i*) simultaneous welding at several locations and thus saving production time and (*ii*) The electrodes being flat will not wear out fast.

5.5.5 Butt Welding

This process is used for joining rods and tubes end to end.

Butt welding is classified as (*a*) Resistance butt welding and (*b*) Flash butt welding depending on how the heating of the ends (to be joined) is carried out.

In resistance butt welding (Fig. 5.13) the ends butt against each other and the contact resistance offered to flow of current heats up the ends. If the heat is high enough, the ends fuse and the bond is formed under pressure.

In flash butt welding an arc is established between the ends and once the ends are heated to welding temperature the ends are forced together to form a weld (Fig. 5.14).

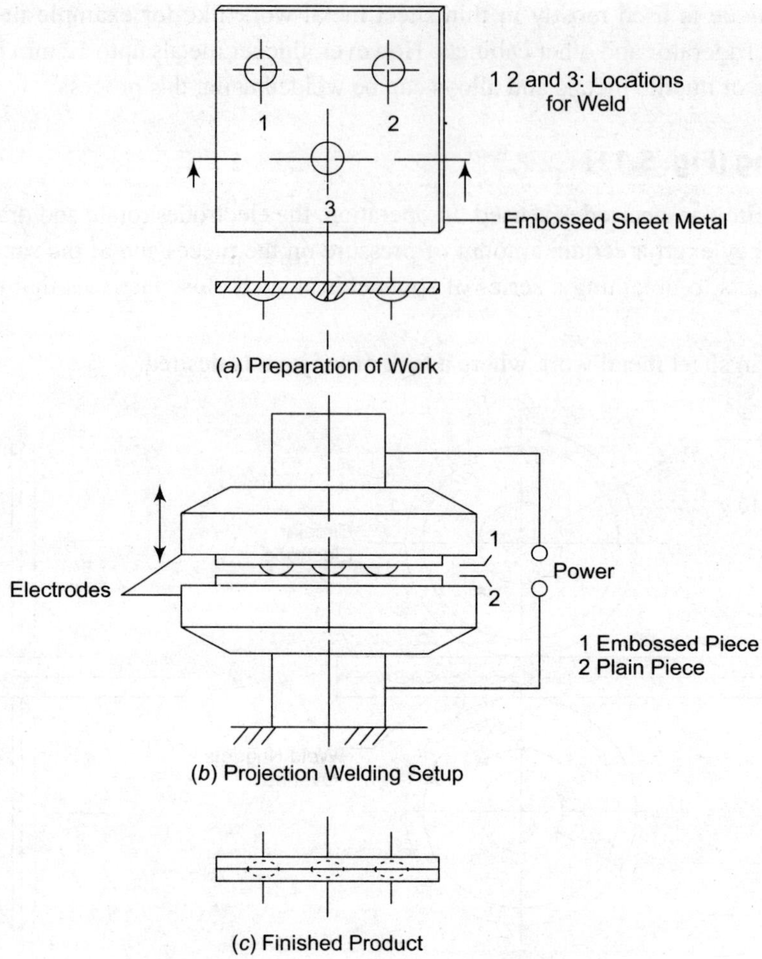

Fig. 5.12 Projection Welding

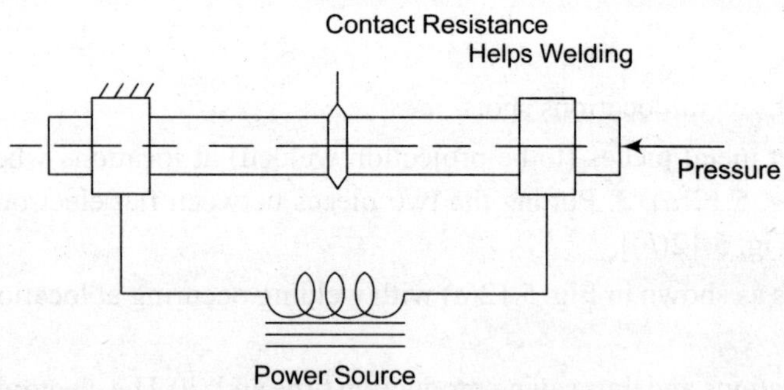

Fig. 5.13 Resistance Butt Welding

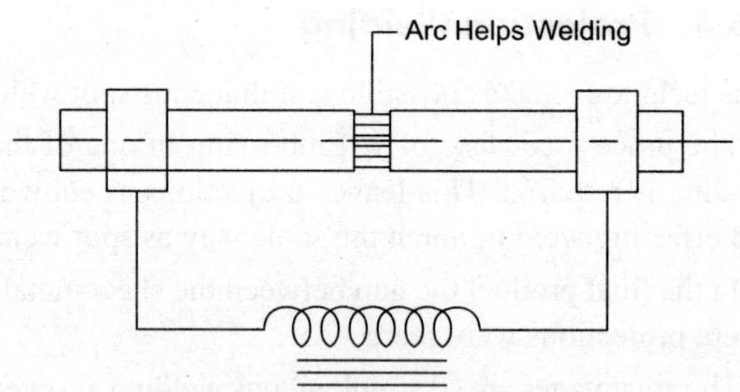

Fig. 5.14 Flash Butt Welding

5.6 THERMIT WELDING

This is a special type of welding used for joining heavy sections at site. For example, the rails on the rail road are joined using this technique.

In this method, a mould is built around the two ends to be joined using refractory material (Fig. 5.15). A mixture of finely divided aluminium and magnetic iron oxide called 'Thermit mixture' is kept in a crucible hanging over the mould. The thermit mixture is ignited using a magnesium ribbon or highly inflammable powder having barium peroxide. Soon chemical reaction takes place as per equation,

$$8Al + 3Fe_3O_4 \rightarrow 4Al_2O_3 + 9Fe$$

and superheated molten iron results. The molten iron is made to flow into the mould and fuse with the parts to be joined.

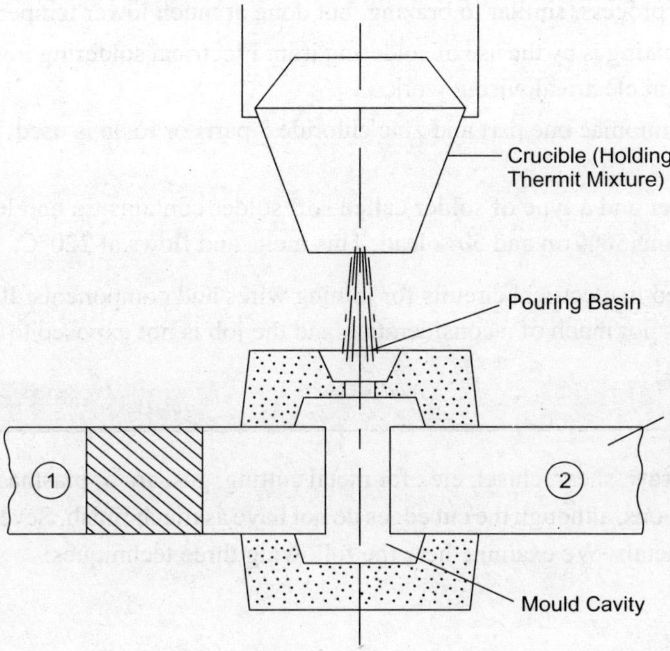

Fig. 5.15 Thermit Welding

5.7 BRAZING

Brazing is also a metal joining process, but it differs from welding in respect of the following.

- Strength — Strength of the brazed joint is less compared to welded joint.
- Brazing temperature — It is less than in welding.
- The base metal is not heated to fusion temperature.

Since the base metal does not melt, a filler metal is a must and this is provided by a brazing alloy. When this alloy is melted and introduced into the gap, the capillary action of the base metal draws in the brazing metal and the joint is made.

Brazing Procedure

1. The ends to be brazed are thoroughly cleaned mechanically or chemically to remove oil, dirt, grease and oxides.
2. The ends are heated to a temperature at which the brazing filler metal can melt and be drawn into the base metal.
3. Suitable flux (Example: Borax) is applied to the surfaces to be joined to remove any oxide present.
4. Brazing alloy is introduced into the joint in the form of rods as in gas welding or it is preplaced in the form of ring, washer or shim as per requirement, which melts to make the joint as the base metal is heated up.

Heating Methods: One of the usual methods is to use the welding torch. Other methods include induction heating (at 50 to 10,000 Hz.), furnace heating and heating by dipping into bath of molten filler metal.

Brazing metals: Several brazing alloys are available. The most common is the copper-zinc combination (60 : 40).

Advantages: The one important advantage of brazing is that dissimilar metals are easily joined. Also brazing is useful in joining very thin sections of metals. The process is rapid and gives neat joint.

Application: Typical applications are: (*i*) brazing of carbide tip to a cutting tool and (*ii*) brazing of copper tubings in refrigerators pneumatic and hydraulic circuits.

5.8 SOLDERING

Soldering is another metal joining process, similar to brazing, but done at much lower temperatures (Less than 430°C).

The most common method of heating is by the use of soldering iron. Electrical soldering irons are available in different wattages and are used extensively in electrical wiring work.

Flux used is a mixture of sal-ammoniac one part and zinc chloride 3 parts or rosin is used. For electrical wiring rosin-cored solder is always preferred.

The filler metal is called a solder and a type of solder called soft solder contains tin and lead in varying proportions, the most common combination being 50% tin and 50% lead. This melts and flows at 220°C.

Application: Soft soldering is used in electrical circuits for joining wires and components. It is also used in light-gauge sheet metal work where strength is not much of a consideration and the job is not exposed to high temperatures.

5.9 METAL CUTTING

Apart from using hacksaw, power saw, shear, chisel, etc., for metal cutting, gas, arc or plasma are also used and are found to be advantageous in certain situations, although the cut edges do not have as much finish. Several techniques are available for using gas and arc for cutting metals. We examine here the following three techniques:

1. Oxygen lance cutting
2. Oxyacetylene flame cutting
3. Arc cutting

5.9.1 Oxygen Lance Cutting

This method is highly useful in cutting heavy sections of steel.

The oxygen lance is simply a piece of black pipe about 6 mm in diameter, fitted at one end with a valve to which a hose carrying oxygen is connected.

The oxygen lance cutting depends on the fact that heated steel gets oxidized quickly and by directing a jet of oxygen on the heated steel the oxidation is almost instantaneous. This, together with the fact that the melting point of the oxide is somewhat lower than that of the steel, makes cutting of steel using oxygen jet very easy.

To start the cut, the starting point is preheated to a bright red heat, the open end of the lance is brought into contact with the spot and the oxygen valve is opened. The metal gets oxidized fast and the iron oxide together with molten iron formed, is blown out by the oxygen jet. In the process, the end of the lance burns brilliantly and provides the necessary heat for continued cutting action.

Applications

(*a*) Piercing holes in heavy sections of steel.

(*b*) Cutting open frozen metal from the tap holes in blast and open hearth furnaces.

(*c*) Oxygen mixed with iron and aluminium powder, used in lance, enables cutting into concrete, fire bricks, cast iron, bronze and other heat resisting materials.

5.9.2 Oxyacetylene Flame Cutting

In this, a cutting torch is used which resembles a welding torch; but the tip of the cutting torch has a hole at the centre with a set of holes surrounding it. Through the central hole a stream of oxygen issues and oxyacetylene mixture comes out from the holes around. When the torch is lighted and used, the oxyacetylene flame preheats the metal and the oxygen oxidizes it in a narrow gap. The slag and molten metal formed are blown away by the oxygen jet. Metals up to 750 mm thick have been cut.

It must be remembered that this method is almost exclusively used to cut ferrous metals and in particular steel.

Applications: This cutting technique is very much used in ship building industry, fabrication of heavy structural items etc.

5.9.3 Arc-Cutting

This method depends on the high temperature attainable using arc and this enables melting and cutting away almost any metal. Cutting of stainless steel and non-ferrous metals is easily done. But the quality of cutting is poorer compared with gas cutting.

5.10 DEFECTS IN WELDED JOINTS AND TEST FOR SOUNDNESS

Defects that may occur in welded joints are: Blow holes, cracks, inclusions and imperfect fusion.

The testing methods used are:

Non-destructive tests: Magnetic particle test, X-ray or Gamma ray test, Ultrasonic test, etc.

Destructive tests: Nick break test, Tensile test, etc.

QUESTION BANK

2-Marks Questions

1. Name the various metal joining processes.
2. What is welding? What are its applications?
3. List the equipment used for gas welding.
4. List the types of gas flames used in welding and give their applications.
5. What is the procedure for lighting a welding torch?
6. What are the safety precautions to be taken in gas welding?
7. What are the advantages of gas welding?
8. What is arc welding?
9. Name the basic equipment used in arc welding.
10. What are the power sources used in arc welding?
11. Why do you use flux coated electrodes?
12. What is atomic hydrogen arc welding?
13. What is resistance welding?
14. List the various types of resistance welding processes.
15. What is brazing? Give examples of applications.
16. What is soldering? Give examples of applications.
17. How do you cut metal using oxyacetylene gas?
18. What are the various forms in which edges are prepared for welding?
19. What are the methods used for testing soundness of welded joints?

10-Marks Questions

1. Explain the method of submerged arc welding with the help of a neat sketch.
2. Explain the procedure for spot welding. Show the set up with a neat sketch.
3. Explain the procedure for thermit welding. Show the set up with a neat sketch.
4. What is Electroslag welding process? Explain with the help of a neat sketch how the process is carried out.

Choose the Correct Answer/Answers

1. The temperature developed in oxyacetylene gas welding is
   (*a*) 500° C (*b*) 1900° C (*c*) 3500° C
2. The pressure in an oxygen cylinder is
   (*a*) 14 Mpa (*b*) 140 Mpa (*c*) 1.4 Mpa
3. The pressure in an acetylene cylinder is
   (*a*) 10 Mpa (*b*) 5 Mpa (*c*) 1.7 Mpa
4. Oxidizing flame is used for welding:
   (*a*) Steel (*b*) Aluminium (*c*) Bronze
5. Reducing flame is used for welding:
   (*a*) Steel (*b*) Cast iron (*c*) Monel
6. Wearing goggles protects the eyes from
   (*a*) Infra red rays (*b*) X-rays (*c*) Ultra violet rays
7. Straight polarity means
   (*a*) making electrode negative and work positive
   (*b*) making work negative and electrode positive
   (*c*) using alternating current
8. Flux covering of electrode in shielded metal arc welding provides
   (*a*) a gaseous shield around the weld
   (*b*) removes any oxide film already existing
   (*c*) raises the temperature of weld metal
9. TIG welding is used because
   (*a*) it is a simple and cheap process
   (*b*) it provides quality welds
   (*c*) very high temperature is available for welding
10. Atomic Hydrogen arc welding is used because
    (*a*) it can weld heat-resistant metals
    (*b*) hydrogen provides a protective shield — an advantage
    (*c*) very high temperature is available for welding
11. Spot welding is used in
    (*a*) butt welding of parts
    (*b*) joining overlapping pieces of sheet metal
    (*c*) sheet metal parts welding
12. Thermit welding is used in
    (*a*) rail road work
    (*b*) fabrication of the components of nuclear plant
    (*c*) joining thin sheets of metal
13. Seam welding is applied in
    (*a*) joining of thick plates
    (*b*) fabrication of sheet metal boxes
    (*c*) sheet metal work where leakproof joint is desired

14. Brazed joint

(*a*) has greater strength than a welded joint
(*b*) has the same strength as soldered joint
(*c*) has strength in between that of welded and soldered joints

15. Soft solder (50% Tin + 50% Lead) melts at

(*a*) toom temperature (*b*) 950° C (*c*) 220° C

16. Arc welding requires

(*a*) very high voltage (*b*) 220 V (*c*) low voltage

17. Gas welding is

(*a*) cheaper (*b*) slow (*c*) useful in welding thick sheets

18. Dissimilar metals can be joined by

(*a*) shielded Metal Arc Welding (*b*) brazing
(*c*) soldering

Answers

1. (*c*) **2.** (*a*) **3.** (*c*) **4.** (*c*) **5.** (*c*) **6.** (*c*) **7.** (*a*) **8.** (*a*) and (*b*) **9.** (*b*)
10. (*a*) and (*b*) **11.** (*b*) and (*c*) **12.** (*a*) **13.** (*c*) **14.** (*c*) **15.** (*c*) **16.** (*c*) **17.** (*a*) and (*b*)
18. (*b*) and (*c*).

Fill in the blanks

1. For welding thick sheets the edge preparation should be of the form ....
2. The inert gas used in TIG welding is ....
3. Brazing makes .... joint compared with soldering.
4. .... flame is used for surface hardening.
5. Dissimilar metals can be joined by ....
6. Quality weld in Aluminium and Magnesium alloys is best done using ....
7. .... welding process is economicaly viable for welding very thick plates of 100 mm and above.
8. The most common brazing alloy is Copper-Zinc combination in ratio ....
9. .... polarity is advantageous in welding mild steel with greater speed ensuring sound weld.
10. In Atomic-Hydrogen arc welding .... provides the protective shield.

Answers

1. U or V or double-U
2. Argon/Helium
3. Stronger
4. Carburizing
5. Brazing
6. TIG
7. Electro-slag
8. 60 : 40
9. Straight
10. Hydrogen

Chapter-6

Machining

Machining is yet another manufacturing process and it means shaping of objects/components by removing excess material from a given job/workpiece. Here we confine our study to machining of metals only.

Machining is employed for the following reasons.

(*i*) For the purpose of proper fitting and accurate orientation, many of the castings and forgings require machining before they are put into use.

(*ii*) Some small precision components can be made only by machining.

(*iii*) Thread forming (screw cutting) — both external and internal is easily done by machining.

(*iv*) Production of accurate profiles, grooves, keyways, spirals etc., require machining.

Machine tools are used for machining. They employ cutting tools to remove excess material from the given job. There are different types of machine tools which come under the broad classification of

(*i*) General purpose

(*ii*) Special purpose and

(*iii*) Automatic screw cutting machine tools.

The latest addition to this list is the revolutionary Computer Numerical Control (CNC) machine tools, which have a distinct role to play in the future production activities.

The general purpose machine tools are:

(*i*) Lathe

(*ii*) Drilling Machine

(*iii*) Shaping Machine

(*iv*) Planing Machine and

(*v*) Milling Machine

Examples of special purpose machine tools are:

(*a*) Special lathes like capstan, turret and copying lathes.

(*b*) Boring Machine

(*c*) Broaching Machine

(*d*) Production Milling Machine

(*e*) Production Drilling Machine etc.

Automatic machine tools, also called Automatic screw cutting machines (or simply automats), are used for mass production of essentially small parts using a set of pre-designed and job-specific cams.

Under CNC machine tools, we have CNC turning centre, which does all the work of a lathe and CNC machining centre, which does milling, drilling etc., with provision for automatic tool changing and tool wear correction built into it.

6.1 LATHE

In general, machine tools are used to produce

(*i*) cylindrical surfaces,

(*ii*) flat surfaces and

(*iii*) curvilinear surfaces.

Lathe is primarily used for producing cylindrical surfaces, although it can produce flat surfaces, curvilinear surfaces and also produce internal features like holes. With attachments, it can do grinding, milling, gear cutting, etc., and that is how lathe is often called a versatile machine tool.

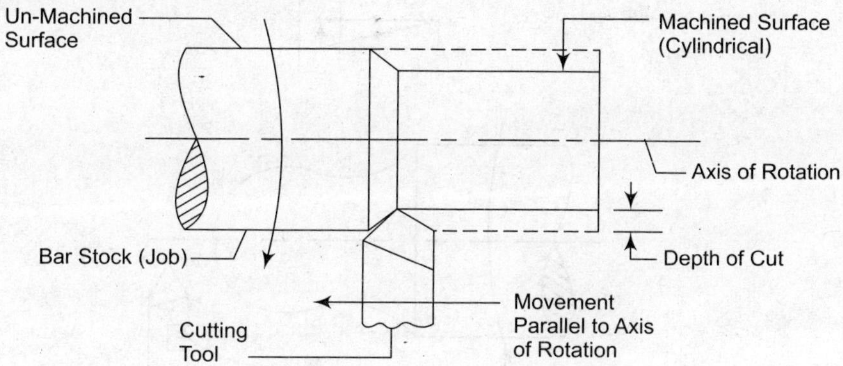

Fig. 6.1 (*a*) Principle of Turning

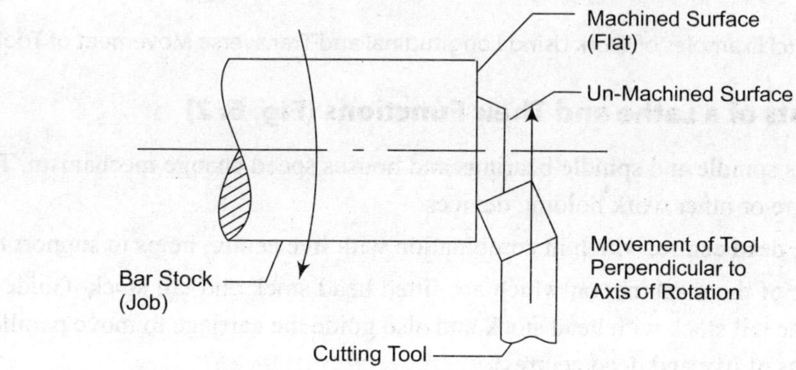

Fig. 6.1 (*b*) Principle of Facing

Working principle: In Fig. 6.1(*a*) is shown a job rotating about its axis. The cutting tool is traversed (moved) parallel to the axis of rotation. As the cut takes place, a cylindrical surface is produced. This operation is called turning. If, however, the cutting tool is traversed in a direction perpendicular to the axis of rotation as shown in Fig. 6.1(*b*), a flat surface is produced. In this case, the operation is called facing. Combination of the tool movements in the two directions produces various forms as shown in Fig. 6.1(*c*).

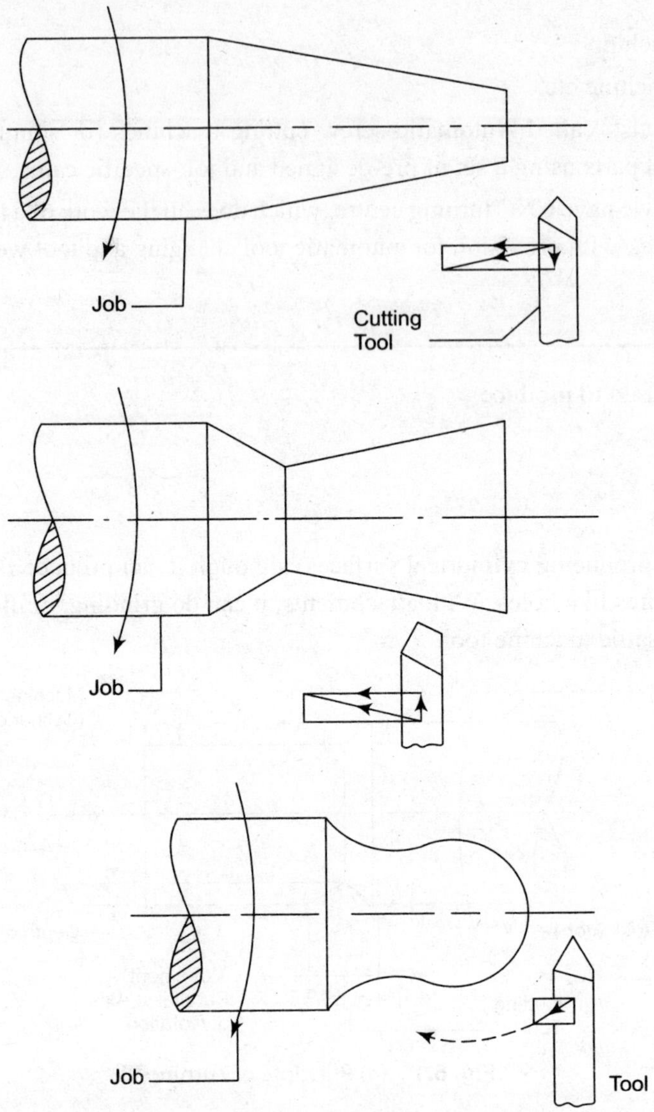

Fig. 6.1 (c) Examples of Work Using Longitudinal and Transverse Movement of Tool

6.1.1 Main Components of a Lathe and Their Functions (Fig. 6. 2)

(a) *Head stock:* It supports spindle and spindle bearings and houses speed change mechanism. The spindle is hollow and holds the live centre or other work holding devices.

(b) *Tail stock:* It holds the dead centre, which in combination with live centre, helps to support long jobs.

(c) *The bed:* It is the base of the machine, on which are fitted head stock and tail stock. Guide ways are formed on the bed, which align the tail stock with head stock and also guide the carriage to move parallel to the axis of lathe (line joining the centres of live and dead centres).

(d) *The carriage:* The carriage moves over the bed and it consists of two parts: Saddle and Apron. The saddle carries the cross slide over which is fitted the compound rest. Cross slide gives the transverse movement to cutting tool. The apron carries the control devices like longitudinal and transverse feed hand wheels etc.

(e) *Compound rest:* This carries the tool post on a swivel base. This arrangement helps cutting tool movement at an angle to the lathe axis and is useful in taper turning.

(f) *The lead screw:* This is used in thread cutting. It gets the drive from the spindle through change gears. (Ref. Fig. 6.6)

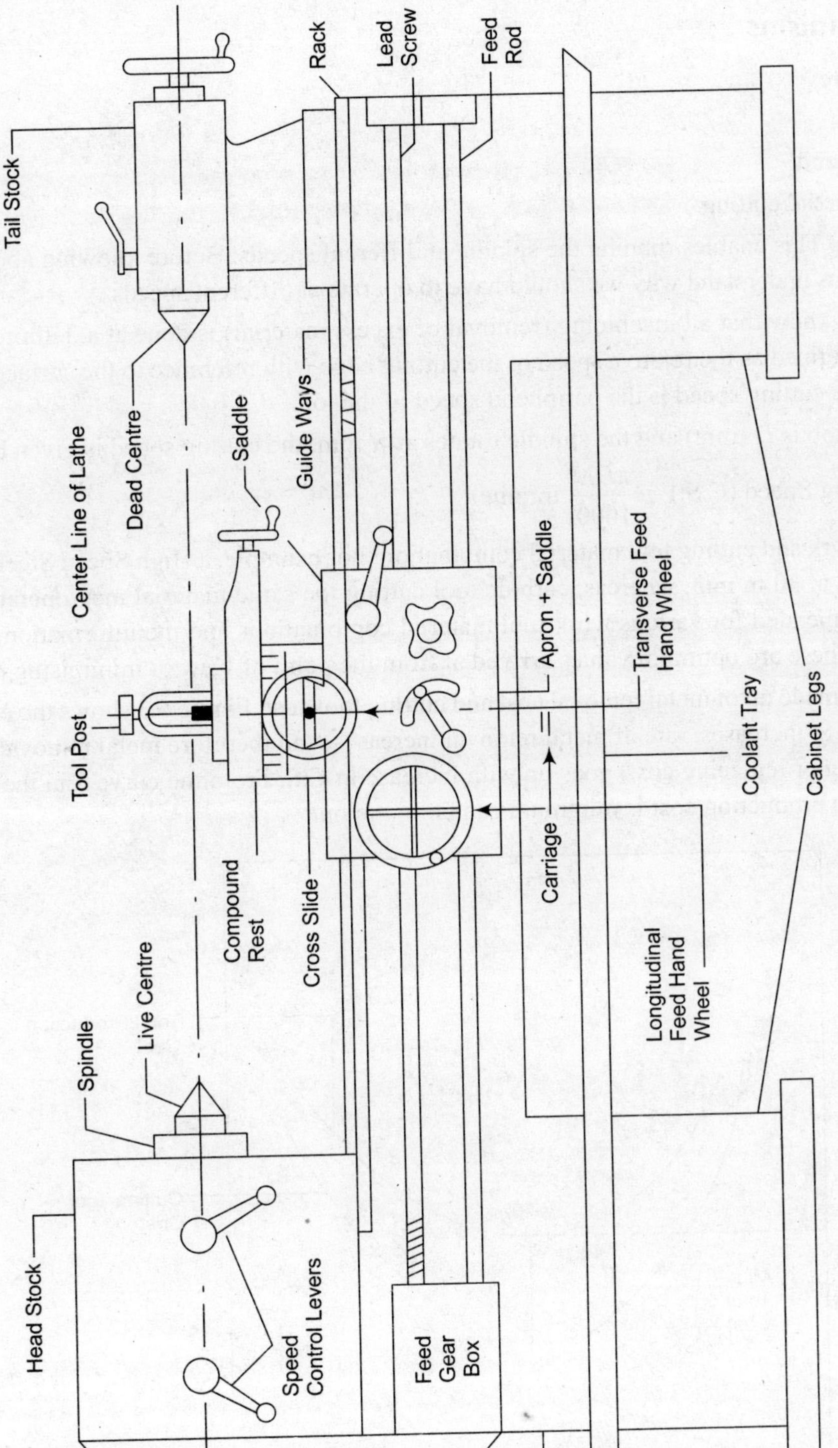

Fig. 6.2 Lathe and Its Main Components

(g) *The feed rod:* This helps in the uniform movement of tool in longitudinal and transverse directions. This also gets the drive from the spindle. The drive is given through feed gear box.

(h) *Legs:* Support the bed and provide proper height for operating the lathe.

6.1.2 Lathe Mechanisms

The main mechanisms are

(a) Speed mechanism,

(b) Feed mechanism and

(c) Mechanism for thread cutting.

(a) *Speed Mechanism:* This enables running the spindle at different speeds. Before knowing about some of the speed change mechanisms, let us understand why we should have to operate at different speeds.

First of all, we should know that all machining (removal of excess material) is done at a definite speed called cutting speed. Cutting speed is defined as the relative speed of the cutting edge with reference to the surface of the job being cut. In the case of turning, the cutting speed is the peripheral speed of the job.

If the diameter of the job is D (mm) and the spindle rotates at N rpm, the cutting speed is given by

$$\text{Cutting Speed (C.S.)} = \frac{\pi DN}{1000} \text{ m/min} \quad \text{...(1)}$$

C.S. is based on the work and cutting tool material combination. For example, a High Speed Steel (H.S.S.) tool cutting mild steel should operate at 30 m/min, whereas, carbide tool cutting the same material may operate at 100 m/min. The cutting speeds are recommended for various work tool material combinations and the information is made available in machinist's hand book. These are optimum values arrived at from the point of view of minimising production cost.

The production cost is made up of metal removal cost and cutting tool cost. Figure 6.3 shows the effect of C.S. on these two parameters. As the C.S. increases, rate of metal removal increases, and therefore metal removal cost decreases. But, the tool cost (resharpening or replacing cost) goes up with increase in C.S. From the curve 3 in the figure, the optimum cutting speed at which the production cost is minimum can be found out.

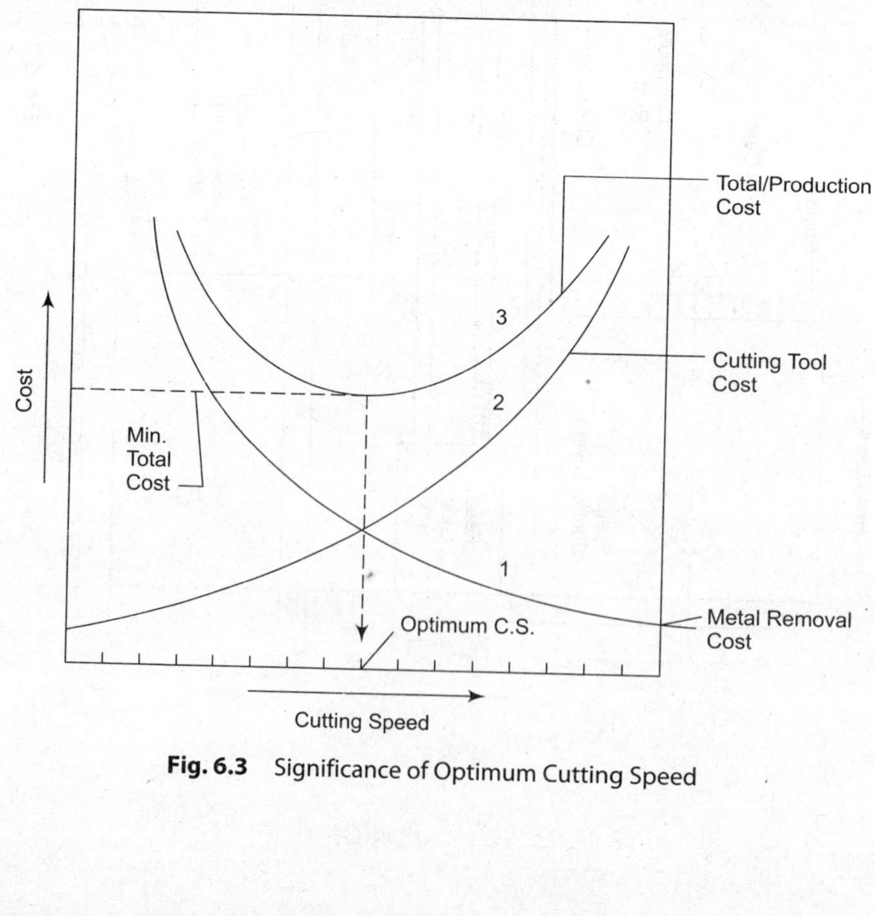

Fig. 6.3 Significance of Optimum Cutting Speed

Once this optimum C.S. is fixed the spindle speed is calculated using equation (1).

The other factor necessitating spindle speed change is in form-turning (for example thread cutting using a thread cutting tool having cutting edge of the same form as the thread). Here a lower C.S. is to be used as otherwise chatter (vibration of cutting tool, job, etc.) may result.

Also in finishing operation, higher C.S. is advantageous in bringing about a better surface finish.

In view of the above factors, mechanism to change the spindle speed becomes essential.

Types of speed change mechanism:

Mechanical	: (*i*) Step cone pulley drive
	: (*ii*) Gear drive
Hydraulic drive	: Uses hydraulic pump and motor
Electrical drive	: Uses variable speed motor with suitable controls

Figure 6.4 shows a few simple examples of speed change mechanisms.

(*b*) *Feed Mechanism:* Feed refers to the tool movement during metal cutting, either in longitudinal or transverse direction. Larger feed rate improves metal removal rate (MRR) but the surface finish diminishes. Feed of tool is brought about either manually or automatically.

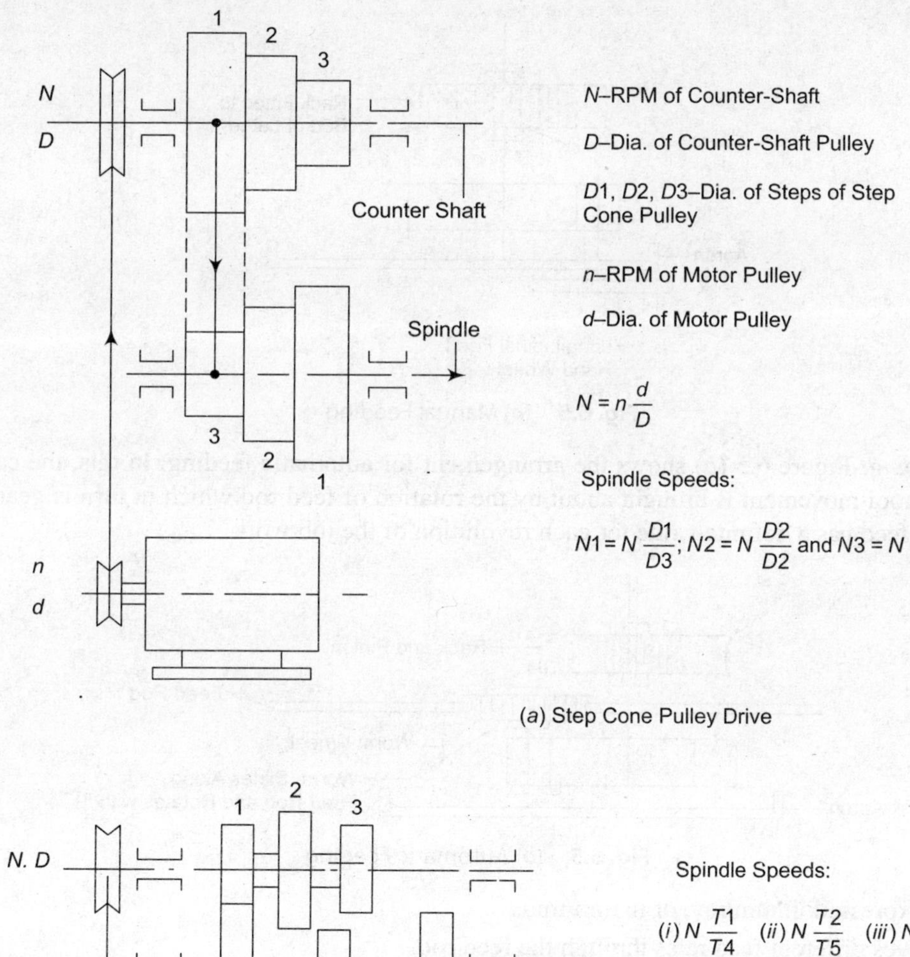

(*a*) Step Cone Pulley Drive

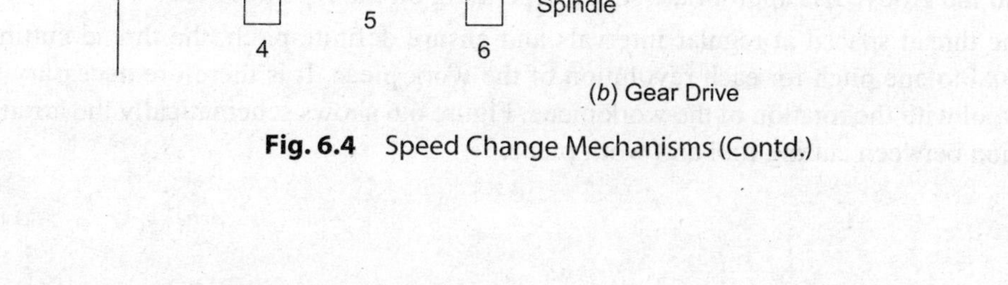

(*b*) Gear Drive

Fig. 6.4 Speed Change Mechanisms (Contd.)

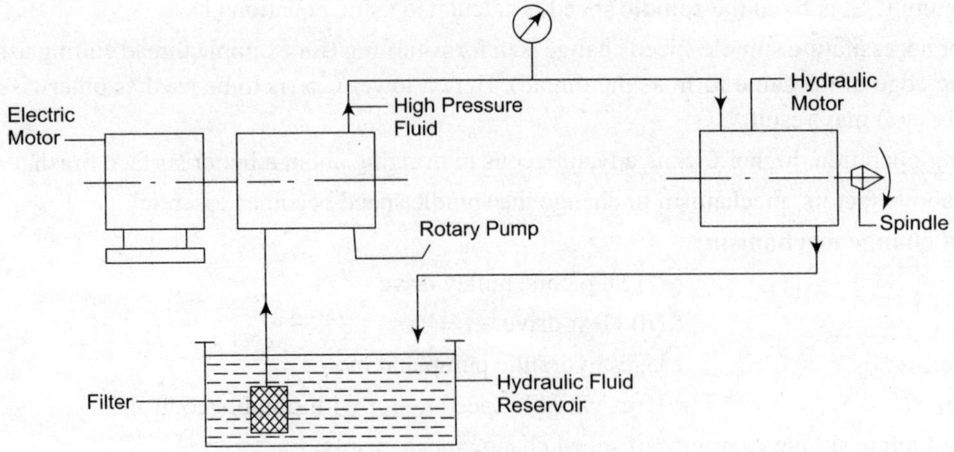

Fig. 6.4 Speed Change Mechanisms

(*i*) *Manual feeding:* Figure 6.5 (*a*) shows a simple sketch of manual feeding. In manual feeding, there is no correlation between the tool movement and the rotation of the job.

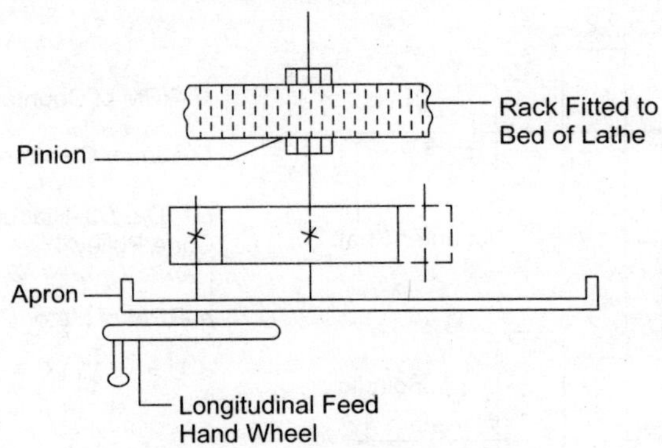

Fig. 6.5 (*a*) Manual Feeding

(*ii*) *Automatic feeding:* Figure 6.5 (*b*) shows the arrangement for automatic feeding. In this, the carriage movement and hence the tool movement is brought about by the rotation of feed rod which in turn is geared to the spindle. Therefore, the feed has a definite value for each revolution of the job/work.

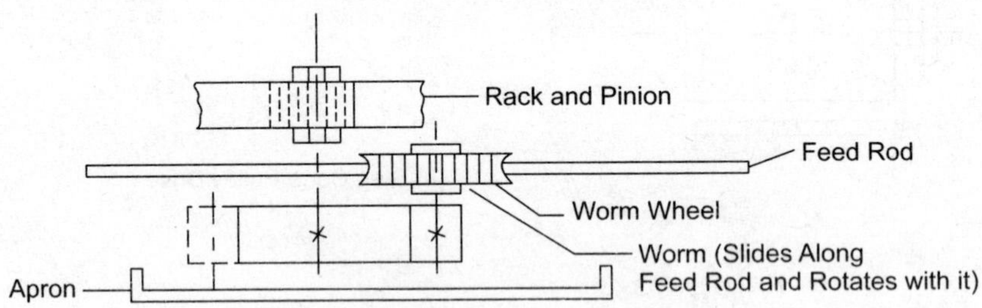

Fig. 6.5 (*b*) Automatic Feeding

Feed is therefore expressed in mm/rev. or in mm/min.

A feed gear box gives different feed rates through the feed rod.

(*c*) *Mechanism for thread cutting:* Screw thread is nothing but a helical groove cut on a cylindrical piece. The helix has a definite pitch and the groove has appropriate shape depending on the type to thread.

In order to have the thread spaced at regular intervals and ensure definite pitch, the thread cutting tool must move through a distance equal to one pitch for each revolution of the work piece. It is therefore necessary to relate the linear motion of the cutting tool with the rotation of the work piece. Figure 6.6 shows schematically the arrangement for getting this co-ordinated motion between cutting tool and work piece.

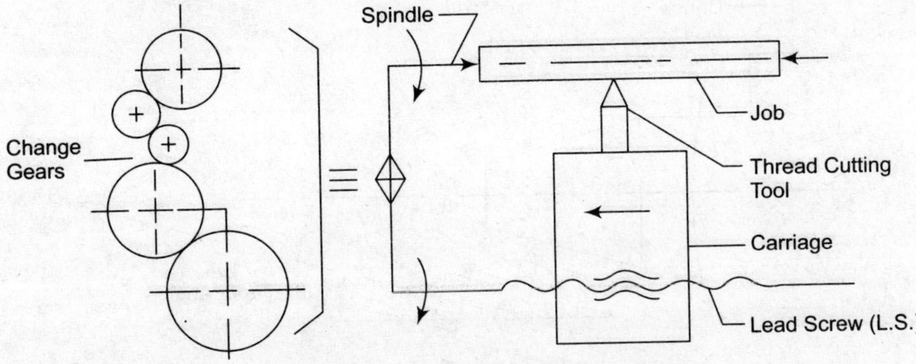

Fig. 6.6 Thread Cutting Mechanism

6.1.3 Machining Parameters

Any machining operation involves the following parameters and must be chosen properly.

(*a*) Cutting speed (C.S.)

(*b*) Feed

(*c*) Depth of cut

Selection of C.S. has been discussed in the previous section and so also feed.

By depth of cut, we mean the amount of tool advancement towards the job before each cut is taken (Ref. Fig. 6.1). Depth of cut is large for rough operation and is small for finishing operation. Larger depth of cut enables larger metal removal rate.

6.1.4 Basic Lathe Operations

(*a*) Turning:

(*i*) Plain turning, (*ii*) Step turning and (*iii*) Taper turning

(*b*) Facing

(*c*) Drilling and Boring

(*d*) Knurling

(*e*) Thread cutting

(*f*) Chamfering

(*g*) Parting off

Let us consider a few of the operations in detail.

Taper turning: Methods available are:

1. Swivelling of compound rest,
2. Off-setting tail stock,
3. Use of FORM TOOL,
4. Taper turning attachment and
5. Co-ordinated longitudinal and transverse feed of tool.

Swivelling of compound rest: Let us take an example (Fig. 6.7). The compound rest is swivelled about the vertical axis through the taper angle.

The tool is fed against the rotating job using hand wheel on compound rest. This makes the tool traverse in a direction inclined to the axis of rotation and thus a tapered surface is produced. The required dimension is obtained by successively giving depth of cut (movement of tool, in this case, perpendicular to axis of rotation using cross feed hand wheel).

This method is applicable for short tapers with relatively large taper angles.

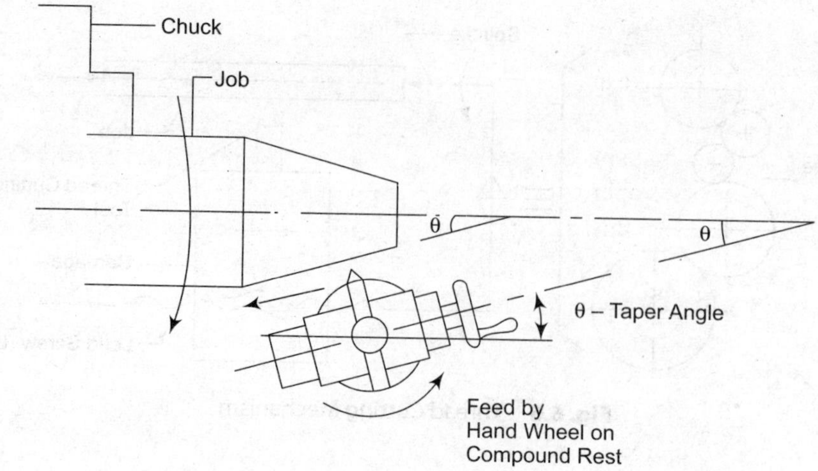

Fig. 6.7 Taper Turning by Swivelling of Compound Rest

Off-setting tail stock: In this method, tool traverses parallel to axis of lathe as in plain cylindrical turning, but the axis of rotation of the job is inclined to the lathe axis at an angle equal to the taper angle. The inclination is obtained by off-setting the tail stock as shown in Fig. 6.8. The feed is given using longitudinal feed hand wheel. The depth of cut is given by moving the tool in steps perpendicular to axis of lathe using the cross slide hand wheel.

The required off-set is calculated as follows:

Referring to Fig. 6.8,

$$\sin\theta = \frac{X}{L} \text{ and } \tan\theta = \frac{(D-d)}{21}$$

For small angles, $\sin\theta \approx \tan\theta$. Therefore,

$$X = \frac{L}{1}\frac{(D-d)}{2}$$

This method is employed for producing long tapers with relatively small taper angles.

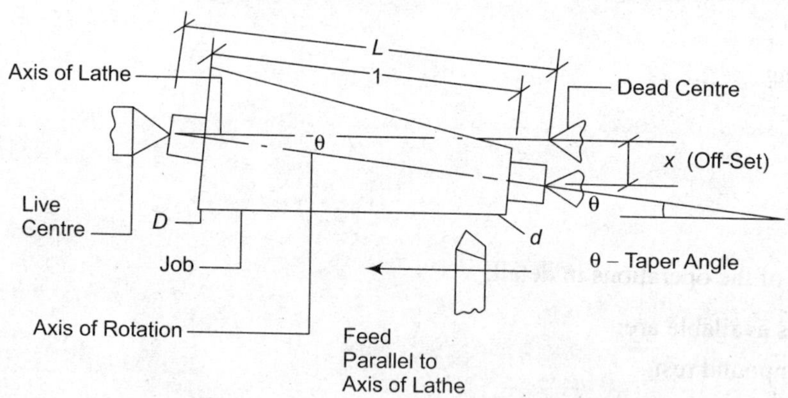

Fig. 6.8 Taper Turning by Off-Setting Tail Stock

Taper turning attachment: In this method, the job rotates about the axis of lathe and the tool is made to follow an inclined path by having an attachment to lathe. The attachment consists of a guide bar which can be set at the given taper angle and a slide which moves over it (Fig. 6.9).

For taper turning, the cross slide is disconnected from its cross feed screw and is connected to the slide on the guide.

As the longitudinal feed wheel is operated and the carriage is traversed, the cross slide is guided by the guide bar and hence the tool moves in a direction inclined to the axis of lathe producing the required taper.

Co-ordinated tool movement in the longitudinal and transverse directions: Although a skilled operator may use this method and do a good job, there can be no repeatability and consistency in quality. A CNC (Computer Numerical

Control) lathe on the other hand, run by a definite programme enables an unerring co-ordination of the movement of the tool in longitudinal and transverse directions, which makes possible obtaining tapers of consistent quality.

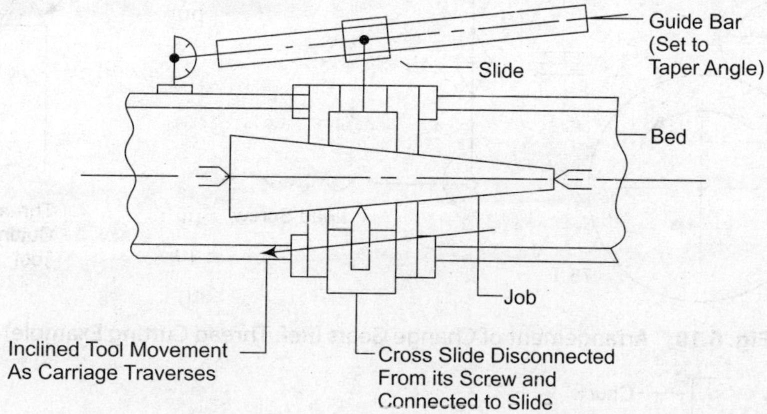

Fig. 6.9 Taper Turning Attachment

Thread Cutting: A screw is characterised by (*i*) pitch and (*ii*) shape of thread/groove. The pitch of a thread is defined as the distance between corresponding points on adjacent threads. In section 6.1.2, the thread cutting mechanism has been discussed. In order to cut a thread of specified pitch, it is necessary to have suitable gearing between the spindle and the lead screw.

(Remember that for every one revolution of job, tool has to move by a distance = one pitch, p)

The calculation for gearing (change gears) is as follows:

Let the gearing introduced have a ratio x. For one revolution of the spindle/job, the lead screw rotates by $1x$ revolution. Therefore, the carriage and hence the tool moves by a distance = $1xP$, where P is the pitch of Lead Screw (L.S.).

For obtaining the specified pitch on job, this distance ($1xP$) should be equal to 'p' the pitch of job.

Therefore
$$p = 1xP$$

or
$$x = \frac{p}{P}$$

i.e.,
$$\frac{\text{No. of teeth on driver}}{\text{No. of teeth on driven}} = \frac{\text{Pitch of the job}}{\text{Pitch of L.S.}}$$

Usually it is written as

$$\frac{\text{Driver}}{\text{Driven}} = \frac{p}{P}$$

Let us take an example of cutting thread of 1 mm pitch on a lathe having L.S. of 3 mm pitch.

The gearing required is given by

$$\frac{\text{Driver}}{\text{Driven}} = \frac{p}{P} = \frac{1}{3} \quad \text{or} \quad \frac{1}{3} \cdot \frac{25}{25} = \frac{25}{75}$$

Change wheel arrangement is shown in Fig. 6.10.

The gear on the spindle has 25 T and that on L.S. 75 T. (Note: Lathe is supplied with a set of change gears, for example, 25 T to 100 T in steps of 5 teeth is one set).

The required shape of thread is obtained by using a FORM tool *i.e.*, a cutting tool of appropriate shape. For example a metric V-thread has a tool with cutting edge shaped as shown in Fig. 6.10.

Drilling and Boring: Figure 6.11(*a*) shows the arrangement for drilling. The job is held in a chuck and drill is fitted in the tail stock. The drill is fed (axially moved) by rotating the hand wheel on the tail stock.

The boring operation is shown in Fig. 6.11(*b*). This operation is intended for enlarging an already drilled hole. A boring tool is held in the tool post and is fed by using the longitudinal feed hand wheel or using automatic feed.

Knurling: This is an operation meant to give a serrated surface to handles and knobs to facilitate gripping by hand. For example, the winding knob of a wrist watch has knurl on it.

The knurling operation is shown in Fig. 6.12.

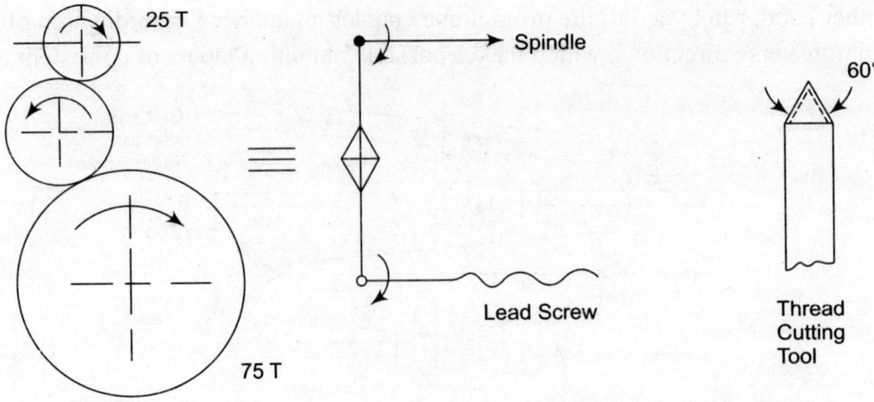

Fig. 6.10 Arrangement of Change Gears (Ref. Thread Cutting Example)

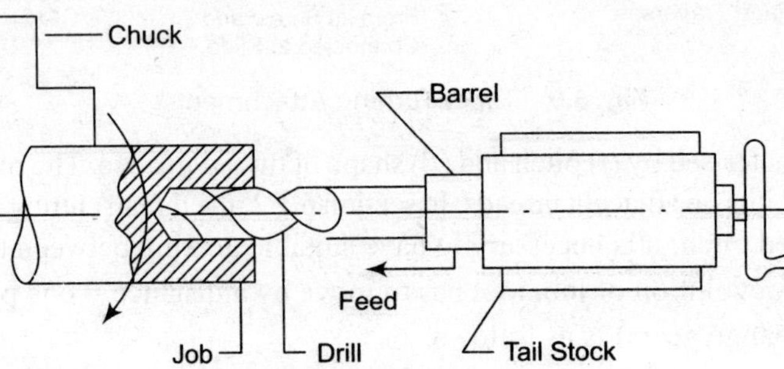

Fig. 6.11 *(a)* Drilling on Lathe

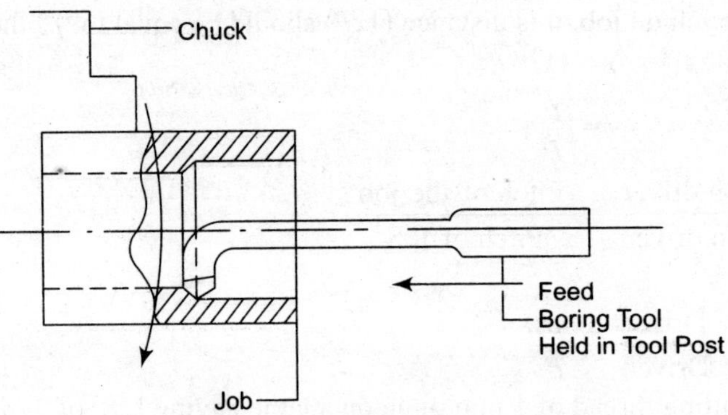

Fig. 6.11 *(b)* Boring Operation

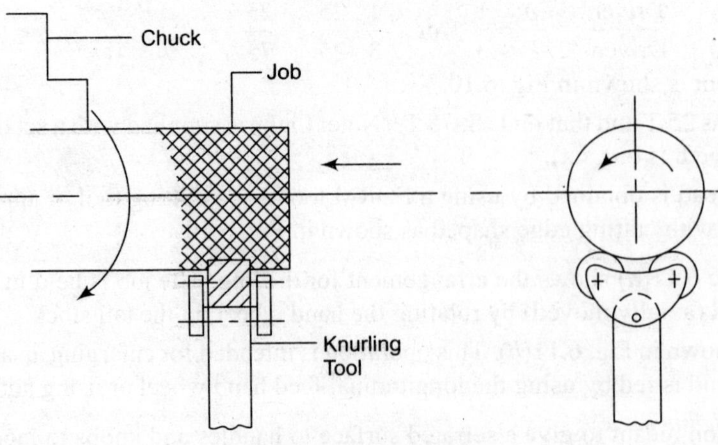

Fig. 6.12 Knurling

Chamfering: The operation is meant for removing sharp corners. Figure 6.13 shows the operation.

Parting-off: Figure 6.14 shows the set-up. Here the machined component is parted-off from the bar stock using parting-off tool.

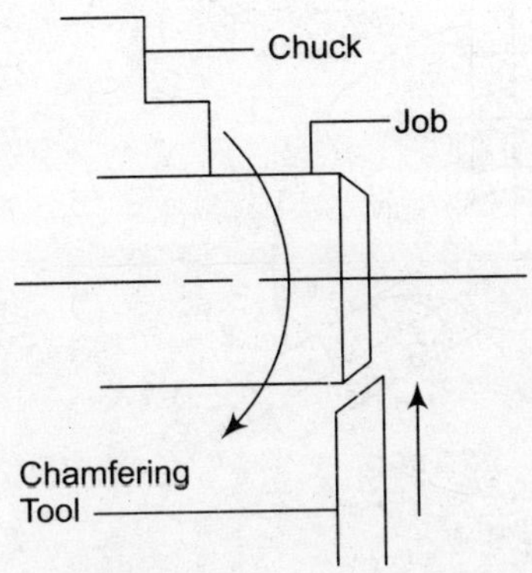

Fig. 6.13 Chamfering

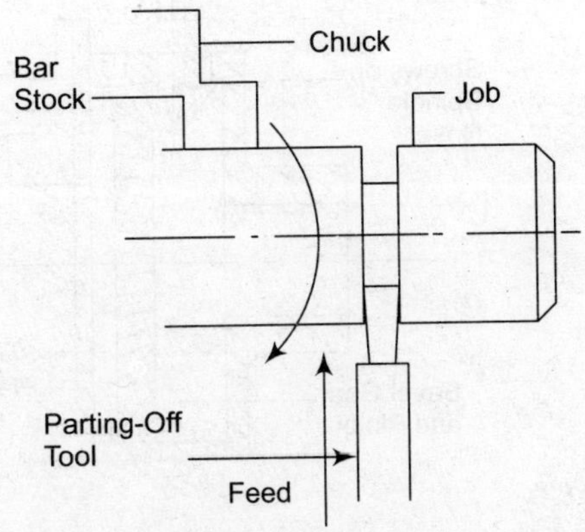

Fig. 6.14 Parting-Off

6.1.5 Work/Job Holding Devices

Following are some of the work holding devices used on a lathe:

(*a*) Chuck: (*i*) 3-jaw self-centering type

(*ii*) Independent 4-Jaw type

(*b*) Collects

(*c*) Holding between centres

(*d*) Mandrel

(*e*) Face plate

(*f*) Special chucks like magnetic, pneumatic etc.

When several devices are available, the production engineer has to select the right one in a given situation. This requires the knowledge of the characteristics of the devices.

The characteristics of work holding devices are:

1. Gripping ability *i.e.*, how well the device grips the job.
2. Quickness of operation *i.e.*, how fast you can grip the job.
3. Accuracy in centering the job. Lathe operations require centering of job. *i.e.*, the axis of the job should coincide with the axis of the lathe.
4. Manner of holding *i.e.*, whether the device holds the job by external surface or internal surface.

(*a*) Chucks:

(*i*) *Three-jaw self-centering chuck:* Figure 6.15 shows a 3-jaw self-centering chuck. The three jaws move in or out simultaneously with a fair degree of accuracy. They are actuated by a scroll plate, which in turn is rotated by means of a bevel gear arrangement. A chuck key engages a slot in the head of a bevel pinion and helps rotate the scroll plate.

The operation of this chuck is fast because of the self centering feature. However, odd jobs cannot be held in this chuck. Also the accuracy may be lost over a period of time and requires repair or replacement.

(*ii*) *Independent 4-jaw chuck:* (Fig. 6.16) This has four jaws, each being operated individually by a screw and nut arrangement. Gripping efficiency is better compared with 3-jaw chuck and also very high accuracy of centering is obtainable. However, it takes time to centre and grip jobs. Another advantage is that odd jobs can be held.

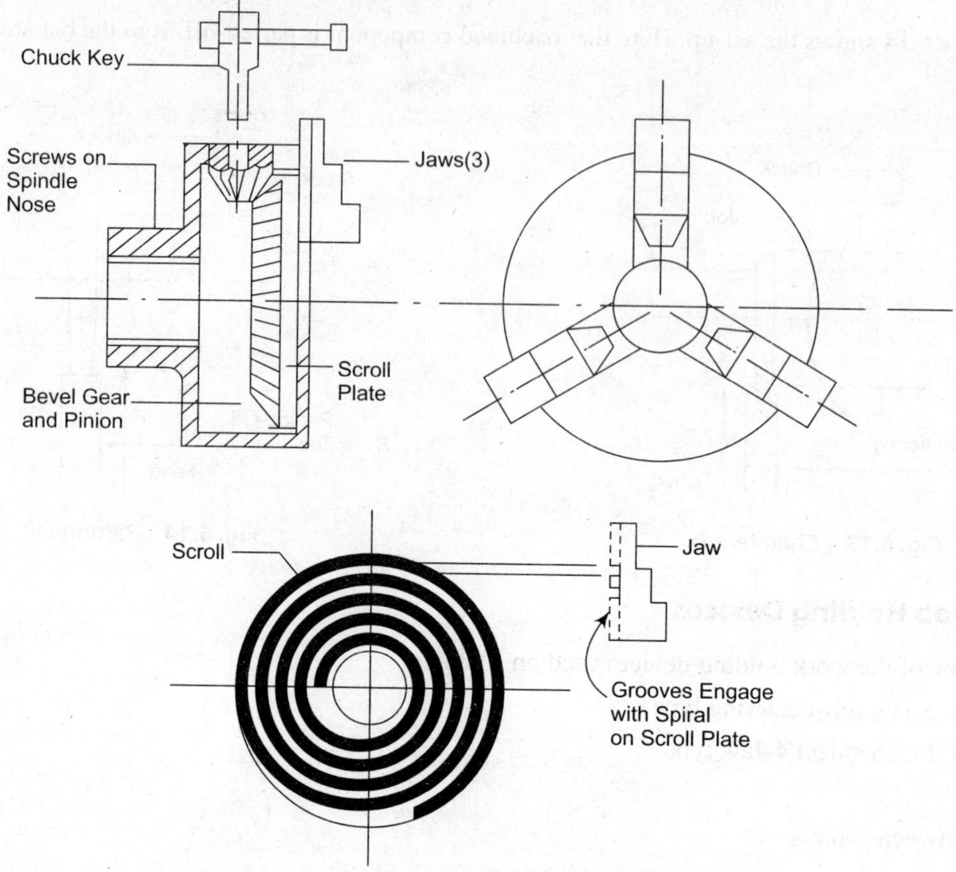

Fig. 6.15 3-Jaw Self-Centering Chuck

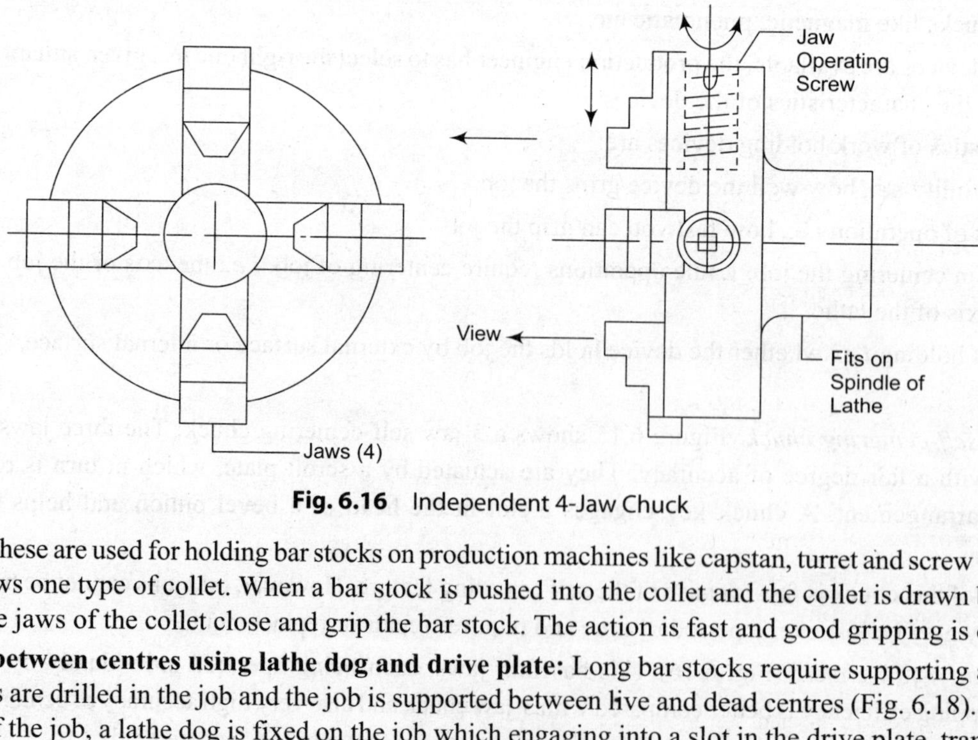

Fig. 6.16 Independent 4-Jaw Chuck

(*b*) **Collets:** These are used for holding bar stocks on production machines like capstan, turret and screw cutting automats. Figure 6.17 shows one type of collet. When a bar stock is pushed into the collet and the collet is drawn backwards in the tapered bush, the jaws of the collet close and grip the bar stock. The action is fast and good gripping is obtained.

(*c*) **Holding between centres using lathe dog and drive plate:** Long bar stocks require supporting at both ends. For this, centre holes are drilled in the job and the job is supported between live and dead centres (Fig. 6.18). In order to bring about rotation of the job, a lathe dog is fixed on the job which engaging into a slot in the drive plate, transfers the motion from the spindle to the job.

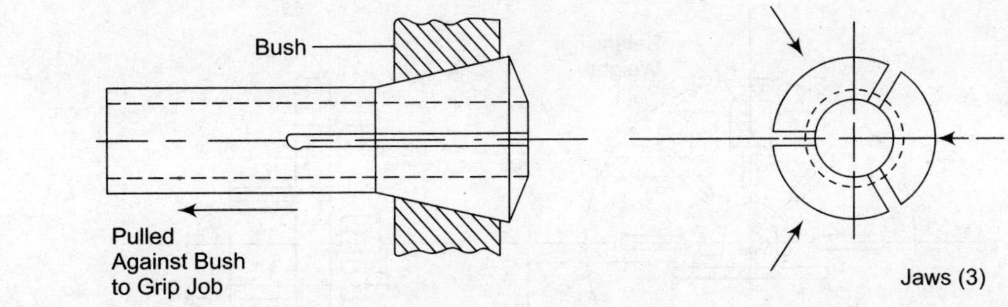

Fig. 6.17 Collet Chuck

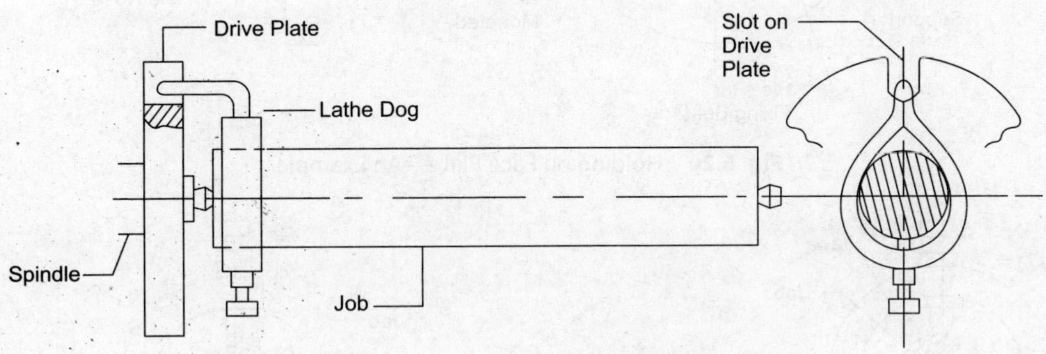

Fig. 6.18 Holding Between Centres

(*d*) **Mandrel:** This type of device is used for holding jobs with pre-drilled or bored holes. Different types and sizes of mandrel are available. Figure 6.19 shows a Solid Mandrel, which having a slight taper, enters the hole in the job making a force-fit. Thus the job becomes one with mandrel and when the mandrel is rotated between centres, the job also rotates.

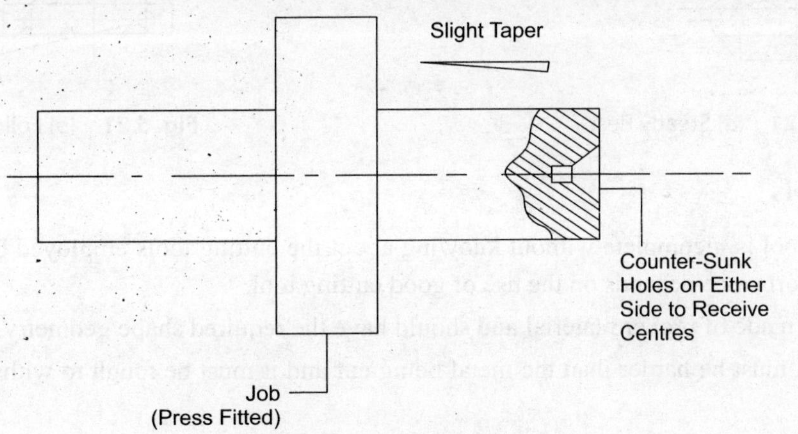

Fig. 6.19 Solid Mandrel

(*e*) **Face plate:** This, fitted to the spindle, enables mounting odd jobs, which cannot be held using other types of holding devices.

Figure 6.20 shows the mounting of a bearing block on face plate for machining its bore.

Steady rest and follower rest: Steady rest is a device used to support long slender (large L/d ratio) jobs and prevent them from bending under the action of cutting force. See Fig. 6.21 (*a*). The jaws of the rest are adjusted to contact the work piece before machining is commenced. The steady rest is located at a suitable place and clamped to the bed of the lathe.

Follower rest [Fig. 6.21 (*b*)] is also used to support slender jobs, but it is fitted to the carriage and it travels along with the carriage. As the cut takes place the support is shifted and is always directed against the tool force.

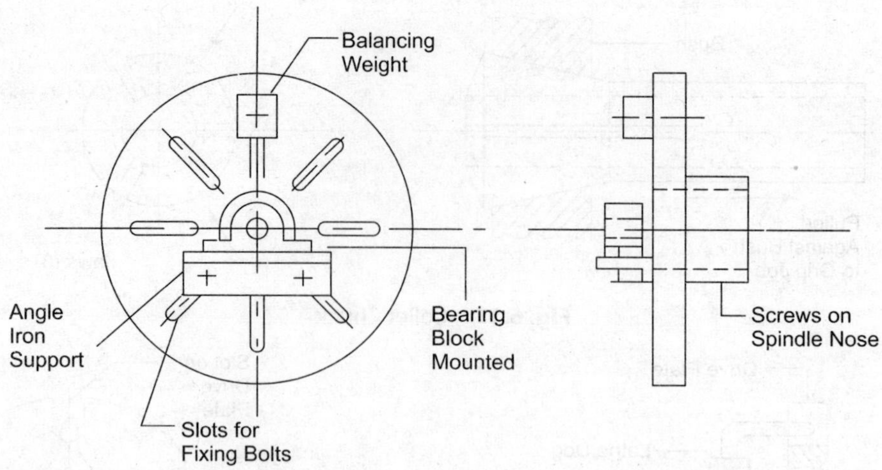

Fig. 6.20 Holding on Face Plate—An Example

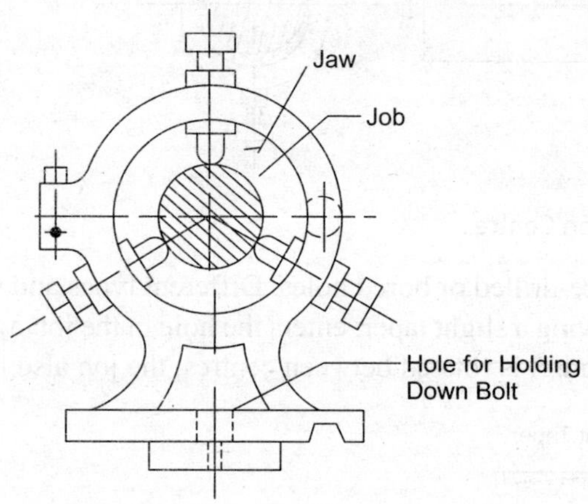

Fig. 6.21 (*a*) Steady Rest

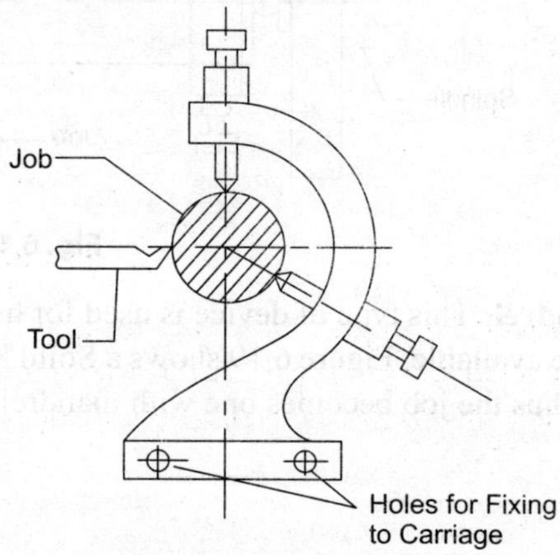

Fig. 6.21 (*b*) Follower Rest

6.1.6 Cutting Tools

Study of any machine tool is incomplete without knowing about the cutting tools employed by them. However good a machine tool is, its performance depends on the use of good cutting tool.

Cutting tool must be made of proper material and should have the required shape/geometry.

Cutting tool material must be harder than the metal being cut and it must be tough to withstand shock and vibration during machining.

High carbon steel, high speed steel, carbides and diamonds are some of the cutting tool materials.

Shape or geometry of cutting tool: The cutting edge of a tool is formed by the intersection of two planes. See Fig. 6.22 (*a*). The plane on which the cut material (chip) flows over, is called the face and the other one is called the flank.

The angle formed by the line drawn perpendicular to the machined surface and the face is called **Rake angle.** The angle between the machined surface and the flank is called **clearance angle/Relief angle.**

Good rake angle makes the tool sharp, but too much will weaken the cutting edge. Clearance angle is necessary to prevent tool flank rubbing against the machined surface but it is kept a minimum so that the cutting edge is not weakened.

The rake and clearance angles of turning tools are shown in Fig. 6.22 (*b*).

Cutting tool must be held in the tool post rigidly and with minimum over hang (Ref. Fig. 6.23) to avoid machine tool vibration and ensure machining accuracy.

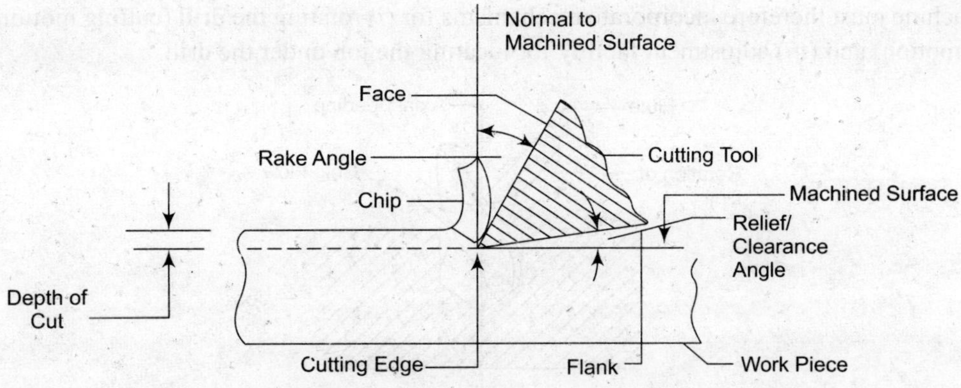

Fig. 6.22 (*a*) Geometry of Cutting Tools

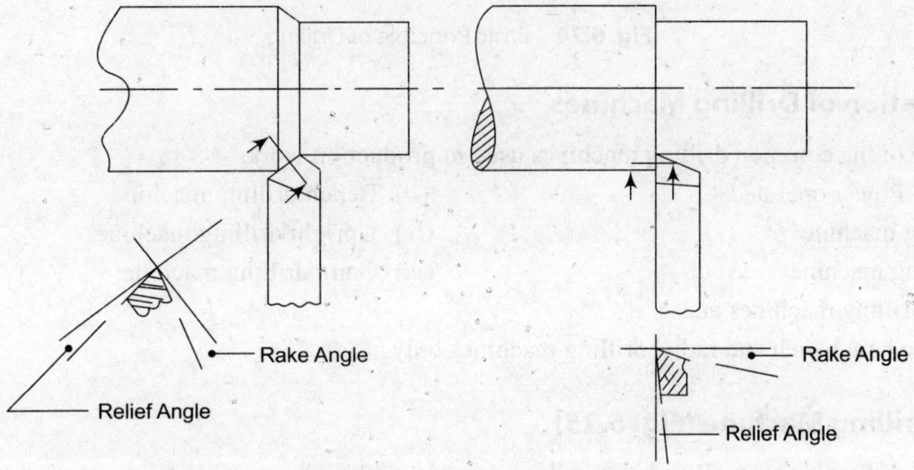

Fig. 6.22 (*b*) Geometry of Cutting Edges as Seen on Turning Tools

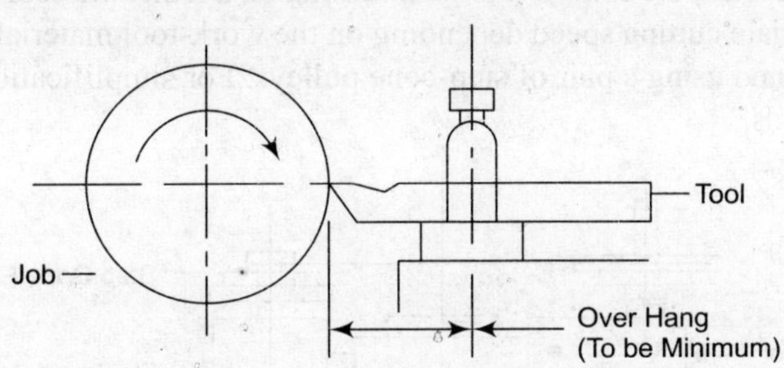

Fig. 6.23 Fixing Tool in Tool Holder

6.1.7 Specification of Lathe

Lathe is specified by the largest cylindrical job that could be mounted and machined. This would mean specifying the maximum distance between centres and the swing over bed (twice the distance from the lathe centre to the top of guide ways).

6.2 DRILLING MACHINE

Drilling, by definition, is originating a hole in a job and machine tool used for this purpose is called Drilling Machine.

6.2.1 Working Principle of Drilling

Refer Fig. 6.24. The operation involves rotating the drill about its axis and feeding it axially. The cutting edges (at the end of the drill) remove the material and the cut material (chip) comes out through the flutes (grooves) cut on the drill.

Any drilling machine must therefore incorporate mechanisms for (*i*) rotating the drill (cutting motion), (*ii*) moving the drill axially (feed motion) and (*iii*) adjustment facility for locating the job under the drill.

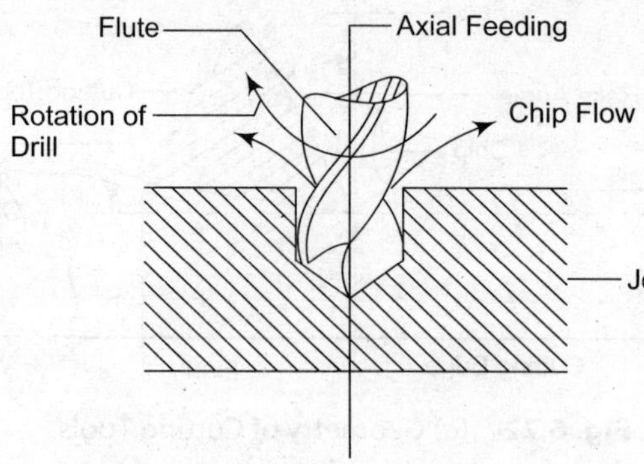

Fig. 6.24 Basic Principle of Drilling

6.2.2 Classification of Drilling Machines

Following are some of the common drilling machines used in production work.

(*i*) Hand drill —Power operated
(*ii*) Bench drilling machine
(*iii*) Pillar drilling machine
(*iv*) Upright drilling machine
(*v*) Radial drilling machine
(*vi*) Gang drilling machine
(*vii*) Production drilling machines etc.

We shall examine here bench and radial drilling machines only.

6.2.3 Bench Drilling Machine (Fig. 6.25)

It is so named because the size is small and is usually mounted on a bench or a table for convenience in operation.

The electric motor provides drive for the spindle. As with turning on a lathe, the economics of drilling dictates that the spindle be rotated to give appropriate cutting speed depending on the work-tool material combination. In the drill shown, different spindle speeds are obtained using a pair of step-cone pulleys. For simplification only three steps are shown, by which we get three different speeds.

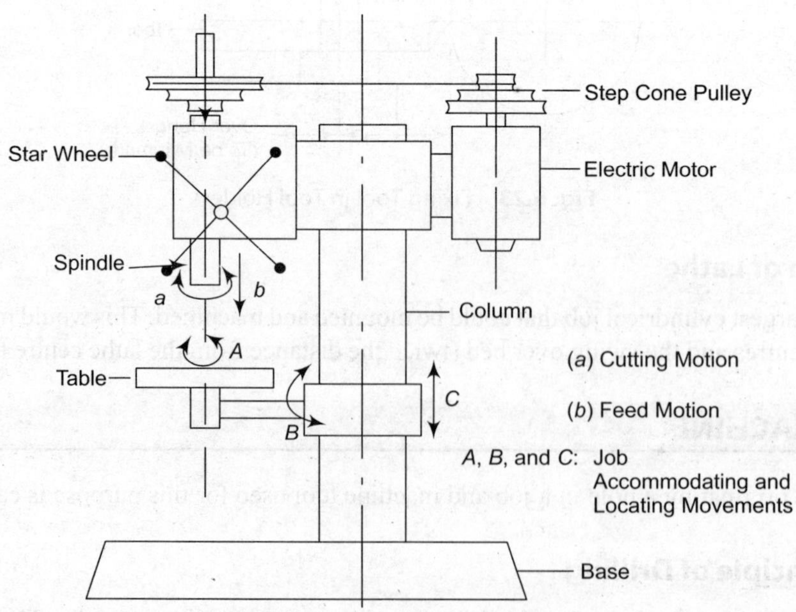

Fig. 6.25 Bench Drilling Machine

The job is fixed to the table and the drill is fed axially using the star wheel. The mechanism for feeding the drill is shown in Fig. 6.26. The quill has rack teeth cut on it and when the pinion, engaging with the rack, is rotated using the star wheel, the quill moves up or down. The downward motion of the quill moves the spindle downwards and the drill fitted to the spindle gets the feed motion.

Before drilling, the job must be located in such a way that the drill axis comes directly over the point where drilling is to be done. To facilitate this, the table is made rotatable about its axis (Ref. Fig. 6.27) (motion *A*) and the table itself can be swung around the axis of the column (motion *B*). To illustrate the method, consider locating point *P* under the axis of the drill. The table is first rotated about its axis so that point *P* moves along path *A* and comes to path *B*. Now, by swinging the table, point *P* moves along path *B* and is brought under the axis of the drill.

The movement *C* is for accommodating jobs of varying heights and all the three movements—*A*, *B* and *C* once adjusted, are locked before drilling operation proceeds.

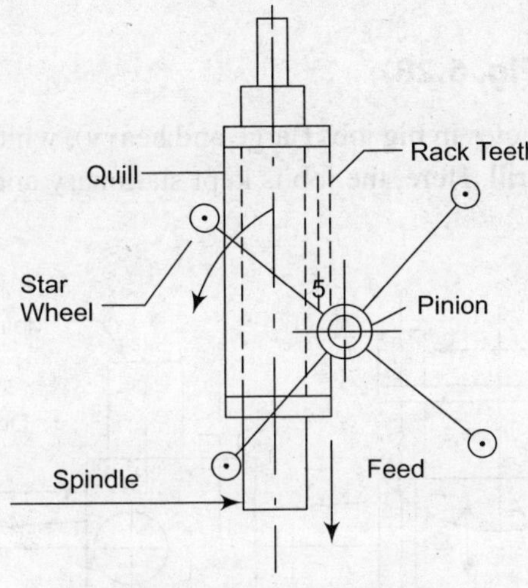

Fig. 6.26 Drill Feed Mechanism

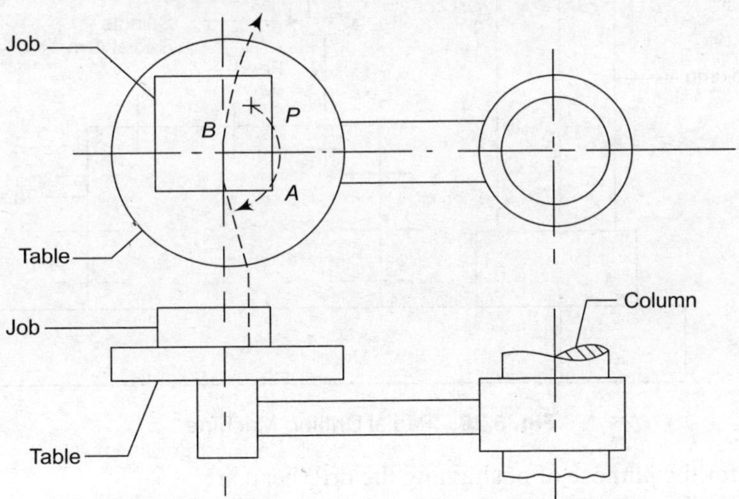

Fig. 6.27 Locating A Job Under Drill—An Example

6.2.4 Cutting Parameters

(*i*) *Cutting speed:* In drilling, the cutting speed is taken as the peripheral speed of the drill.

$$\text{Cutting speed (C.S.)} = \frac{\pi d N}{1000} \text{ m/min.}$$

where d = Diameter of the drill (mm) and

N = rpm of drill

(*ii*) *Feed:* It is the axial movement of the drill expressed in mm/rev. of drill or as mm/minute.

(*iii*) *The depth of cut:* It is equal to half the diameter of the drill used.

6.2.5 Procedure in Drilling

The drill locations are first marked out using a scriber and centre-punched. Then the job is taken to the drilling machine, clamped and drilled.

6.2.6 Specification

A bench drilling machine is specified by the largest diameter of job in which machine can drill a centre-hole.

6.2.7 Radial Drilling Machine (Fig. 6.28)

This has been specially designed to drill holes in big jobs (large and heavy), which cannot be easily moved about to bring the drill locations under the axis of the drill. Here, the job is kept stationary and the drill head is moved to the required location.

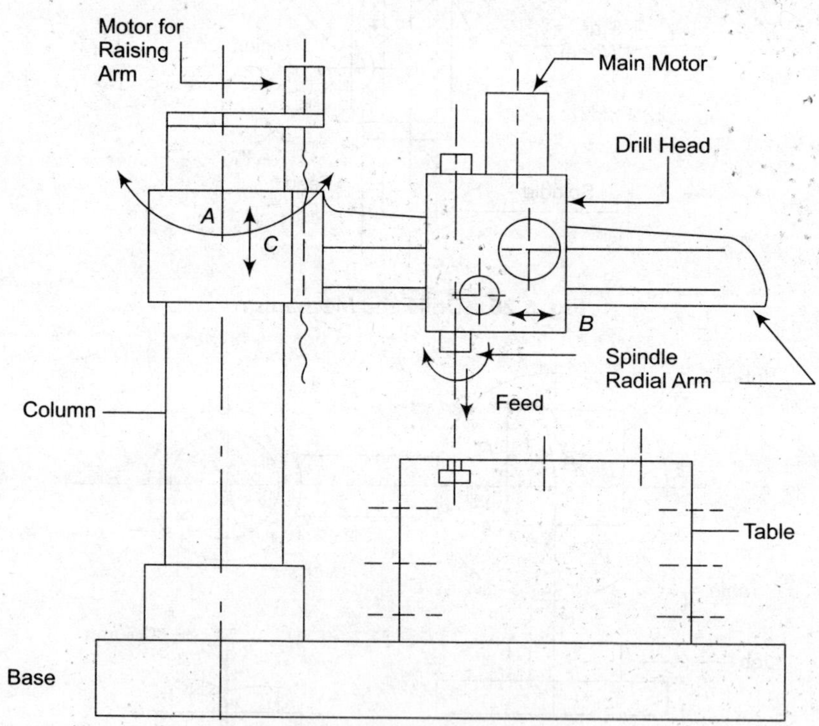

Fig. 6.28 Radial Drilling Machine

The various movements for the purpose of positioning the drill head are:

(*i*) Swing of the radial arm about the axis of the column (movement A)

(*ii*) Movement of the drill head along the radial arm (movement B)

Movement C is for height adjustment of the arm to accommodate job.

All the three movements are locked, once the drill head is positioned.

The drill head carries the main motor (to provide drive for the spindle), speed and feed gear boxes, the spindle and spindle bearings.

The job is mounted on the box table or directly over the base.

6.2.8 Operations Done on a Drilling Machine

Although the drilling machine is designed to originate holes, it is employed to do many other allied operations such as (*i*) Reaming, (*ii*) Tapping, (*iii*) Boring, (*iv*) Counter boring, (*v*) Spot facing, (*vi*) Counter-sinking, etc.

Figure 6.29 shows the scheme of these operations.

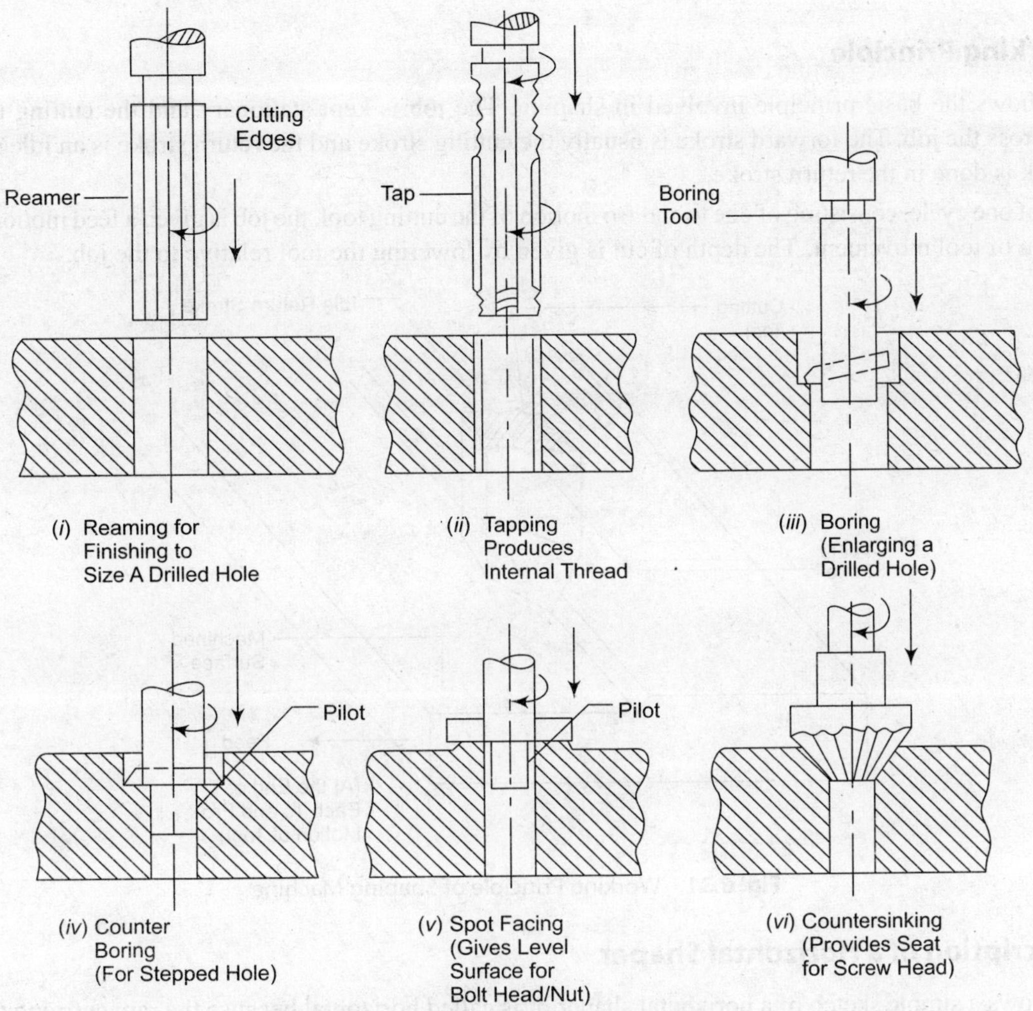

Fig. 6.29 Operations on Drilling Machine

6.2.9 The Drill

Some of the types of drills used are:

(*i*) Twist drill

(*ii*) Straight fluted drill

(*iii*) Core drill

(*iv*) Gun drill, etc.

The twist drill is the most common type of drill used and its main features are shown in Fig. 6.30.

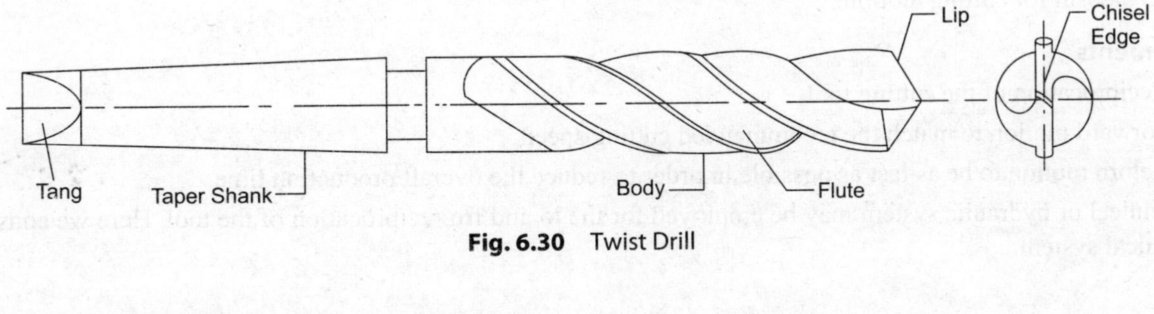

Fig. 6.30 Twist Drill

6.3 SHAPING MACHINE (SHAPER)

Shaper is a machine tool primarily intended for producing flat surfaces. However, key way cutting in shafts and bores, machining form surfaces of curvilinear contours, gear cutting, etc., are also done on this machine.

Shapers are used for machining smaller jobs. When large size jobs are to be machined a planer is used.

6.3.1 Working Principle

Figure 6.31 shows the basic principle involved in shaping. The job is kept stationary and the cutting tool is made to reciprocate across the job. The forward stroke is usually the cutting stroke and the return stroke is an idle stroke, because no useful work is done in the return stroke.

At the end of one cycle, consisting of one to and fro motion of the cutting tool, the job is given a feed motion perpendicular to the direction of tool movement. The depth of cut is given by lowering the tool relative to the job.

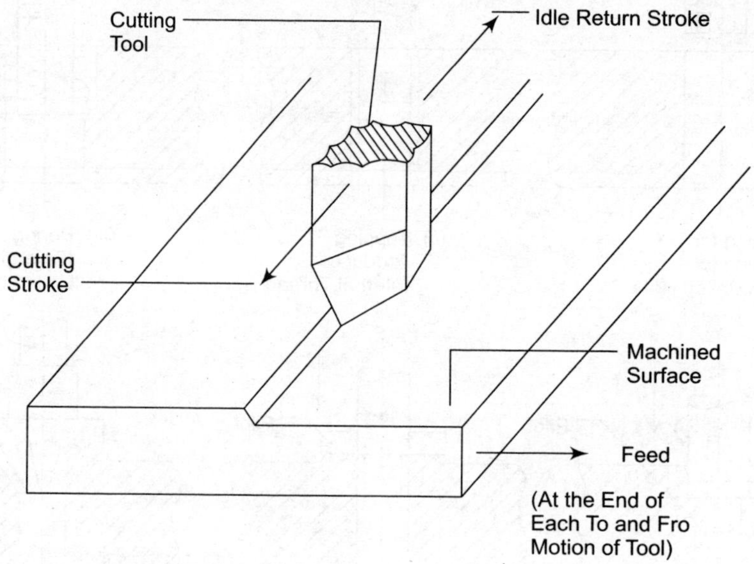

Fig. 6.31 Working Principle of Shaping Machine

6.3.2 Description of a Horizontal Shaper

Figure 6.32 shows a simple sketch of a horizontal shaper. It is called horizontal because the ram carrying the cutting tool reciprocates in a horizontal plane.

From the figure, it can be seen how the working principle described in the previous section has been translated into practical means by incorporating the following mechanisms.

(*i*) Mechanism for to and fro motion of cutting tool (cutting motion)

(*ii*) Mechanism for feed motion and

(*iii*) Mechanism for giving depth of cut.

Let us make a brief study of these mechanisms.

(*i*) Mechanism for cutting motion:

Requirements

(*a*) Reciprocation of the cutting tool

(*b*) Forward motion to match the recommended cutting speed

(*c*) Return motion to be as fast as possible in order to reduce the overall production time.

Mechanical or hydraulic system may be employed for the to and fro reciprocation of the tool. Here we consider only a mechanical system.

For the fast returning of the tool, two mechanical systems are available:

(*a*) Crank and slotted lever mechanism and

(*b*) Whitworth quick return motion mechanism

Let us now study just one mechanism viz., Crank and slotted lever mechanism in detail.

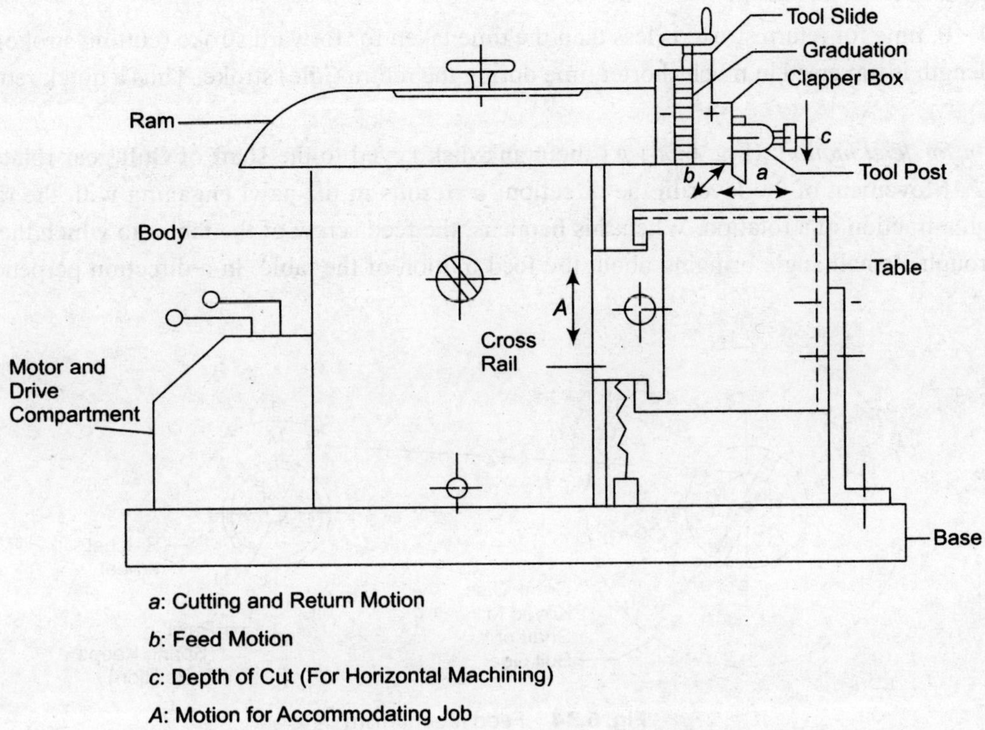

Fig. 6.32 Shaping Machine (Shaper)

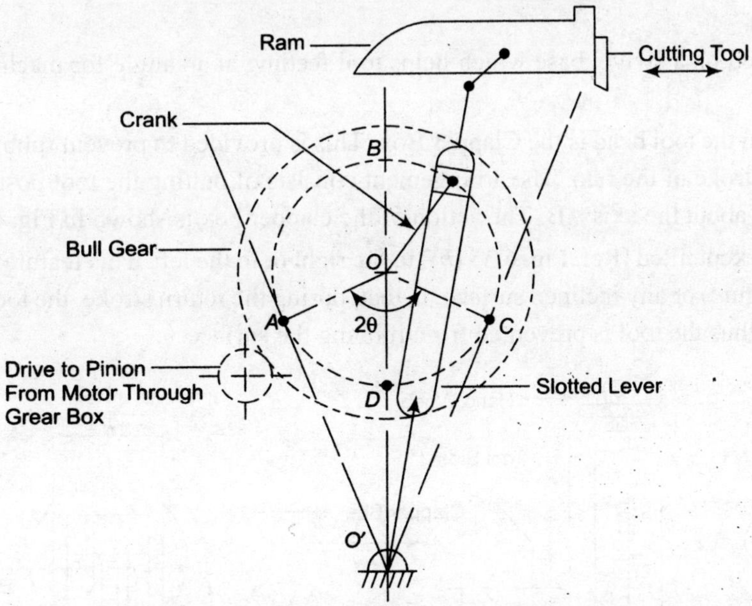

Fig. 6.33 Crank and Slotted Lever Mechanism

Crank and slotted lever mechanism: Figure 6.33 shows the line diagram of the mechanism. The crank (Bull gear) driven off by an electric motor rotates at a constant speed (ω). As the crank rotates, the slotted lever oscillates about the fulcrum *O*′. This makes the ram reciprocate and therefore the cutting tool fixed to it.

As the crank rotates clockwise through the arc *ABC* the ram moves forward and cutting stroke takes place. Further movement of the crank through the arc *CDA* results in return stroke.

Taking the ratio,

$$\frac{\text{Time for cutting stroke}}{\text{Time for return/idle stroke}} = \frac{(360 - 2\theta)/\omega}{2\theta/\omega} = \frac{180 - \theta}{\theta}$$

Since $\theta < 180 - \theta$, time for return stroke is less than the time taken for forward stroke (cutting stroke). In other words, the same stroke length is traversed in much shorter time during the return (idle) stroke. Thus a quick return motion of the tool is obtained.

(*ii*) *Mechanism for feed motion:* (Fig. 6.34) As the crank disk keyed to the shaft of Bull gear rotates, the lever *MN* oscillates about *F*. Movement of the lever in the direction '*a*' results in the pawl engaging with the ratchet wheel and rotating it through a fraction of a rotation. When this happens, the feed screw of the table (to which the ratchet wheel is fixed) rotates through a small angle bringing about the feed motion of the table, in a direction perpendicular to that of the ram.

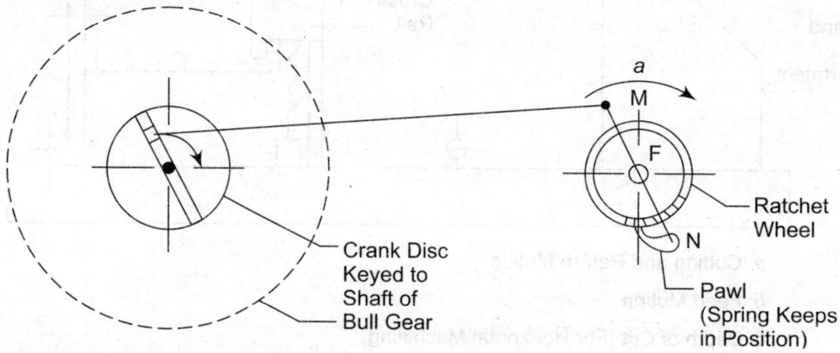

Fig. 6.34 Feed Mechanism

(*iii*) *Mechanism for giving depth of cut:* Refer Fig. 6.35. The depth of cut is given by lowering the tool slide on the tool head using the hand wheel.

The tool head is mounted on a swivel base which helps tool feeding at an angle for machining inclined surfaces like for example, a dove tail.

The additional feature in the tool head is the Clapper Box. This is provided to prevent rubbing of tool on the machined surface during the return stroke of the ram. The arrangement consists of putting the tool post on the central piece of the clapper box, which hinges about the axis AB. The action of the clapper box is shown in Fig. 6.35 (*a*).

Clapper box can also be kept tilted [Ref. Fig. 6.35 (*b*)] to the right or to the left. This feature helps in machining vertical surface (as shown in the figure) or any inclined surface, in that, during the return stroke, the tool lifts in a plane away from the machined surface and thus the tool is prevented from rubbing the surface.

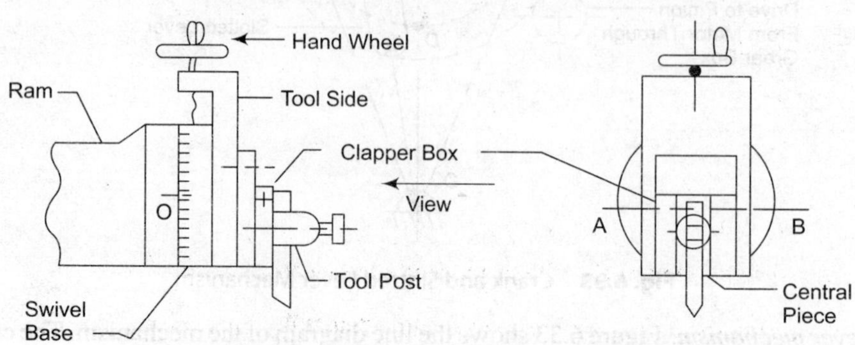

Fig. 6.35 Tool Head and Clapper Box

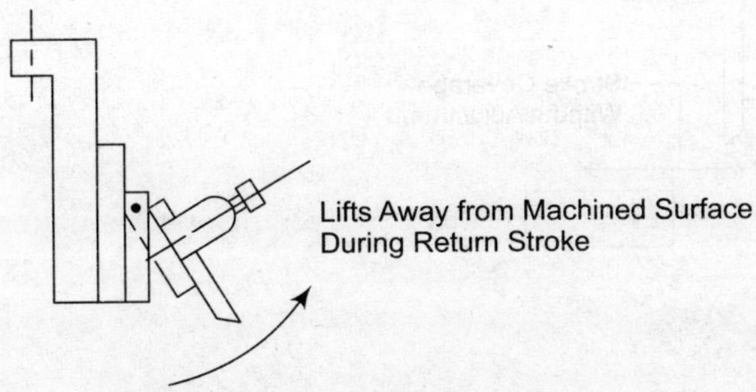

Fig. 6.35 (*a*) Action of Clapper Box

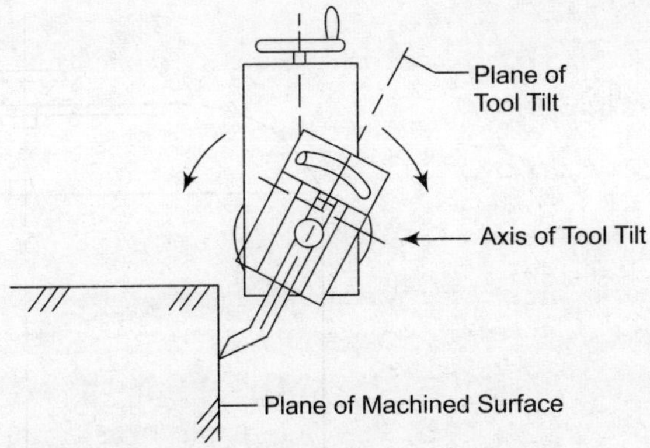

Fig. 6.35 (*b*) Machining a Vertical Surface

Adjustments to be made before machining:

(*i*) Ram speed setting

(*ii*) Stroke adjustment

(*iii*) Ram position adjustment

(*i*) *Ram Speed:* As with other machining processes, we must ensure an optimum cutting speed in the shaping process also. With the crank and slotted lever mechanism, the velocity of the ram is zero at the beginning and at the end of the stroke and is a maximum at the middle of stroke. The cutting speed that must be maintained must therefore match with the average speed of the ram.

The speed of ram is varied using a gear box in between the electric motor and the Bull gear (which also serves as a crank for the slotted lever mechanism)

The speeds available are in steps and the one nearest to the optimum speed must be chosen and set up.

(*ii*) *Stroke adjustment:* Depending on the length of job to be shaped, stroke length must be adjusted (Fig. 6.36). This is done by varying the radius of the crank. For this, the slide in the slotted lever is radially moved inward or outward by the arrangement shown in Fig. 6.36. Nearer the slide block to the centre, smaller is the stroke and vice versa.

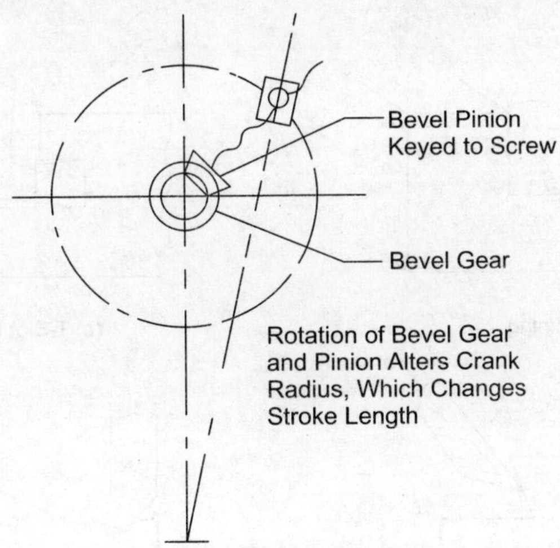

Fig. 6.36 Stroke Adjustment

(*iii*) *Position adjustment:* When a job is mounted on the table of the shaper, the reciprocation of the ram may not occur over the length of job properly. See Fig. 6.37. In order to adjust the position of the ram properly, the ram fixing screw is loosened, the ram moved to the required position and the screw tightened again.

6.3.3 Shaper Operations

Figure 6.38 shows some of the operations done on a shaper. They are:

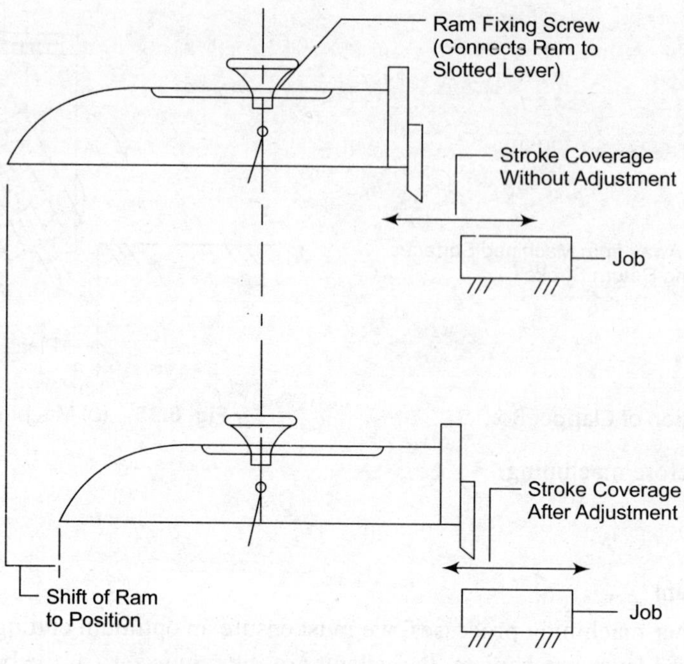

Fig. 6.37 Ram Position Adjustment

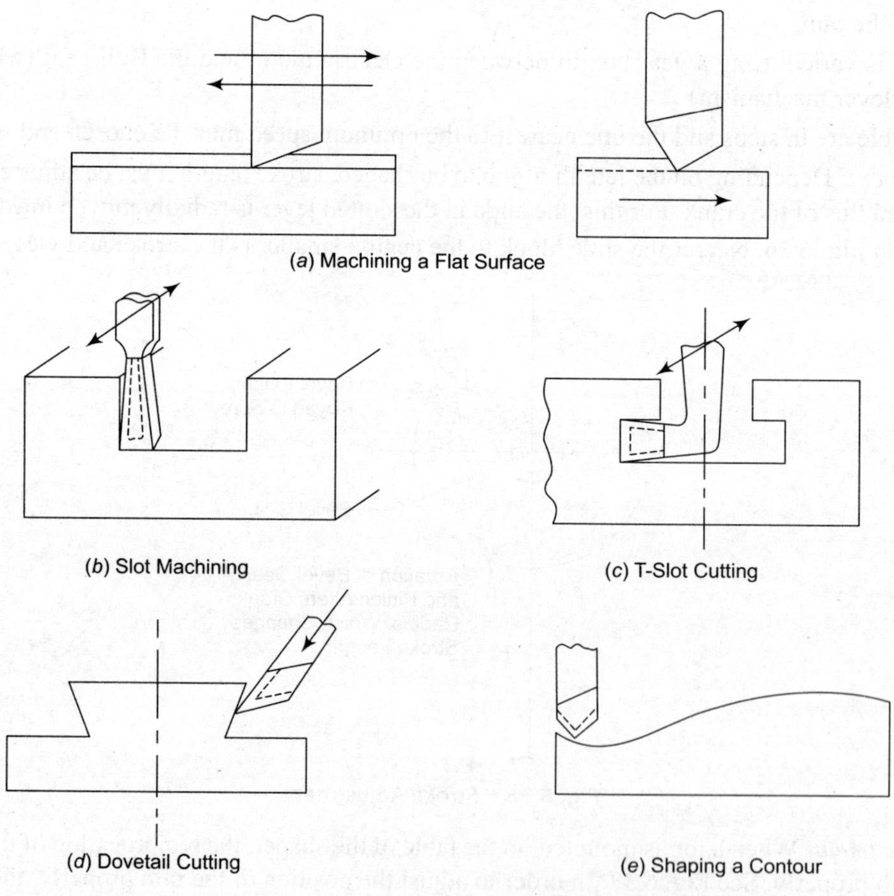

Fig. 6.38 Operations on a Shaper

(*a*) Machining a flat surface
(*b*) Machining a slot
(*c*) Cutting a T-slot
(*d*) Machining a dove tail
(*e*) Shaping a contour

6.3.4 Classification

Shapers are classified based on the plane in which the ram reciprocates. Thus we have horizontal and vertical shapers. The vertical shaper is otherwise called a slotting machine (slotter). The horizontal shapers are further classified as pushcut and drawcut shapers. The pushcut type has been examined in section 6.3.2. In the drawcut type, the cutting takes place in the return stroke. A universal shaper has a table that could be tilted.

6.3.5 Specification

The size of a shaper is given by the maximum stroke length of the ram in mm.

6.4 PLANING MACHINE (PLANER)

Planing machine is primarily intended for machining large flat surfaces. It is also used for machining long grooves, slots, vertical surfaces and contours.

The working principle is similar to that of a shaper, in that the relative motion between the cutting tool and the job is linear. The basic differences between shaper and planer are:

(*i*) In a shaper, the cutting tool moves against a stationary job giving the main motion (cutting motion), whereas in a planer the job moves against a stationary cutting tool.

(*ii*) The feed motion in a shaper (for horizontal machining) is given to the job, whereas in a planer the motion is given to the cutting tool, which moves across the job a small distance every time a to and fro motion of the job is completed.

6.4.1 Types of Planer

Following are the main types of planers.

(*i*) Double housing planer

(*ii*) Openside planer

(*iii*) Universal planer (cuts on both strokes)

(*iv*) Pit type planer (for very large jobs)

We shall examine here a double housing planer which is very commonly used.

6.4.2 Double Housing Planer

Figure 6.39 shows a simplified diagram of a Double Housing Planer. This has two columns (housings) on either side of the table. The cross rail moves vertically up or down over the guideways provided on the housings. This motion helps in accommodating jobs of different heights on the table. One or two tool heads traverse along the guide ways provided on the cross rail. Side tool heads are sometimes provided, which move over in a vertical direction along the guideways on the housings.

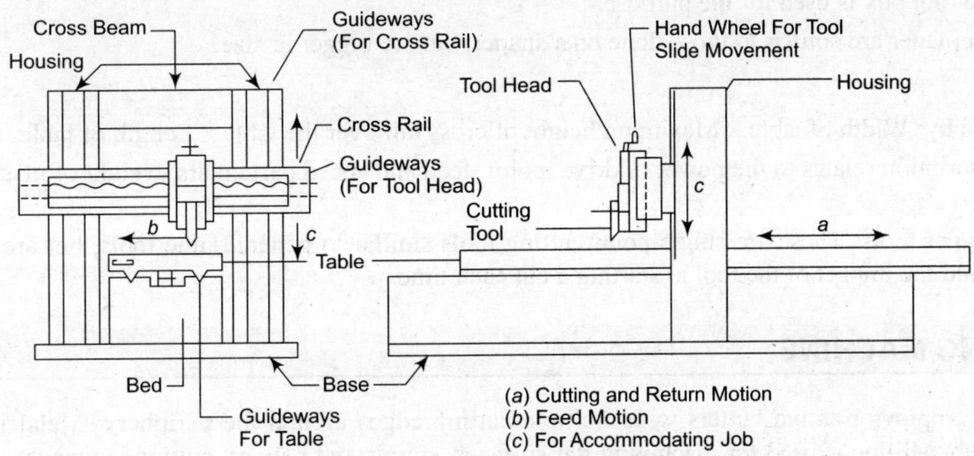

Fig. 6.39 Double Housing Planer

The table reciprocates along guideways machined on the bed. The table reversal is brought about by a simple mechanical arrangement shown in Fig. 6.40. Power is transmitted through cross belting during cutting stroke and through open belt during idle stroke.

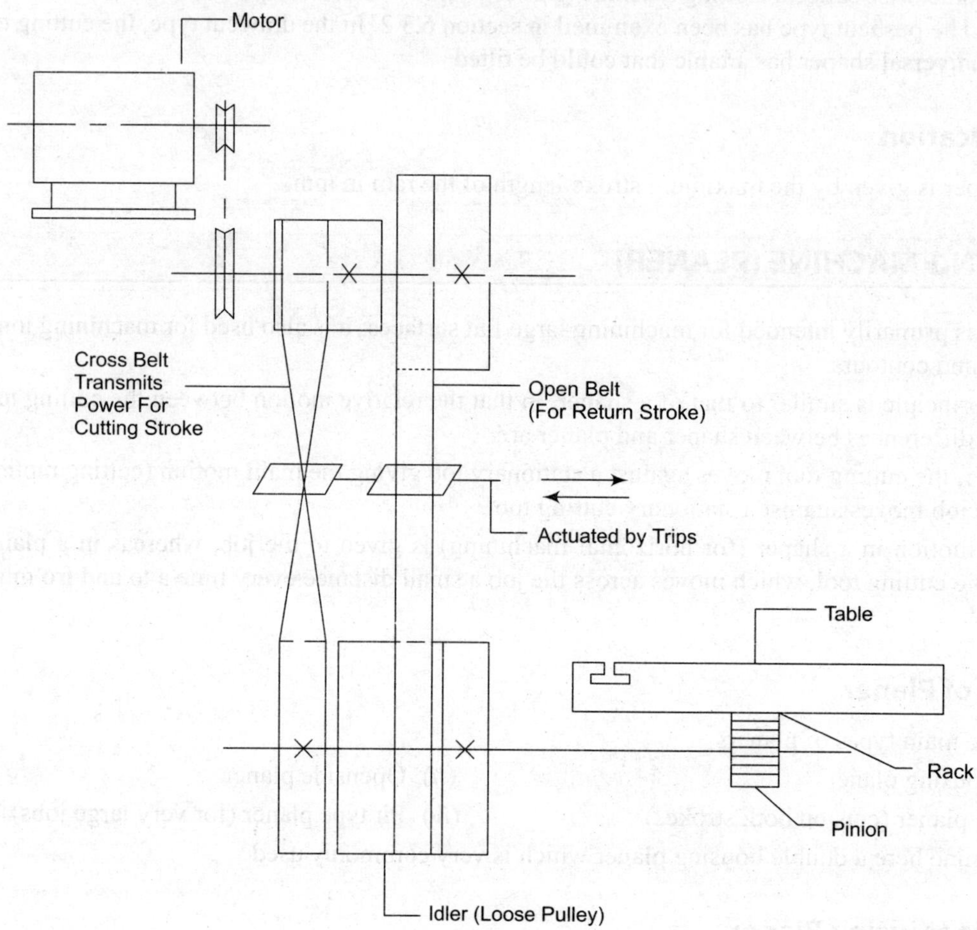

Fig. 6.40 Table Drive with Quick Return

In the alternative, a hydraulic or electric drive is used to advantage, but the cost will be higher. In an electric drive a variable speed reversible D.C. motor is used. The reversing of the table is simplified by incorporating limit switches, which make the necessary change over.

As in the shaper, the cutting tool is mounted in a clapper box to lift it clear of the work on the idle stroke. Sometimes, a solenoid—lifted tool box is used for the purpose.

Jobs done on a planer are similar to those done on a shaper, but are bigger in size.

Specification

Planer is specified by: Width of table × Maximum height of cross rail over the table × Length of table.

Additional description relates to the power of drive motor, feed and speed particulars, weight of machine, floor space required etc.

Planer and Shaper tools: These are single point cutting tools similar to general lathe tools, but are heavier in cross section to withstand the impact of the tool in starting a cut each time.

6.5 MILLING MACHINE

Milling machine employs rotating cutters with multiple cutting edges around the periphery. Metal Removal Rate is therefore very high. Milling is used for machining flat surfaces, spirals and helixes, cutting keyways, slots and grooves and producing cams, gears and complex contours.

6.5.1 Working Principle

Refer Fig. 6.41. The job to be machined is fed against a rotating cutter and each cutting edge removes a small amount of material (chip) in the form of a 'Comma'. The rotation of the cutter provides the cutting motion. This type of milling is called upmilling and is most commonly employed.

Sometimes, for specific advantage, for example machining a thin job, down milling is employed. Here the feed direction coincides with the direction of cutting motion.

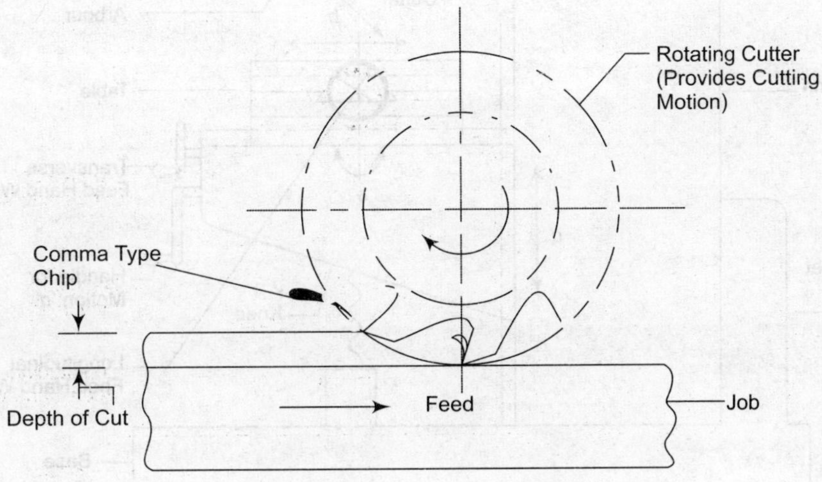

Fig. 6.41 Principle of Milling

6.5.2 Classification of Milling Machines

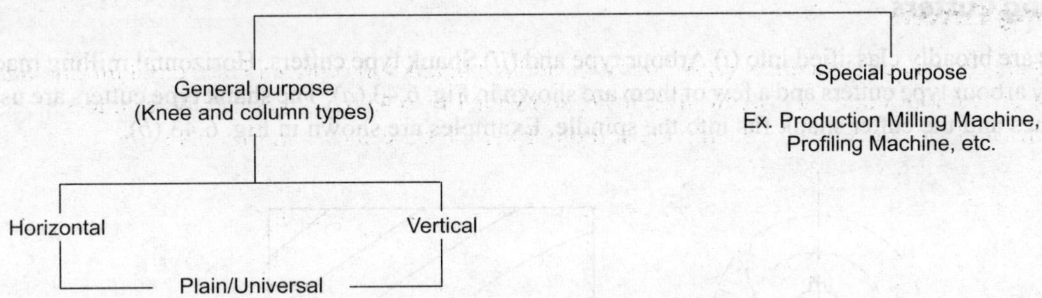

6.5.3 Column and Knee Type Horizontal Universal Milling Machine

Figure 6.42 is a diagrammatic view of the milling machine showing important parts and motions. The machine is called horizontal because the cutter rotates about a horizontal axis.

It is universal because the table of the machine can be swivelled about a vertical axis, which helps in helical milling, cam milling etc.

The machine consists of a rigid base on which stands the column. The column houses the drive motor and speed mechanism in addition to providing support for the spindle and spindle bearings. The knee is capable of being moved up or down over the guide ways formed on the front of the column. This motion helps in accommodating jobs of varying heights and also giving depth of cut.

The table consists of three parts. Bottom one is the cross slide which can be moved towards or away from the column (transverse motion). It is guided by the guideways machined on the top of the knee. Over this part comes the middle one, which is a swivel base. The top part (actual table) is mounted on the swivel base and has longitudinal motion in the guideways machined in the swivel base.

The cutter is mounted on an Arbour, which fits into the spindle of the machine. The arbour is supported at the other end by means of an end support fitted to the over arm.

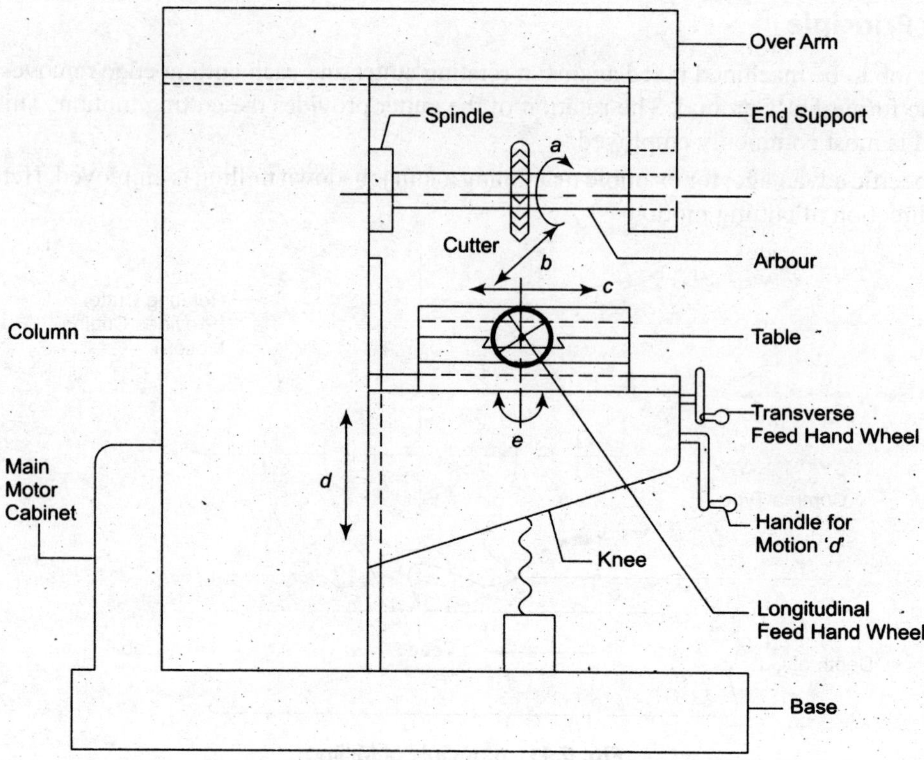

Fig. 6.42 Column and Knee Type Horizontal Universal Milling Machine

6.5.4 Milling Cutters

Milling cutters are broadly classified into (*i*) Arbour type and (*ii*) Shank type cutters. Horizontal milling machines having arbours employ arbour type cutters and a few of them are shown in Fig. 6.43 (*a*). The shank type cutters are used on vertical milling machines and the cutter shank fits into the spindle. Examples are shown in Fig. 6.43 (*b*).

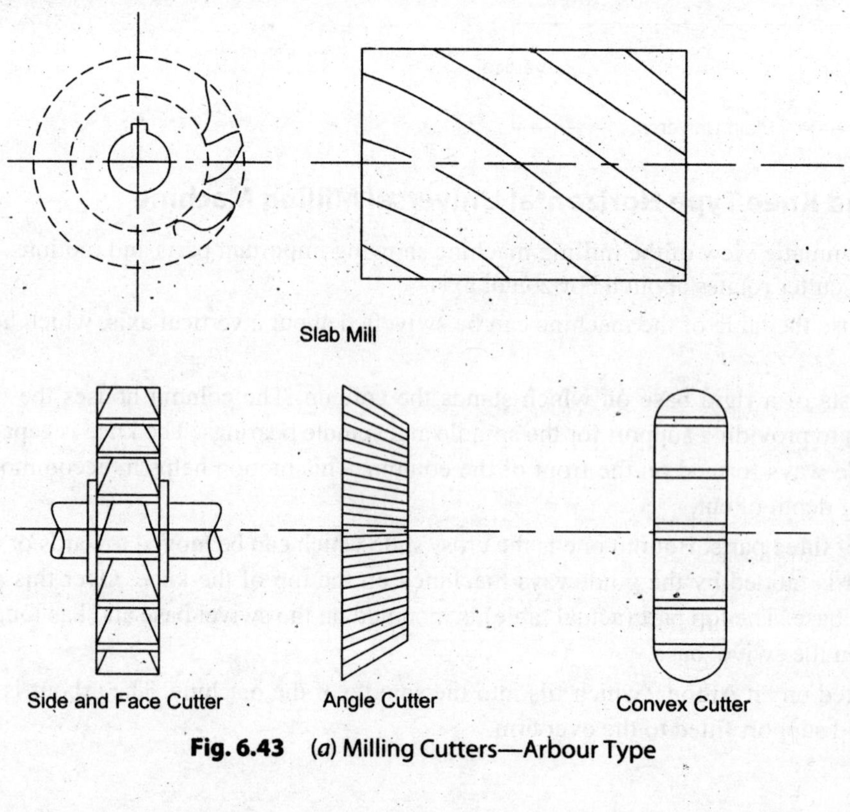

Fig. 6.43 (*a*) Milling Cutters—Arbour Type

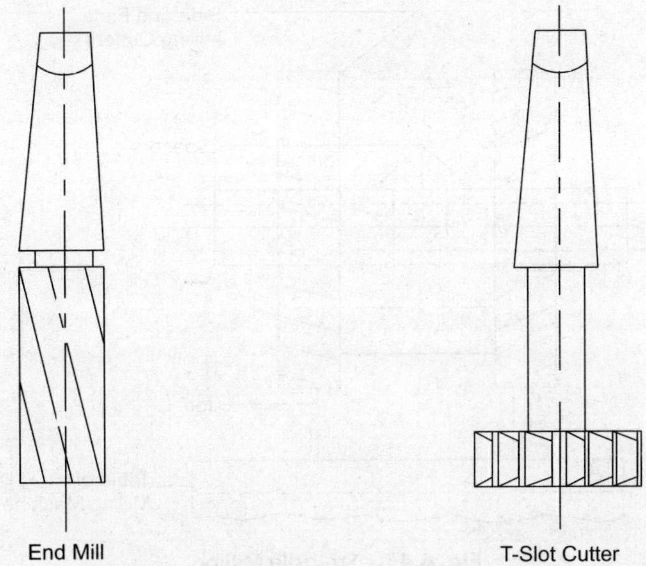

Fig. 6.43 (*b*) Milling Cutters—Shank Type

6.5.5 Milling Techniques

Basic principle remaining the same, techniques employed are varied to obtain specific advantages.

Following are the techniques used in milling in order to improve productivity:

(*i*) Slab Milling
(*ii*) Straddle Milling
(*iii*) Abreast Milling
(*iv*) String Milling
(*v*) Gang Milling etc.

(*i*) **Slab Milling:** Figure 6.44 shows the arrangement in Slab Milling. The cutter, called Slab mill is wide enough to cover the entire width of job and as the job passes underneath the cutter once, (or twice for finishing) the whole machining is complete. The production rate is very high and the process is used for machining flat surfaces economically.

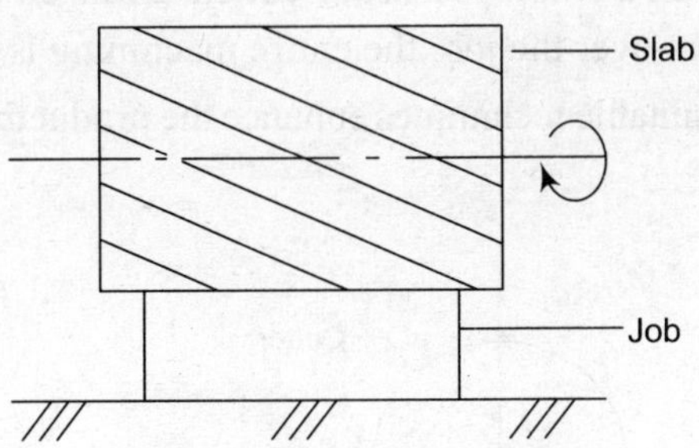

Fig. 6.44 Slab Milling

(*ii*) **Straddle Milling:** Figure 6.45 shows the arrangement for milling a job to the required width 'x' by employing two side and face milling cutters. The cutters are spaced at a distance equal to the width of the job. This arrangement makes possible machining in one pass to the required width with consistent dimensional accuracy. In-process inspection is therefore not required and hence production time is reduced considerably.

(*iii*) **Abreast Milling:** In this, smaller jobs are placed side by side (Fig. 6.46) and a slab mill machines all of them simultaneously.

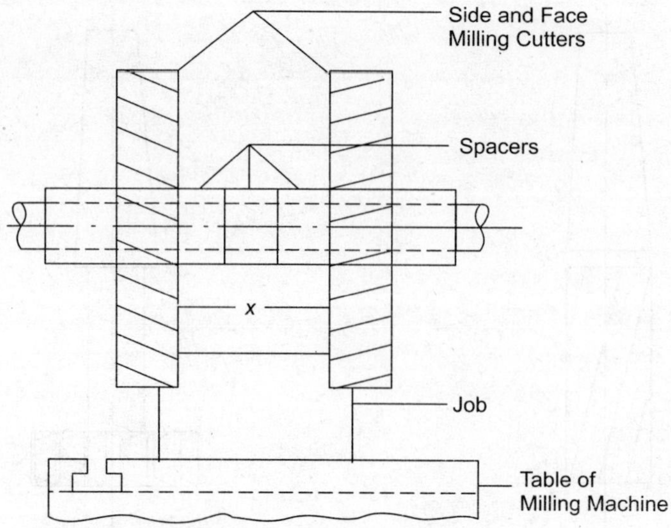

Fig. 6.45 Straddle Milling

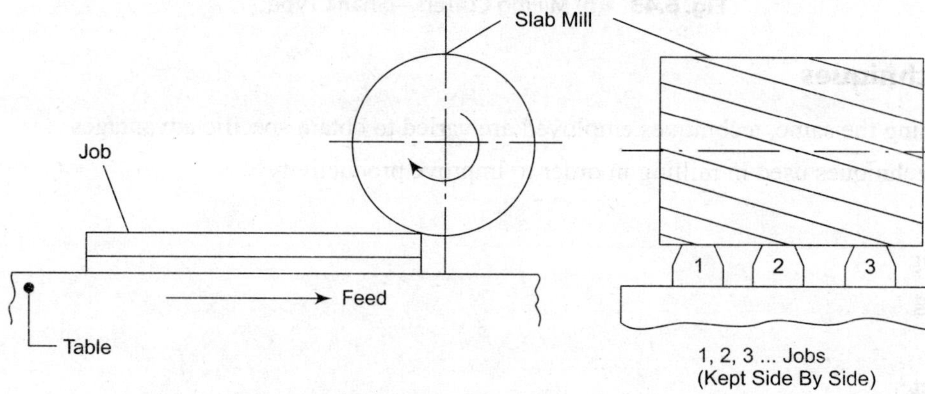

Fig. 6.46 Abreast Milling

(*iv*) **String Milling:** (Fig. 6.47): Smaller components may be arranged one behind the other in a string like fashion and machining done as the table traverses from one end to the other.

(*v*) **Gang Milling:** (Fig. 6.48): This method employs a set or cutters arranged to conform to the required shape of the contour and in one pass of the cutters over the job, the entire machining is done.

Given above are some examples of how suitable techniques enhance the productivity and thus bring down the production cost.

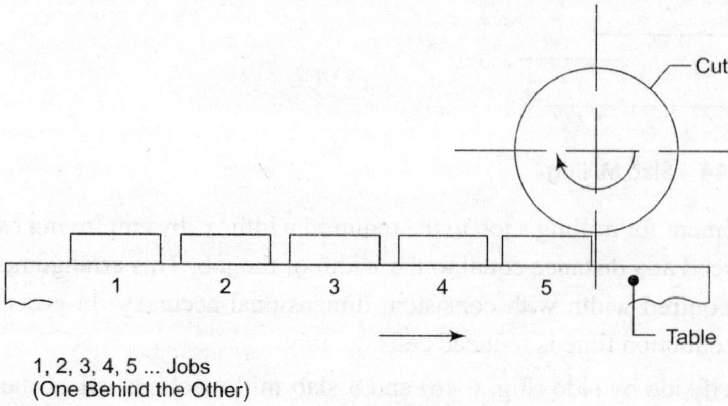

Fig. 6.47 String Milling

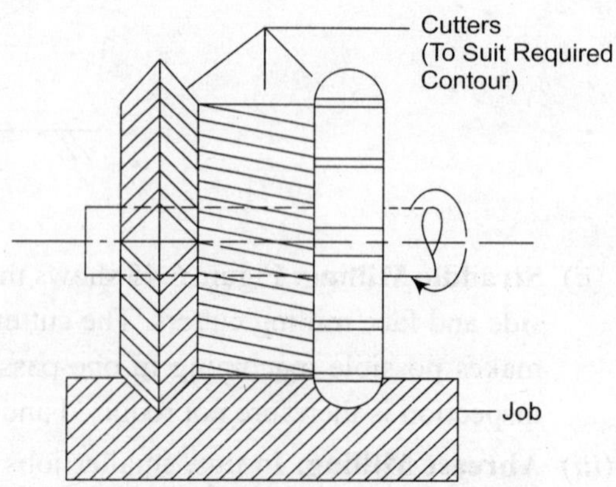

Fig. 6.48 Gang Milling

6.5.6 Milling Machine Attachments

Several attachments are designed to make milling a versatile machine tool. Key way, spiral and helix milling are just a few examples of the operations done on the milling machine using the attachments. Among the many attachments, Dividing Head (Indexing Head) is the most important and interesting attachment, using which, gears-Spur, bevel and helical, hexagon and other shapes are easily milled.

Dividing Head: Two types are available. One is the Mechanical and the other Optical (used for precision work).

Dividing head is used for spacing features at definite angular intervals on a job. It is also used in spiral milling, cam milling etc. Let us study here the Mechanical type.

Figure 6.49 shows schematically the mechanism of a dividing head. The crank is used to rotate the spindle of the dividing head through worm and worm wheel arrangement. A standard dividing head has a transmission ratio of 40 : 1 *i.e.*, when the crank is rotated 40 times, the spindle of the dividing head rotates once. Th usage of the dividing head is best explained by taking a few examples.

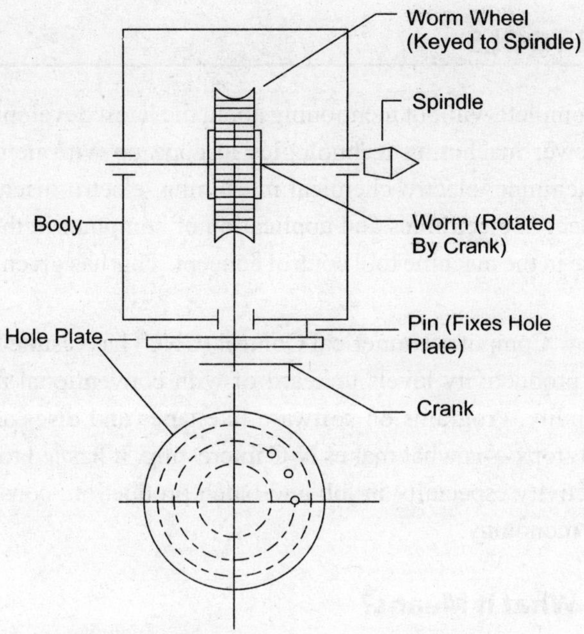

Fig. 6.49 Schematics of Mechanical Dividing Head

Example 1: To mill a hexagon on a bolt head (Fig. 6.50).

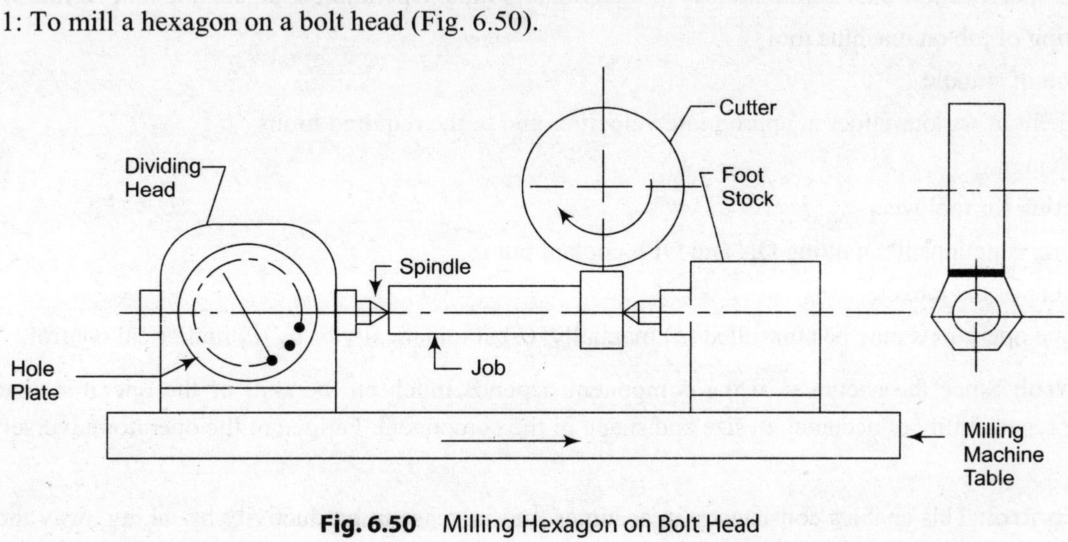

Fig. 6.50 Milling Hexagon on Bolt Head

This requires the bolt (held between the spindle of the dividing head and the foot stock centre) to be rotated about its axis 1/6 of a rotation each time a face is machined. This is done by rotating the crank by 40/6 of a revolution *i.e.*, 6 and 2/3 rev. In order to precisely rotate through fraction of a crank rotation, a hole plate is used (Standard hole plates are made available with each Dividing Head). A hole plate consists of several hole circles. In the present example, after rotating the crank through full six turns, for the remaining 2/3 rotation, we may choose a plate with 18 hole circle and rotate the crank through 12 holes (hole spaces) on this circle. Thus the job is rotated by l/6th of a revolution.

In general, if N features are to be equi-spaced around a job, indexing (amount of rotation of the crank) required is 40/N and the fraction of crank rotation is set using hole plates.

Example 2: To find indexing required for spacing 76 teeth on a gear.

Here $N = 76$

Indexing = 40/76 = 10/19 of crank rotation

Taking a hole plate having a 19 hole circle, the crank needs to be rotated through 10 hole spaces on this circle.

6.6 MODERN MACHINE TOOLS

Introduction to machine tools is incomplete without mentioning about the latest developments in this field. Newer machine tools have been built to absorb newer machining technologies to cope up with newer and tougher materials. Newer technologies include ultrasonic machining, electro-chemical machining, electro-discharge machining, laser machining, etc. While this is so, the advancement in electronics and application of computer in the machine tools have brought in a significant and revolutionary change in the machine tool control concept. This has given birth to an entirely new generation of machine tools.

Numerical Control (N/C) and now Computer Numerical Control (CNC) has resulted in totally new class of automated machine tools with quality linked productivity levels unheard of with conventional machine tools. The essence of the Numerical Control lies in its flexibility. Programs on software like tapes and discs as against rigid programming with Hardware like cams, templates and stops —is what makes N/C interesting. It has led to several advantages like reduction in 'Lead Time', increase in productivity especially in job and batch production, consistency in accuracy and above all elimination of the human skill and monotony.

6.6.1 Numerical Control—What it Means?

Controlling a machine tool implies operating the machine tool in the most appropriate manner to enable producing components to the required specification with minimum loss of material and time. Operation of a machine tool includes:

(*i*) Mounting of job on machine tool
(*ii*) Rotation of spindle
(*iii*) Movement of various slides at appropriate velocities and to the required limits
(*iv*) Tool changing
(*v*) Correcting for tool wear
(*vi*) Auxiliary functions like putting ON and OFF coolant pump
(*vii*) Releasing of the job, etc.

All the above operations may be controlled (*a*) manually, (*b*) automatically or by (*c*) numerical control.

Manual Control: Since the accuracy of the component depends much on the skill of the operator, manual control cannot ensure repeatability of accuracy in size and shape of the component. Fatigue of the operator adversely affects the productivity.

Automatic Control: This enables consistency in accuracy and increase in productivity by taking away the skill factor and reducing the physical exertion of the operator. But this type of control is characterised by Rigid Programming in the sense that hardwares like cams, templates, stops, *etc*., are used for programming. Any change in size or shape of the component requires redesigning and making of new cams, altering of the settings of stops, etc., an exercise consuming lot of time and money.

Numerical Control: This type of control on the other hand is characterised by Flexible Programming. Flexibility in programming refers to the ease with which program can be altered or modified. In this, the programming is done on punched tape/punched card/magnetic tape and the like. This method enables altering the program with least time and expense in order to modify a component or even to make an entirely new component. Consistency in accuracy is ensured as with automatic control. N/C is best suited for job and small lot production.

6.6.2 Program on Tape

Yes. Whatever one has to do to produce a component, like setting the spindle in motion, giving depth of cut, feeding the tool to obtain a particular contour, etc. — are all automatically done by the machine itself as per the instructions now contained in a punched tape. The information (program) is put into the tape in coded form, which the control unit of the machine can understand. See Fig. 6.51, which shows the schematics of Numerical Control.

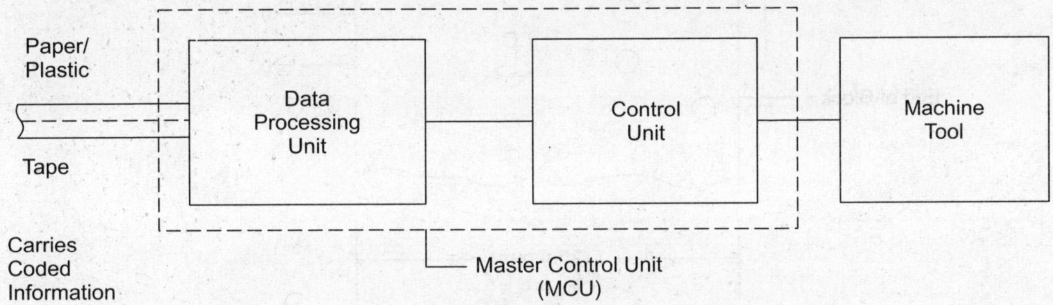

Fig. 6.51 Schematics of Numerical Control of Machine Tool

Punched tape: Although there are many other forms of putting the information into the control unit like magnetic tape, floppy disc, etc., we shall examine here a punched paper tape. The punched paper tape (Fig. 6.52) carries information in **coded** form and the first step in making a punched tape is to write out the best sequence of operations for a given job. Next, instruction for each operation is specified in a standard FORMAT. For example, the following is an instruction.

N01 G01 X05000 Y00300 F150 S400 M03 (EB)

Here: N stands for the sequence number of operation

G refers to preparatory function (Ex: G01 is for linear interpolation used for generating a taper or straight cut.)

X is command for movement of tool/slide in X-direction (X05000 means movement by 5000 units of say 0.01 mm each).

Y is command for movement in Y-direction

F for feed (mm per min.)

S for spindle speed (rpm)

M command for miscellaneous function (Ex: M03 is command for spindle rotation in CW direction)

EB End of Block

(**Note:** Cartesian co-ordinate system is used for defining tool/slide movements)

As shown above, the instruction consists of several characters — Numerals, Alphabets and Symbols. Each character is coded by punching holes on the tape (Fig. 6.52) in a row, the number and arrangement of holes having been defined. For example character 'N' is identified by holes in 7th, 3rd and 1st channels. Number '5' is coded as holes in 5th, 3rd and 1st channels.

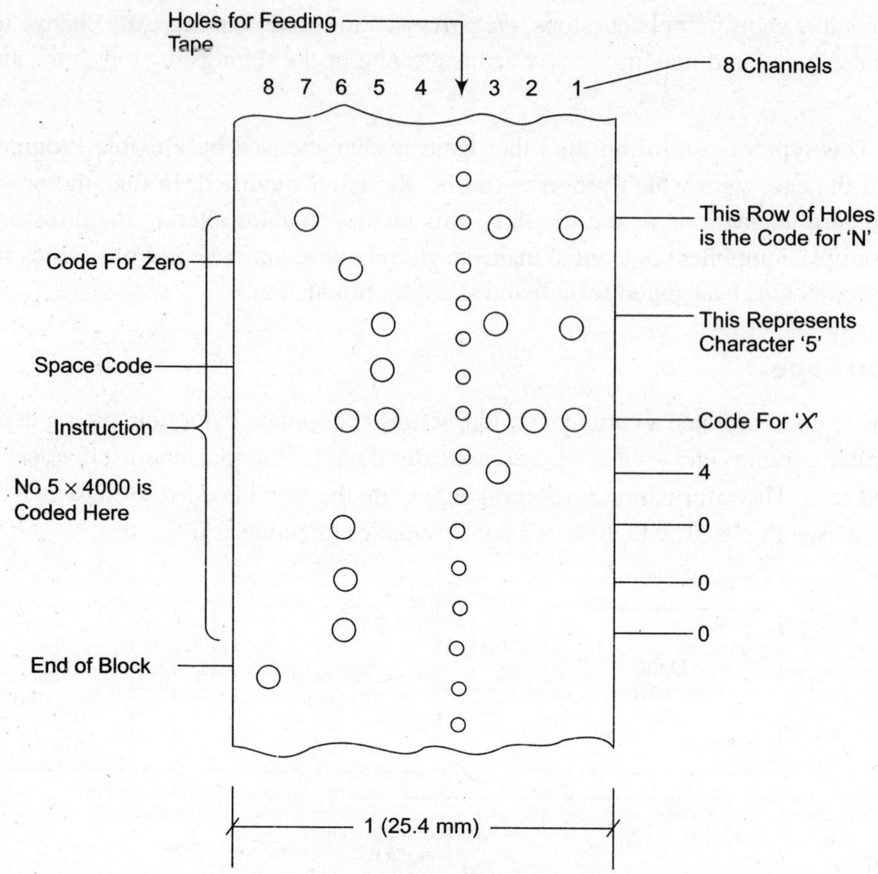

Fig. 6.52 Sample of Punched Tape

A group of characters makes one word. Ex: N05 is a word.

Several words in a standard format make one instruction as above.

The instructions are now transferred to the paper tape by typing on a machine called Flexo-Writer, which in addition to typing the matter, punches and produces the control tape (punched tape).

Thus a software with a program for making a job is prepared.

An example of Part Programming:

Let us take the case of a simple turning operation as per the drawing shown in Fig. 6.53.

The job requires turning a bar stock of ϕ20.00 to ϕ15.00 for a length of 35 mm.

The sequence of operations to be programmed is as follows:

(*a*) Positioning the tool at Start Point (Start Point is 5 mm away from the job).

(*b*) Movement of tool in *X*-direction by 7.5 mm to give depth of cut – 2.5 mm (Sign: –ve) (Sign of co-ordinates has been defined as shown in inset).

(*c*) Rotation of spindle.

(*d*) Tool feed in *z*-direction by 40 mm at the given feed rate (Sign: –ve) [from (2) to (3)]

(*e*) Withdrawal of tool along *X* and *Z* directions (both +ve) in order to clear the tool from the job [from (3) to (4)].

(*f*) Return of tool to position (1).

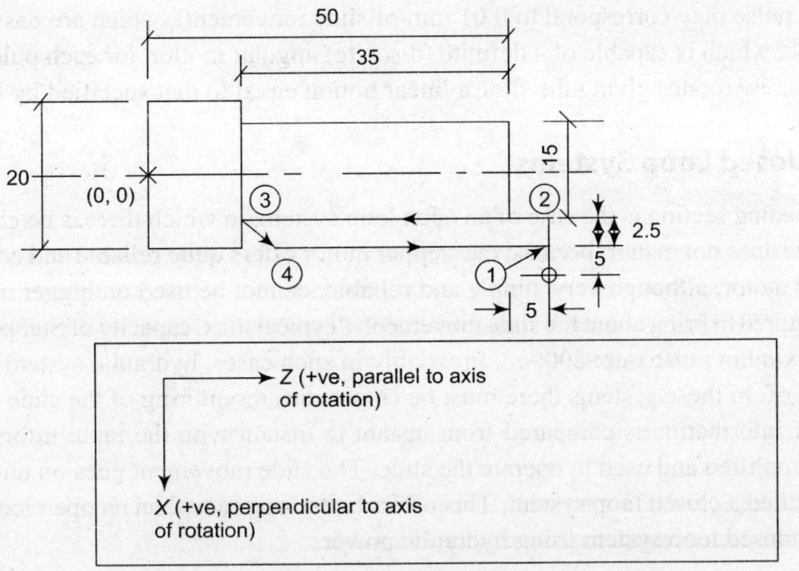

Fig. 6.53 Job for Turning Operation on N/C Lathe

The Program:

N01 G00 X1500 Z5500 F0	:	(*a*) G00 is command for positioning tool at start point (F0 is command for rapid traverse)
N02 G01 X-750 F0 M03	:	(*b*) and (*c*) movement of tool along *X*-axis and rotation of spindle
N03 Z-4000 F150	:	(*d*) feed in *Z*-direction (mm/min.)
N04 X250 Z250	:	(*e*) tool withdrawal
N05 Z3750	:	(*f*) Return of tool to original position.

(**Note:** *a*,*b*,*c*,... refer to sequence of operation).

6.6.3 Slide Control on an N.C. Machine

Refer Fig. 6.54. Here, based on the decoded information from the tape reader, the machine control unit (MCU) sends message to a pulse generator about the magnitude of the slide movement. The pulse generator in turn, generates a definite

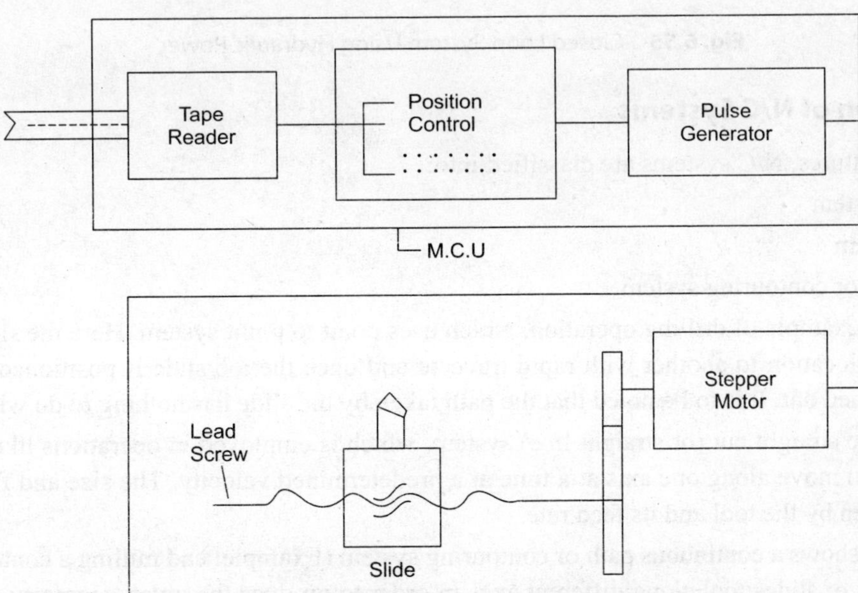

Fig. 6.54 Slide Control in a N/C Machine Tool

number of pulses (each pulse may correspond to 0.01 mm of slide movement), which are passed on to a Stepper motor. The stepper motor is one, which is capable of a definite (discrete) angular motion for each pulse received, and when this motor rotates, the lead screw rotates giving the slide a linear notion equal to that specified by the program.

6.6.4 Open and Closed Loop Systems

The example in the preceding section is the case of an open loop system, in which there is no check on the actual position reached by slide. But this does not matter, because the stepper motor offers quite reliable and consistent results. However, the system with stepper motor, although very simple and reliable, cannot be used on bigger machines, where very high torque and power are required to bring about the slide movement. (Typical max. capacity of Stepper motor: Torque: 200 Nm, Power: 4.5 kW, and maximum pulse rate: 8000/s). Invariably in such cases, hydraulic system or electrical system using D.C. motor is made use of. In these systems there must be continuous monitoring of the slide position using 'Feed back devices'. The feed back information is compared from instant to instant with the input information and the difference called 'error signal' is amplified and used to operate the slide. The slide movement goes on until the error becomes zero. This type of system is called a closed loop system. This offers better accuracy than an open loop system. Figure 6.55 is a schematic diagram of a closed loop system using hydraulic power.

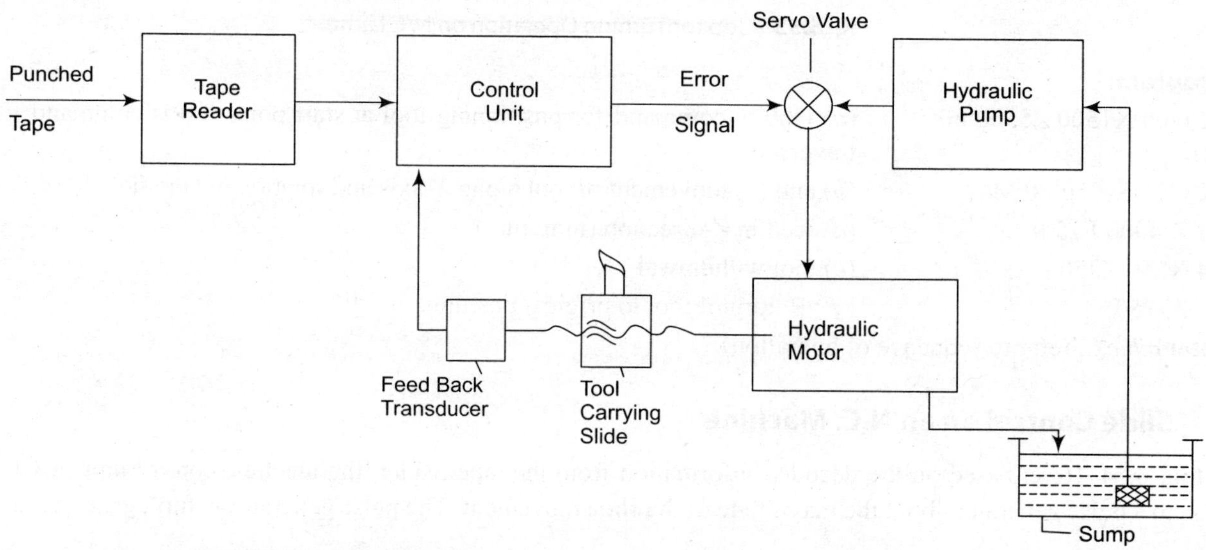

Fig. 6.55 Closed Loop System Using Hydraulic Power

6.6.5 Classification of N/C Systems

Based on application features, N/C systems are classified into:

(*a*) Point to point system

(*b*) Straight cut system

(*c*) Continuous path or contouring system

Figure 6.56 gives an example of drilling operation, which uses point to point system. Here the slide movement takes place from one drilling location to another with rapid traverse and once the job/slide is positioned under the drill, the drilling operation is carried out. It is to be noted that the path taken by the slide has nothing to do with machining.

Figure 6.57 shows the straight cut (or straight line) system, which is employed in operations like turning on a lathe. Here the slide/tool has to move along one axis at a time at a predetermined velocity. The size and finish of the job thus depends on the path taken by the tool and its feed rate.

In contrast, Fig. 6.58 shows a continuous path or contouring system (Example: end milling a contour), which calls for simultaneous movement of slides/tool along different axes in order to produce the given geometry and size of job. The movement of the slides and their velocities must be so coordinated that the tool takes specific path relative to the job.

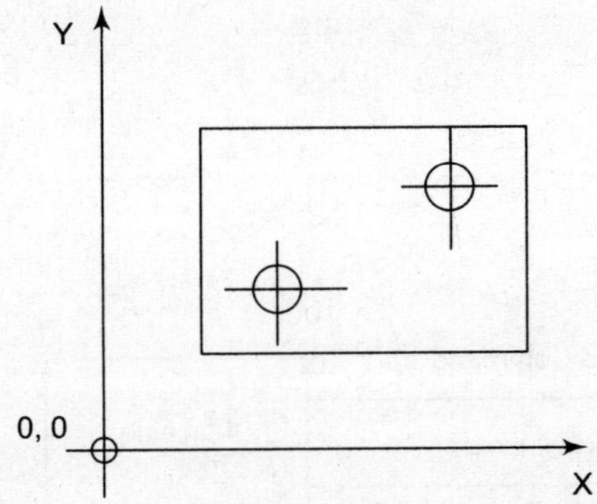

Fig. 6.56 Point to Point System (Ex: Drilling)

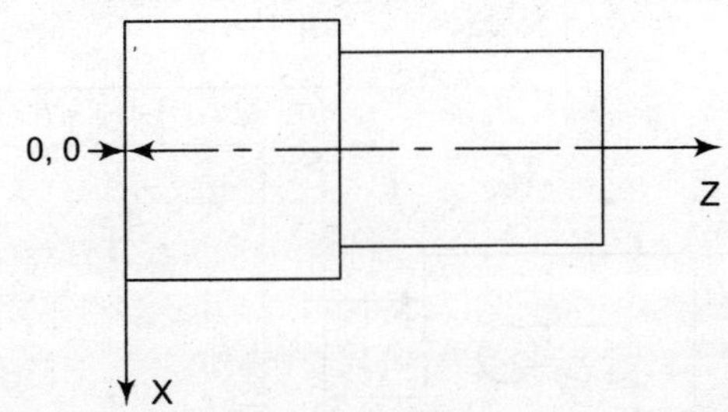

Fig. 6.57 Straight Cut System (Ex: Turning)

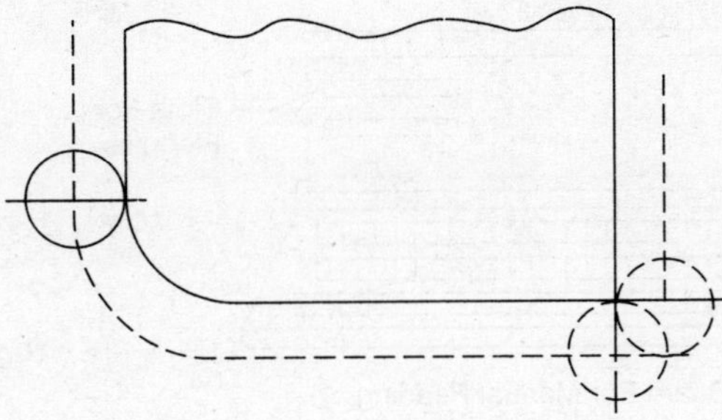

Fig. 6.58 Contouring System (Ex: End Milling a Contour)

6.6.6 Computer Numerical Control (CNC)

In section 6.6.2. we have seen how a part program is written and transferred to a control tape. This method of manually preparing a control tape can be used efficiently for both point to point and straight cut NC systems, where calculation of tool paths is fairly simple.

But in the case of programming for a contouring system, tool path calculations are involved and are therefore best handled by a computer. The computer, in addition to programming does part of the control function, which in a NC system is relegated to Master Control Unit (MCU).

A machine tool having a dedicated computer to help prepare the program and control some or all of the operations of the machine tool is called a CNC machine tool. The schematics of a CNC system is shown in Fig. 6.59.

6.6.7 APT (Automatically Programmed Tool)

Once a computer is incorporated into the N.C. System, it is but natural to exploit the whole capability of the computer like its ability for fast computing, logic and its vast memory. Naturally, control concept has been simplified by framing a computer language. Just like Fortran and other computer languages, we have APT (Automatically Programmed Tool) for machine tool control.

The programmer (part programmer) can now use some simple terms (some defined by him and others pre programmed) and write a program indicating (rather than calculating and writing) the cutter path relative to pre-defined job surfaces. Based on the information given in the program, the computer calculates the cutter path and helps in preparing the control tape.

6.6.8 Turning and Machining Centres

The development of Machine Tools in the wake of the application of computer has resulted in two distinct types of machine tools: (*a*) Turning Centre and (*b*) Machining Centre

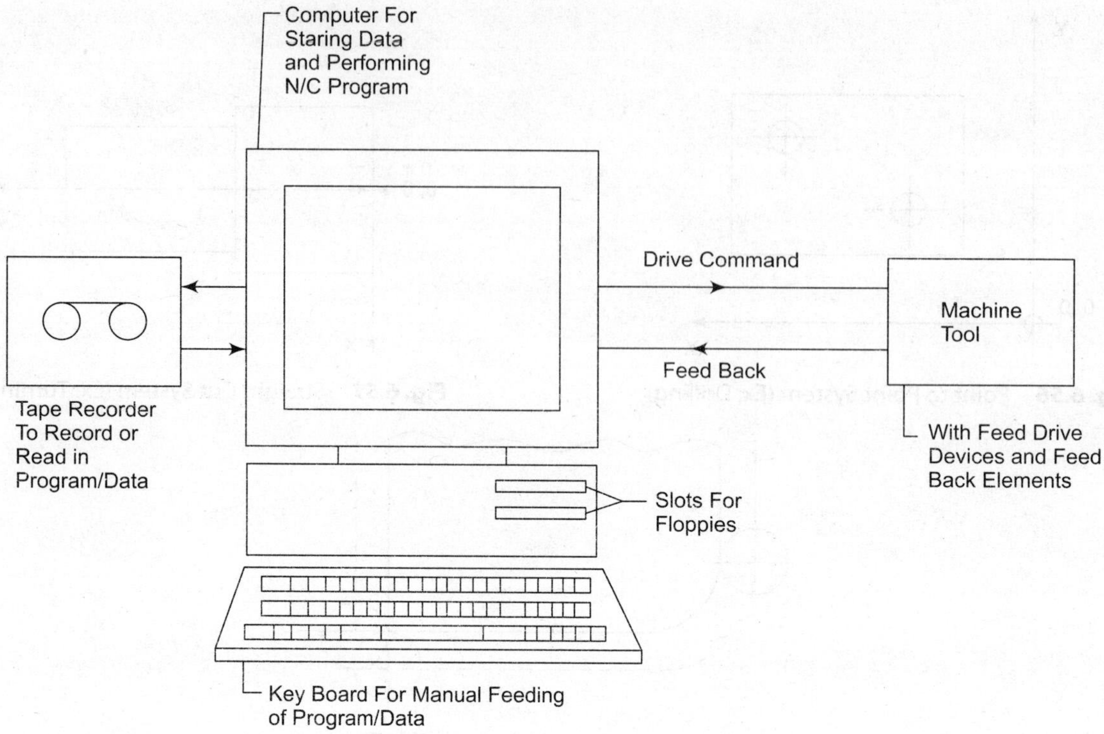

Fig. 6.59 Schematics of CNC System

The Turning centre is capable of all operations done on a conventional lathe and the Machining centre is capable of doing jobs of milling, drilling and shaping.

Both machines run on computer programs and the present day machines have added facilities like simulation of the working of the program on the monitor (VDU), zooming for examining intricate details of the part, checking dimensions etc. Once the program is found satisfactory the control is passed on to the machine tool.

The other important features of the CNC Machines are:
(*i*) Automatic tool changing facility and (*ii*) Automatic tool wear correction.

6.6.9 Advantages of N/C

1. Offers high degree of accuracy, quickness and reliability.
2. Flexible programming permits quick' changes in dimensions and geometry of job.
3. Programs can now be stored in compact form in the form of paper rolls, magnetic tapes and floppies.
4. The human error is totally eliminated in the production process and therefore the scrap due to rejects is almost nil.
5. Programming cost is negligible compared with programming with hardware as used on conventional automats.
6. The machine utilisation time is vastly improved due to reduction in ideal time.
7. No special fixtures and jigs are called for.
8. Reduced 'Lead time' due to reduction in setting time and possibility of quick change over from one job to another.
9. CNC machines may be programmed to give the production time and thus cost estimation becomes simpler and quicker.

6.6.10 Adaptive Control

A further sophistication in N/C, is the addition of 'Adaptive Control'. It increases the productivity of machining process by automatic on-line determination of the speed and feed and varying them to suit the cutting conditions as encountered during the cutting process. This is highly useful in maximising productivity in the process of machining extremely hard materials like titanium and stainless steel.

6.6.11 CAM and CIM

CAM stands for Computer-Aided Manufacturing and CIM stands for Computer Integrated Manufacturing system.

In section 6.6.6 we have learnt about the application of computer to NC machines. The use of computer has helped in particular (*a*) tool path calculation and (*b*) control of machine tool. Now a step further in the application of computers is the use of interactive graphics. This has made NC part programming a very fascinating and easy task.

The application of interactive graphics (IG) in the first place, has made possible creating, analysing and optimising a product/component design and thus has aided in the design of the product itself. This is how the activity CAD (Computer-Aided Design) has come into being. With the Interactive Graphics made use of in part programming, computers have stepped into helping machining activity as well. Thus we have CAM (Computer-Aided Manufacturing).

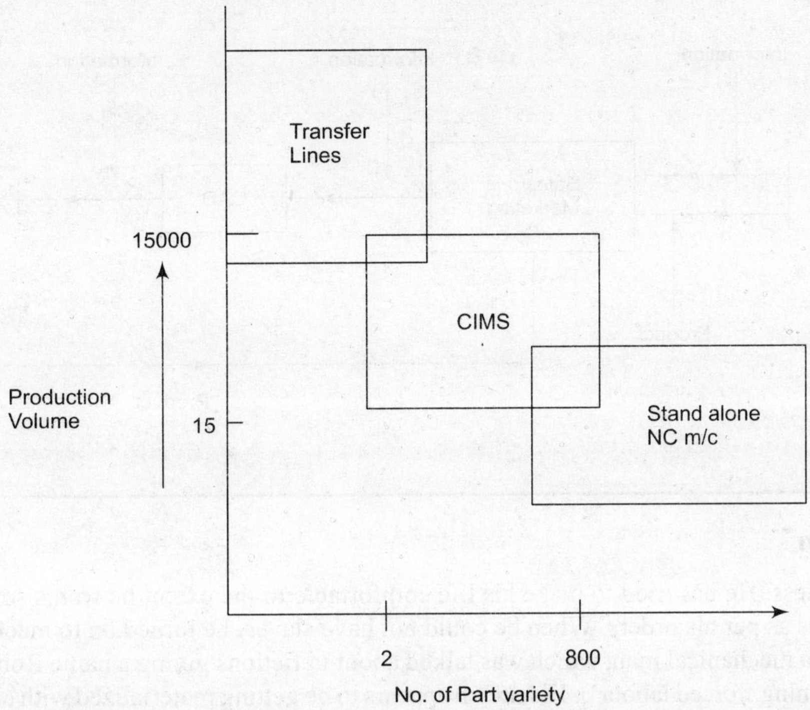

Fig. 6.60 CIM vs Transfer Lines and Stand alone NC machines

Using the same geometric data defining the part during CAD process, the tool path is prepared to help machining process. The output is a listing of APT program, which in turn is used to prepare NC punched tape.

CAD/CAM incorporating the IG system besides helping generate APT program, offers several other advantages. Different views of the part like Isometric, Oblique and Perspective can easily be formed to visualize how the part looks like. The cross-section of the part can be displayed and magnified to examine problem areas in machining.

CAD/CAM approach using graphic terminal also helps in finding alternative methods of fixturing the part.

CAD/CAM system is built with a tool library with various tools used in the shop. Selection of tools may be made using this facility.

CIM (Computer Integrated Manufacturing system): There has been a constant endeavour in the manufacturing industry to improve productivity and the quality of the manufactured goods. CAD/CAM technology has provided a powerful tool to achieve these objectives. CNC machine tools, robots, improved inspection techniques — are some of the innovations brought about by this new technology using the present day powerful digital computers. Computers also help in vast data processing, concerning production planning, production scheduling, inventory listing, cost estimation and host of other management activities.

The CAD /CAM integration together with an automated material handling system connecting together a group of NC machines goes to make a Computer Integrated Manufacturing system (CIM). This provides the most automatic form of production.

CIM is designed to cater to the middle level production both in terms of number of items and number of varieties to be produced as shown in Fig. 6.60.

CIM systems are designed exclusively for machining activity. In some cases, assembly and inspection activities are also included.

CIM offers advantages like (*a*) Increased machine utilization, (*b*) Greater productivity, (*c*) Reduced labour (*d*) Reduced manufacturing lead time (*e*) Lower inspection costs, etc. —all leading to utmost economy in production.

In the ultimate effort towards total automation of a factory, CIM, to day incorporates Business functions as well, in addition to Engineering functions.

Business functions include Marketing and Sales, Accounting, Pay roll preparation, Customer billing etc.

Thus the CIM system provides a total integration of information with automated manufacturing as shown in the Block diagram given below.

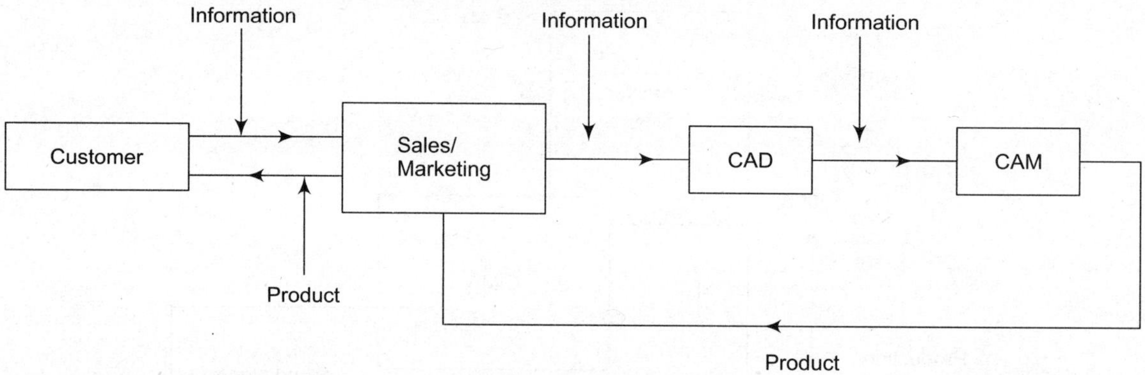

6.7 ROBOTICS

6.7.1 Introduction

Man's ambition is limitless. He has tried to make his life comfortable to the extent he wants someone to be subservient to him and do all his jobs as per his orders. When he could not have slaves, he turned on to machines. He visualized that one day he could create a mechanical man, which was talked about in fictions giving a name Robot (a name derived from Czech.word Rabota meaning 'forced labour'). His dream appears to be getting materialized with already-existing automats develop one step further and correct themselves. Naturally the concept of 'artificial intelligence' ushered in. With base in Mechanical engineering other branches like electrical, electronics, computer and other branches of engineering contributed towards this goal and a new branch of engineering emerged called 'Robotics'. Earlier we used a title Mechatronics indicating integration of Mech. and electronics engineering. Now Robotics as a separate discipline has made in-roads into practically every field of our daily life, like doing domestic chores, industrial production, aerospace, defence, education, entertainment etc., and the list is endless. Pilot-less aircrafts (Drones), Driverless cars are a few interesting examples. With various articulations of a human being successfully replicated in the hardware of a robot, we have now a more and more sophisticated Robotic man. Remember, however, that except for 'Humanoid'(Fig. 6.61), which looks like a human in form and structure, others need not be so.

Now confining our attention to manufacturing activity, robotics may be defined as an integrated advanced automated production system with in-built artificial intelligence. Some examples of robots engaged in manufacturing are: machine loading, welding, spray-painting, material handling and transporting:

6.7.2 Advantages

(*a*) Elimination of fatigue and stress of the working personnel.

(*b*) Maintaining consistent quality of work output eliminating human errors.

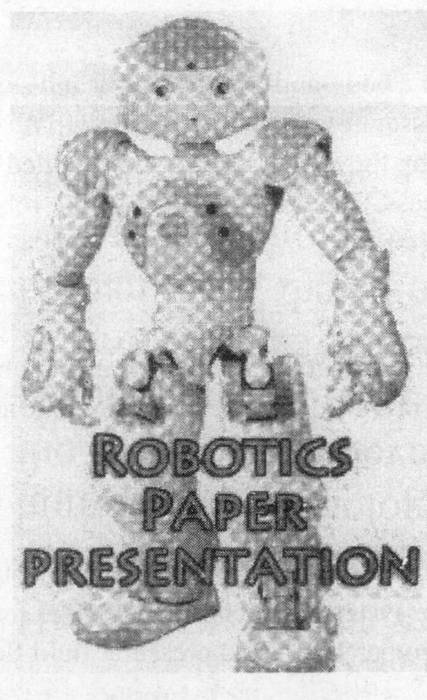

Fig. 6.61 Humanoid

(*c*) Capable of working in hazardous environments, like toxic chemical industries, assembly or diffusing of bombs and explosives, deep sea exploration, nuclear reactor *etc*., thus enabling and ensuring human safety.

(*d*) permits design flexibility with far less lead time.

(*e*) Pattern recognition with computer vision helps track objects.

(*f*) Speed of production is greatly increased.

6.7.3 Classification

We may classify robots into following categories, based on the type of configuration used for the movement of links.

(*a*) Cartesian, (*b*) Cylindrical and (*c*) Spherical co-ordinate systems.

For illustration, Fig. 6.62 shows Robot using Cartesian Co-ordinate system. Here we have three mutually perpendicular movements of the links permitting a work space of prismatic nature. Figure 6.63 shows cylindrical system. In this cylindrical configuration, the work envelope is a portion of a cylinder with two perpendicular axis movements together with a circular motion.

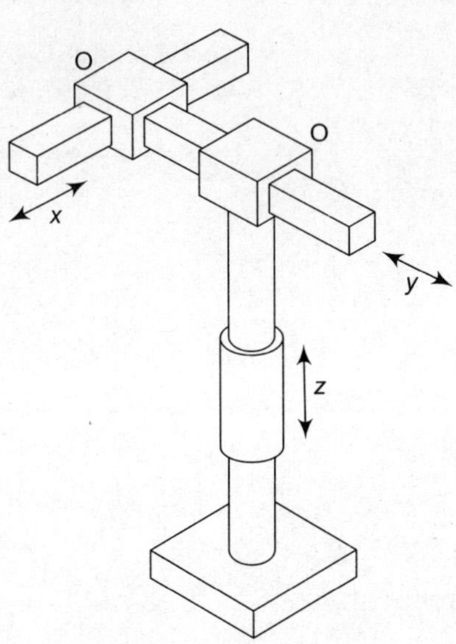

Fig. 6.62 Cartesian Coordinate Configuration Robot

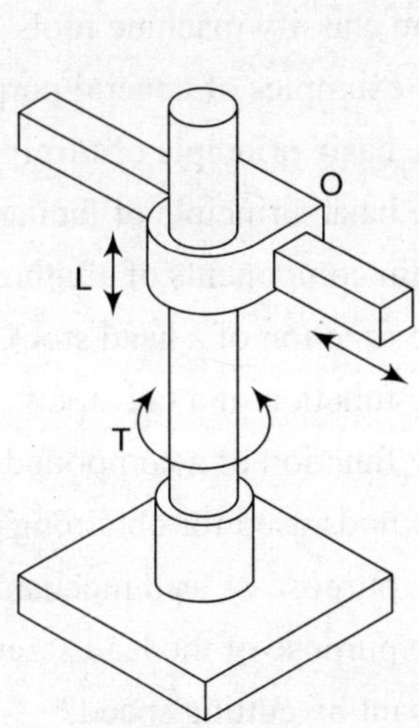

Fig. 6.63 Cylindrical Coordinate Configuration Robot

6.7.4 Components of a Robot

Robot consists of a manipulator consisting of a body-and-arm assembly and a wrist assembly. There are usually 3 degrees of freedom associated with body-and-arm assembly and similarly the wrist. And at the end of manipulator's wrist is a device related to the task that is to be done by the robot. The device is called 'End effector', like for example a gripper for holding a tool for performing some process. The body-and-arm of the robot is used to position the end effector and the robot wrist is used to orient the end effector.

6.7.4.1 Drive Systems

Robotic drives may be electric, hydraulic or pneumatic. Electric drives may be servo or stepper motors. They are more precise compared to hydraulic and pneumatic drives. Hydraulic and pneumatic are of piston and cylinder type and hydraulic is more robust and can take larger loads.

6.7.4.2 Sensors

Robots act on the feed back given by sensors. For example a 'Tactile sensor' (related to perception of touch) helps adjusting robotic grip on held objects. Computer vision systems, which rely on image sensors, help the robot see, identify and give feed back. Other sensors like for example, temperature, fluid pressure, fluid flow *etc.* are also used where required.

6.7.4.3 Power Source

Mostly 'lead-acid battery' is used as a power source. This is safe and has relatively long shelf life, but is rather heavy compared to silver- cadmium battery, which is smaller in volume but more expensive. Battery selection depends on safety, cycle lifetime and weight.

6.7.4.4 Tools

Robot uses tools to perform processing operations on the work piece. Wrist manipulates the tool relative to a stationary / slowly moving object. Examples of tools used as end effectors: Spot welding gun, arc welding tool, spray painting gun, rotating spindle for drilling, grinding, etc.

The above material should be good enough as an introduction to the evergrowing field of Robotics.

QUESTION BANK

2-Marks Questions

1. What is a machine tool?
2. How do you classify machine tools?
3. Give a few examples of general purpose machine tools.
4. Explain the basic principle of turning with a simple sketch.
5. Explain the basic principle of facing with a simple sketch.
6. List the main components of a lathe.
7. What is the function of a head stock?
8. What is the function of a tail stock?
9. What is the function of a compound rest?
10. List the methods used for obtaining different spindle speeds.
11. What is the purpose of feed mechanism?
12. What is the purpose of the lead screw?
13. What is meant by cutting speed?
14. What factors constitute production cost?

15. What are the machining parameters?
16. What is meant by feed and depth of cut?
17. List the basic lathe operations.
18. List the different taper turning methods employed on a lathe.
19. List the various job holding devices used on a lathe.
20. What is a self-centring chuck?
21. What is meant by a steady rest?
22. What is meant by a follower rest?
23. What is drilling?
24. List the various drilling machines used.
25. When do you use a radial drilling machine?
26. List the operations done on a drilling machine.
27. How is the drill fed into the work?
28. What are the parts of a twist drill?
29. What is meant by quick return motion?
30. List the operations done on a shaper.
31. What is a planing machine?
32. List different types of planers.
33. List some of the jobs done on a planer.
34. What is meant by milling?
35. Name some of the arbour type milling cutters.
36. Name some of the shank type milling cutters.
37. Name some of the tool materials.
38. List the important milling techniques.
39. What is straddle milling?
40. What is gang milling?
41. What is the purpose of a dividing head?
42. What are the methods available for controlling a machine tool?
43. What is meant by numerical control?
44. What is a punched tape?
45. Give the classification of N/C based on its application to machine tools.
46. What is computer numerical control (CNC)?

10-Marks Questions

1. Make a neat sketch of an engine lathe and name its parts. Explain briefly their functions.
2. Explain briefly the various operations that can be done by a drilling machine.
3. Explain with a neat sketch the construction of a shaper and mention the functions of the principal parts.
4. Name some typical jobs that can be done on a lathe. Explain taper turning using compound rest.
5. Explain the method of thread cutting on a lathe with the help of schematics.
6. Explain with a neat sketch the slide control on a N/C machine.
7. Explain the closed loop type of N/C using hydraulic power. Make a neat sketch.

Choose the Correct Answer/Answers

1. Lathe is primarily meant for (*i*) producing flat surfaces, (*ii*) gear cutting, (*iii*) cylindrical surfaces.
2. Head stock of a lathe houses: (*i*) Speed change mechanism (*ii*) Feed change mechanism (*iii*) Spindle bearings.
3. Tail stock is used for (*i*) turning between centres (*ii*) holding drill in drilling operation (*iii*) holding boring tool for boring a job.
4. Compound rest is useful in (*i*) turning long tapers, (*ii*) turning short tapers, (*iii*) chamfering a job.
5. Feed refers to (*i*) movement of the tool for cutting, (*ii*) advancement of tool towards the job, (*iii*) rotation of the job.
6. Some of the Cutting tool materials are: (*i*) Cast iron, (*ii*) H.S.S., (*iii*) Mild steel, (*iv*) High carbon steel, (*v*) Ceramics.
7. Drilling machine is used primarily for (*i*) boring, (*ii*) slotting, (*iii*) originating a hole.
8. Materials for a drill are: (*i*) Carbon steel, (*ii*) Duraluminum, (*iii*) H.S.S.
9. Distance moved by the cutting tool for one revolution of the spindle/work is called: (*i*) Depth of cut, (*ii*) Feed.
10. In transferring drive from spindle to the job held between centres, the following are used.
    (*i*) Face plate, (*ii*) Drive plate, (*iii*) Lathe dog
11. Process of reaming refers to
    (*i*) boring a hole (*ii*) threading a hole (*iii*) finishing a hole.
12. Cutting speed is expressed in
    (*i*) mm/rev (*ii*) m/min (*iii*) mm/min.
13. Feed is expressed in
    (*i*) mm/rev (*ii*) m/min (*iii*) mm/min.
14. Collets are used for holding
    (*i*) cylindrical jobs (*ii*) irregular jobs (*iii*) bar stock of various shapes.
15. Mandrel is used for holding
    (*i*) bar stock (*ii*) cylindrical jobs (*iii*) jobs with predrilled holes or bored jobs.
16. Face plate is used to hold
    (*i*) cylindrical jobs, (*ii*) odd jobs (*iii*) pre-drilled jobs.
17. The easiest way of centring a cylindrical job on a lathe is to use
    (*i*) 4-jaw chuck (*ii*) face plate (*iii*) self-centring chuck.
18. Rake angle is given to a tool
    (*i*) to facilitate smooth flow of chip
    (*ii*) to make the tool sharp
    (*iii*) to prevent rubbing of tool against the job.
19. Shaping machine is used mainly to
    (*i*) machine flat surfaces
    (*ii*) cylindrical surfaces
    (*iii*) shape gears.
20. Cutting motion in a Shaper is obtained by
    (*i*) horizontal movement of the table
    (*ii*) downward motion of the tool
    (*iii*) reciprocation of the ram.

21. Quick return motion is obtained by

(*i*) crank and connecting rod mechanism

(*ii*) gear train

(*iii*) crank and slotted lever mechanism.

Answers

1. (*iii*) **2.** (*i*), (*iii*) **3.** (*i*), (*ii*) **4.** (*ii*) **5.** (*i*) **6.** (*ii*), (*iv*), (*v*) **7.** (*iii*) **8.** (*iii*)
9. (*ii*) **10.** (ii), (iii) **11.** (*iii*) **12.** (*ii*) **13.** (*i*), (*iii*) **14.** (*iii*) **15.** (*iii*) **16.** (*ii*)
17. (*iii*) **18.** (*i*), (*ii*) **19.** (*i*) **20.** (*iii*) **21.** (*iii*).

REFERENCES

1. ARCHIE W. CULP JR.: Principles of Energy Conversion, Mcgraw Hill 1985.
2. SUKHATME: Solar Energy, Tata Mcgraw Hill 1988.
3. MALEEV V.L.: I.C. Engines, Mcgraw Hill 1989.
4. HAJRA CHOUDHURY S.K.: Elements of Workshop Technology, Asia Publishing House 1988.
5. BEGEMAN M.I.: Manufacturing Processes, Asia Publishing House.
6. NAGPAL G.R.: Power Plant Engineering, Khanna Publishers.
7. ARRORA and DOMKUNDWAR: A Course in Power Plant Engg., Dhanpat Rai and Sons, Delhi 1988.
8. YORAM KOREN JOSEPH BEN-URI: Numerical Control of Machine Tools, Khanna Publishers, Delhi, 1978.
9. ANANTHANARAYANAN P.N. Basic Refrigeration and Air conditioning, Tata Mcgraw Hill Publishing Co. Ltd. 1982.
10. FRANZO H. CRAWFORD/WILLIAM D. VANVORST, Thermodynamics for Engineers, Harcourt, Brace and World, Inc. N.Y. 1968.
11. MIKELL P. GROOVER AND EMORY W. ZIMMERS JR. CAD/CAM Computer-aided Design and Manufacturing.
12. SRINIVASA REDDY M, SHENOY H.G. AND SUNDERESH S — Elements of Mechanical Engineering 2014.
13. COMPOSITES — Science and Technology by PRASAD R.C AND RAMAKRISHNAN P.

Chapter-7
Engineering Materials

7.1 INTRODUCTION

After studying the various production processes, it would be appropriate to learn about some of the engineering materials that are processed to get the various products required for our use and comfort.

Nature has provided mankind with many kinds of metals and materials. But here we study some engineering materials like aluminium, copper, iron and steel.

Most of the metals, as available in nature are in the form of chemical compounds like for example, iron in the form of iron oxide, copper as copper sulphate, etc. They exist combined with lot of impurities as well. We call them minerals and mining is resorted to get them in areas, where they are available in plenty. These minerals are processed to get pure metals. Again, metals, as obtained, may not have the qualities we want. Invariably, therefore, they are combined with other metals / materials and thus we have alloys or metal composites.

Alloys are combination of two or more metals made by melting them together, whereas composites, using both metals and other ingredients, are made using powder metallurgy techniques.

7.1.1 Cutting Tools Materials and Engineering Materials

First we shall examine in detail some cutting tool materials mentioned vide 6.1.6, used in production processes. They are: (*a*) Carbon tool steel, (*b*) High speed steel (HSS), (*c*) Cast alloys, (*d*) Cemented carbides, (*e*) Ceramics and (*f*) Diamonds.

Carbon tool steel, after suitable heat treatment, acquires hardness of 50–55 Rc. and withstand cutting temperature upto 200°C. With further increase in temperature, it loses its hardness. Therefore, carbon tool steel is used for low cutting speed tools, such as taps, dies, hacksaw blades, hand drills, wood working tools, etc.

High speed steel (HSS) has the following composition:

Carbon – 1.9 %, Manganese – 0.3%, Chromium – 3.8 % and Tungsten – 8%. Tools made of HSS withstand high temperatures upto 600°C without losing hardness. Hardness is of the order 60 Rc.

Cast alloys consist of chromium, tungsten, columbium and carbon in the matrix of cobalt. They withstand temperatures upto 700°C. Since they are very expensive, small bits are cast and the bits are brazed to steel shanks.

Cemented Carbides: Sintered Tungsten-carbide,Titanium Carbide and Tantalum with suitable percentage of Cobalt are all compacted using powder metallurgy technique and used as tool material. They withstand temperature upto 800°C. Hardness is about 70 Rc .These are highly wear-resistant and permit higher cutting speeds.

Ceramics: Al_2O_3, SiC, BC, Titanium Carbide /Boride are examples of Ceramics. Tools are made by compacting the powders of specific grain size and sintering at high temperatures as high as in pottery work. Cutting speed for steel could be around 500 m/min.

Diamond: It is the hardest material, but expensive. Tool life is the highest.

7.1.2 Ferrous Metals

After having learnt about specific cutting tool materials, let us now try to understand about some of the generally used engineering materials like ferrous and non-ferrous metals. Ferrous metal is by far, the most widely used material. It is used mostly as cast iron or steel.

Cast iron application includes cook-ware, engine blocks and components, water main pipes, machine beds, etc. The advantages are corrosion and wear-resistance, vibration-damping characteristics and easy castability.

Steel is used as mild steel or high speed steel. Mild steel is used in building activity as reinforcement rods, water pipes of smaller size, usually galvanized to prevent corrosion. Use of HSS has been discussed earlier.

7.1.3 Non-ferrous Metals

We shall learn here about some of the non-ferrous metals, the most widely used one being aluminium and its alloys. Following are some of the usages:

Aluminium is widely used in the transport, construction, packaging and electrical sectors.

In transport, aluminium is used for making automobile parts like engine blocks, cylinder heads, transmission housings and body panels. It is also used extensively in aircraft and aero-space industries in the form of sheets.

In construction, aluminium is used for roofing and wall cladding. As extrusions, it is used for windows and doors.

In packaging, aluminium is used for beverage can bodies, as foil for household and commercial wrap and for pharmaceutical products.

In the electrical sector, aluminium is used in the form of wires, normally reinforced with steel to form cables.

Copper is yet another useful non-ferrous metal. Because of its high ductility, it is used in ornamental works in buildings; due to its high conductivity, it is used in electrical conductors. It also finds application in cook-ware, having high thermal conductivity. Copper-nickel alloy clad plates are used in marine application and in de-salination plants owing to the high resistance to corrosion. The alloy has excellent weldability.

Magnesium, manganese, zinc, etc., are other important engineering materials.

7.2 COMPOSITES

With every-day research and development in material science, we have variety of composites used not only in tools but also in automobile, aerospace, medical, defence, sporting, electronics and marine applications.

The composites, assuming prominence these days, we shall get into some details of making the composites. We have already studied about tool materials using composites. The concept of this technology hinges on (*a*) how to increase the strength to weight ratio (*b*) how to reduce cost for a given purpose (*c*) how to increase the toughness and shock-resistance, given the available input. The biggest advantage of composites is they can be tailored to specific requirements.

Composites consist of (*a*) reinforcing material and (*b*) some matrix embedding the reinforcements using different techniques. The reinforcement may be in the form of fibres, particles, whiskers or lamellae/mesh. This gives a material that combines the most useful properties of the constituents. Generally, the properties of a composite are superior to those of the individual constituents.

Reinforcing materials: The first ever composite consisted of glass fibre as reinforcing material embedded in plastics (introduced in 1940). Since then technology has grown and newer reinforcement materials have been developed like carbon fibre, alumina-silica fibre, SiC, boron filaments, etc.

Matrix materials: Epoxy resin or other high temperature plastics, aluminium or other metal or a ceramics such as Silicon Nitride are used as Matrices. Epoxy, a thermoset polymer, provides strength upto a temperature of 200°C. Resins such as polymides provide higher temperature strength up to 300°C. Other thermosetting resins include benzo cyclo butane.

Thermoplastic polymers such as poly ether – ether ketone (Peek) permit faster production, but cannot work at high temperatures.

Metal matrix permits working at elevated temperatures.

Ceramix matrix such as alumina, zirconia silicon carbide permits usage at temperatures up to 2500°C.

Application. Some of the major and interesting uses of composites are given below.

1. Cutting tools.
2. Automobile and aerospace components like tails, wings, fuselages, propellers, race car bodies, etc. Carbon composite is the key material in today's launch vehicles to provide heat shield.
3. Armour—for example Indian Arjun tank, an armoured vehicle, uses modular composite armour with a layer of concrete sandwiched between layers of steel. This provides an impenetrable shield.
4. In orthopedics, for example, carbon-fibre nylon composite is used for making leg brace for handicapped. This weighs one third as much as the aluminium and steel brace used earlier. Again it is much stiffer and stronger.

GLOSSARY (PART II)

Chapter 3 (Casting)

Adhesiveness: Ability to stick to any material.

Binders: Materials used to improve the cohesion of sand particles.

Blow holes: A defect in casting caused usually by trapping of air or gas during the pouring of a mould. This leaves cavities in the casting.

Cast iron: Material got from a cupola by smelting pig iron. Carbon content: 2.5 to 6%.

Casting: A product obtained by pouring molten metal into a mould cavity and allowing the metal to solidify. The product takes the shape of the cavity.

Centrifugal casting: A casting process in which molten metal is filled into the mould cavity by centrifugal force.

Cohesiveness: Ability to stick together with like material.

Coke: A product of destructive distillation of coal. It is devoid of most of the volatile matter.

Cold shut: Two streams of molten metal not fusing at the meeting point.

Collapsibility: Ability of the core to break up after the metal has solidified in the mould enabling easy removal of the core from the casting.

Core: A component made from core sand. Placed inside a mould introduces a cavity of the shape of the core in the casting.

Core print: An extension in a pattern, which forms a seat in the mould for the core to be supported [Fig. 3.19 (c)].

Core sand: Used in core making has some additives giving strength and also helping break up of core once the casting has cooled down.

Crucible: An earthen or graphite pot used for melting metals.

Cupola: A furnace for smelting generally pig iron into cast iron.

Die casting: Casting in metal moulds either under gravity or pressure. Mostly non-terrous metals are die-cast.

Dies: Used in die-casting are made of heat resistant materials. They carry the impression of the casting to be made.

Dowel: A wood or metal pin used to align parts of a pattern or core box.

Draft: The taper on a vertical surface of a pattern to facilitate withdrawal of the pattern from the mould.

Fettling: A process of cleaning castings after being knocked out of the moulds.

Flask: A complete moulding box. It has usually two parts, the cope and the drag.

Flux: A material used for removing impurity. In cupola lime stone is used as flux, which combining with the impurities in the melt shows up as slag floating on the molten metal.

Foundry sand: Used for moulding, consists of sieved river sand, clay and moisture.

Furnace: An insulated enclosure for heating or melting metals. Heat is provided by burning oil/gas or using electricity.

Gating: A system of passages in the mould for transferring molten metal into the cavities of the mould.

Honey combing: A defect on the surface of casting. It is a cluster of small cavities closely placed.

Jolting machine: Machine to compact sand in the mould by vibration (jolting).

Ladle: A container to transfer molten metal from a furnace to a mould.

Moulding: Process of making cavity similar to that of casting using moulding sand.

Moulding box: A metal or wooden box open at both ends used in making moulds. Also called flask, it is used in pairs. Upper or top box is called Cope and lower box is called Drag.

Moulding sand: Same as foundry sand.

Moulding board: A flat wooden base for moulding.

Parting sand: Sieved river sand applied to prevent sticking of foundry sand to pattern and also sticking of the two halves of the mould.

Pattern: It is a replica of casting made of wood or other materials, but with certain allowances made for shrinkage, machining, distortion and draft.

Permeability/porosity: of foundry sand is the ability to pass air/gas/vapour through the interspaces of the sand particles.

Pig iron: Metal obtained by smelting iron ore in blast furnace.

Plasticity: Ability to take up and retain the shape given.

Rammer: A tool for packing the moulding sand into the moulding box.

Refractoriness: Ability to withstand high temperatures.

Riser: Passage left in the mould to enable excess molten metal to flow out.

Runner: Passage in a mould for pouring molten metal into the cavity.

Scab: It is a defect in casting, caused by erosion of mould and filling of the space by molten metal.

Swab: A fine-bristle brush which when moistened cleans fallen sand from a mould cavity.

Shift: A defect in casting due to misalignment of cope and drag or core displacement. This defect causes the two halves of the casting to be displaced at the central line.

Slag: Scum floating on molten metal.

Slick: (Fig. 3.11) A moulding tool used for cleaning and finishing the mould cavity.

Sand slinger: A machine to compact sand in a mould by throwing the sand with a great velocity.

Sprue pin: A moulding tool to form the runner passage.

Squeeze machine: A machine to compact sand in a mould by applying uniform pressure.

Swell: Enlargement of casting due to poor ramming or moulding boxes not weighed adequately.

Trowel: A moulding tool for finishing moulds.

Tumbling barrel: A large barrel used to clean small castings by tumbling them with sand and bits of hard metal.

Tuyeres: Passages in a blast furnace/cupola that supply air for smelting.

Vent wire: A pointed wire to poke into the sand mould to leave openings for passage of air/gases to pass through while the molten metal is getting filled into the cavity of the mould.

Chapter 4 (Forging)

Billet: An iron or steel bar.

Blanking: Cutting out of sheet metal a piece of required shape and size using die and punch.

Bloom: An iron or steel bar bigger in size than billet.

Coining: A cold forging technique for making coins and medallions.

Cold rolling: A finishing operation done on hot rolled items. This improves surface finish and gives better dimensional accuracy.

Drawing: A cold forming operation which enables cup-like jobs to be done.

Drop forging: A machine forging technique, which uses a set of closed impression dies.

Extrusion: A process in which heated metal is forced through a die. The extruded product will have the same cross section as that of the die opening.

Forging: Technique of shaping metal by the application of force on metal rendered plastic by heating. Alternately the metal is brought in plastic state by the application of very high pressure and thus a given shape is retained.

Hammer forging: A machine forging technique using flat faced dies.

Hearth: An equipment to heat jobs for forging using an open fire. Combustion air supplied from a blower increases the rate and intensity of heating.

Press forging: Forging made using a power press. Metal heated to forging temperature is squeezed into the cavity of the die and each forging is completed in one stroke of the ram.

Punching: Making of a hole in a sheet metal of the desired shape and size using punch and die.

Rolling: Forging of rods, sheet, structural items like I-section *etc.*, by pushing a bar of heated metal in between rolls.

Slab: A bar whose thickness is much smaller than the width.

Spinning: A cold forming process which is used to shape sheet metal into hollow parts. A spinning lathe is used for the purpose.

Upset forging: A forging technique used for the production of components like bolts, rear axle of automobiles, *etc.*, which require upsetting operation.

Chapter 5 (Metal Joining)

Arc welding: Welding which uses the heat produced by an electric arc struck between the job and an electrode (consumable or non-consumable).

Brazing: It is joining of similar or dissimilar metals with filler material melting at about 600°C (higher than for soldering). The temperature is, however, less than the fusion temperatures of base metals.

Electron beam welding: Welding using a concentrated beam of electrons impinging on the joint to be welded. An electron gun is used for the purpose and the whole operation is performed in an evacuated chamber. The technique is used for welding important components like the ones in aircraft and nuclear industries.

Friction welding: Usually applied for rod or tube welding depends on the heat generated by friction produced between one end rotating and the other end stationary.

Gas welding: Uses oxy-acetylene or oxy-hydrogen gas to provide heat for welding. A gas torch provides the heating flame and a filler metal is used in the form of a rod.

Laser beam welding: This welding technique uses a Laser beam of very high power density of the order of 10^6 W/cm^2. Unlike electron beam welding, Laser beam welding requires inert gas shield to protect the weld pool, but needs no vacuum.

MIG welding: An arc welding process in which bare electrode is continuously fed to make the weld through a shield of Argon or Helium.

Resistance welding: A process of welding in which heat required for welding is provided by the contact resistance (between parts being welded) offered to the flow of a heavy current.

Seam welding: A resistance welding process in which the welded spots overlap giving a seamless joint.

Shielded metal arc welding: An arc welding process using a consumable electrode covered with a coating of flux. The flux burns as welding proceeds giving out gases which form a shield around the weld zone.

Soldering: A metal joining process for joining similar or dissimilar metals by adding a low melting alloy called Solder. The base metal is not affected.

Spot welding: A resistance welding process to effect weld at specific locations. No filler or flux is used in this process.

Submerged arc welding: An automatic welding process, which uses a bare electrode continuously fed through a blanket of flux. Flux feeder and the electrode feeder together move as the welding proceeds. Thick plates are welded using this process.

TIG welding: An arc welding process which uses a non-consumable electrode (made of Tungsten). Filler material is used in the form of a welding rod and an Inert gas like Argon or Helium provides a protective shield for the weld.

Chapter 6 (Machining)

Machine tool: A machine which uses cutting tools to remove excess material from a given job to produce a component of desired size and shape.

Closed Loop System: The system in which the output is fed back for comparison with input for the purpose of correcting any difference between the command and output function.

Code: A pre-determined number and arrangement of holes in punched tape or magnetic areas on a magnetic tape to represent a character.

Cutting Speed: Relative speed of the cutting edge with reference to a point on the surface being cut.

Depth of cut: Amount of tool advancement after each pass of the cutting tool against the job, usually in a direction perpendicular to the direction of tool movement (feed motion).

Feed: Amount of tool movement per revolution of the spindle and therefore the job. It is expressed in mm/rev. or mm/min.

Linear interpolation: It means obtaining intermediate co-ordinate values on a straight line from the given start and end co-ordinate values.

Machining: It is the removal of excess material from a given job to get a component of desired shape and size.

Manual Control: Manual operation of machine functions by means of the push buttons, switches and hand wheels on the operation control panel.

Numerical Control (N/C): Defined very simply as control by numbers does not give a clear picture of the concept. (N/C apart from its use as a manufacturing aid, finds application in various other fields). With reference to manufacturing, it may be said that the information required for manufacturing, taken out from engineering drawing, is coded and put in a paper or magnetic tape for subsequent use. This coding involves arranging a certain **number** of holes in a certain order (Refer Fig. 6.52). This should be the reason why the control is referred as N/C.

Open Loop System: A control system that has no means for comparing the output with input for control purposes.

Program: A sequence of instructions to be executed by the computer or control to solve a given problem.

SOME USEFUL DEFINITIONS AND DATA

Definitions:

Units of Force, Work, Power, Heat and Pressure:

Force: Dyne is the physical unit of force; it is that force which acting on a mass of 1 g gives it an acceleration of 1 cm per second per second (1 cm/s^2)

Newton (N) is the S.I. unit of force; it is that force which acting on a mass of 1 kg gives it an acceleration of 1 metre per second (1 m/s^2)

$$1 \text{ N} = 10^5 \text{ dynes.}$$

Work: Erg is the physical unit of work; it is the work done when the point of application of force of one dyne is displaced through a distance of one cm in the direction of force.

Joule (J) is the S.I. unit of work; it is the work done when the point of application of force of one Newton is displaced through a distance of one metre in the direction of force.

$$1 \text{ J} = 1 \text{ Nm,}$$
$$= 10^5 \text{ dynes. } 100 \text{ cm.}$$
$$= 10^7 \text{ dyne cm} = 10^7 \text{ ergs. Also } 1 \text{ J} = \text{Ws}$$

Power: Watt (W) is the S.I. unit of power; it is the rate of doing work (Joules per second), J/s.

Heat: Calorie is the physical unit of heat and is defined as the quantity of heat required to raise the temperature of 1 g of water from 14.5 to 15.5°C.

Joule (J) is the S.I. unit of heat or thermal energy.

$$1 \text{ calorie} = 4.187 \text{ J} \approx 4.2 \text{ J}$$

Pressure: Pascal (Pa) is the S.I. unit of pressure

$$1 \text{ Pa} = 1 \text{ N/m}^2$$

Atmosphere: (Atm.) This corresponds to the pressure exerted to a Hg column of 760 mm.

$$1 \text{ Atm.} = 1.0133 \times 10^5 \text{ Pa}$$
$$1 \text{ Bar: (1b)} = 10^5 \text{ N/m}^2 \text{ or } 10^5 \text{ Pa} = 0.1 \text{ M Pa}$$

Data:

Heating values of fuels:
(Approx.)

Wood..........................	18,000 kJ/kg.
Coal...........................	28,000 kJ/kg.
Petrol..........................	47,500 kJ/kg.
Diesel..........................	45,000 kJ/kg.
Hydrogen.....................	142,097 kJ/kg.
CO..............................	10,110 kJ/kg.
Methane.......................	55,529 kJ/kg.
Propane.......................	50,399 kJ/kg.
Natural gas..................	37,000 kJ/kg at 1 atm. and 20 C
Atmospheric air:	$\gamma = 1.4$
(at STP)	$c_p = 1000$ J/kg K
	$c_v = 714$ J/kg K

Fuel air ratio required
for combustion (approx.)

Coal: 1:10

Oils: 1:15

Steam:	Critical pressure = 225 Atm.
	Critical temperature = 374°C
	Critical volume = 0.00317 m^3/kg
I.C. Engine:	Compression ratio: (approx)
	Petrol 7 : 1
	Diesel 21 : 1
	Sparking voltage: 12000–15000 V
	Injection pressure: 150 to 250 Atm.
Solar energy:	Incident on earth – 600 to 900 W/m^2
	Above the atmosphere —1500 W/m^2

Composition, specific gravity and melting point of some metals:

Metal	**Composition**	**Specific gravity**	**Melting point C (approx)**
Cast iron	Carbon : 2 to 6%	7.1	1100–1300
Brass	70% Cu + 30% Zn	8.5	700
Steel (Mild or low carbon)	Carbon : 0.1 to 0.2%	7.8	1400
Steel (Medium Carbon)	Carbon : 0.3 to 0.6%		
Steel (High Carbon)	Carbon : 0.8 to 1.2%		
Stainless steel	18 Cr + 8 Ni	7.9	1500
Bronze	90% Cu + 10% Sn	8.8	800
Gold		19.3	1063
Silver		10.5	960

Elements of Mechanical Engineering

First/Second Semester B.E. Degree Examination, June/July 2014

Time: 3 hrs. **Max. Marks: 100**

Note: 1. *Answer any FIVE full questions, choosing at least two from each part.*

2. *Answer all objective type questions only on OMR sheet page 5 of the answer booklet.*

3. *Answer to objective type questions on sheets other than OMR will not be valued.*

4. *Use of steam tables is not permitted.*

Part – A

1. (*a*) Choose the correct answers for the following: (04 Marks)

(*i*) The centrifugal forces generated by the earth rotation on the far side results in another bulge rise on this side of the earth.

(a) Lunar tides (b) Earthquakes

(c) Volcanoes (d) None of these

(*ii*) The condition of steam in the boiler is always

(a) Dry (b) Wet

(c) Saturated (d) Superheated

(*iii*) Superheater is used

(a) Inside the boiler drum

(b) To convert wet steam into dry steam

(c) In the path of the gases to increase volume of steam

(d) To increase temperature of steam above saturation temperature.

(*iv*) Babcock and Wilcox boiler is ______ boiler.

(a) Fire tube (b) Water tube

(c) Air tube (d) Fioxi tube

(*b*) With the help of a temperature - enthalpy diagram, explain the mechanism of its formation of steam. (10 Marks)

(*c*) Name any five boiler mounting and accessories and state their functions. (06 Marks)

2. (*a*) Choose the correct answer for the following : (04 Marks)

(*i*) In reaction turbine, the pressure drops.

(a) In nozzles (b) In moving blades

(c) In fixed blades (d) In both fixed and moving blades.

(*ii*) Kaplan turbine is

(a) A high head mixed flow turbine (b) An impulse turbine, outward flow

(c) A reaction turbine, outward flow (d) Low head, axial flow.

(*iii*) Delaval turbine is a

(a) Impulse turbine (b) Reaction turbine

(c) Velocity compounded turbine (d) Pressure compounded turbine.

(*iv*) In a gas turbine, if the working substance is continuously recalculated, then it is called as,

(a) Open cycle gas turbine (b) Closed cycle gas turbine

(c) Mixed flow gas turbine (d) None of these

(*b*) Differentiate between open cycle and closed cycle gas turbine with neat sketches. (08 Marks)

(*c*) Sketch and explain the working of a Kaplan turbine. (08 Marks)

3. (*a*) Choose the correct answer for the following : (04 Marks)

(*i*) In a 4 stroke Cl engine during suction stroke

(a) Only air is sucked (b) Only diesel is sucked

(c) Both air and diesel are sucked (d) Either air or diesel is sucked

(*ii*) The inner diameter of engine cylinder is called as

(a) Stroke (b) Clearance

(c) Bore (d) Pitch

(*iii*) In a diesel engine the fuel is ignited by

(a) Spark

(b) Ignitor

(c) Heat resulting from compressing air that is supplied for combustion

(d) Combustion

(*iv*) Piston speed is equal to

(a) Stroke * rpm (b) 2*stroke * rpm

(c) 48 stroke*rpm (d) $\frac{\text{(stroke*rpm)}}{2}$

(*b*) With the neat sketch, explain the working of 4 stroke diesel engine. (08 Marks)

(*c*) The following observations were obtained during a trial on a four stroke diesel engine:

Cylinder diameter = 25 cm; Stroke of the piston = 40 cm

Crank shaft speed = 250 rpm; Brake load = 70 kg

Brake drum diameter = 2m; Mean effective pressure = 6 bar

Diesel oil consumption = 0.1 m^3/min; CV = 43900 kJ/kg

Specific gravity of diesel = 0.78

Find: (*i*) BP (*ii*) IP (*iii*) FP (*iv*) ηmech (08 Marks)

4. (*a*) Choose the correct answer for the following : (04 Marks)

(*i*) The boiling point of ammonia is

(a) 100°C (b) −33.3°C
(c) 33.3°C (d) 0°C

(*ii*) Most commonly used refrigerant in vapour absorption refrigeration system is,

(a) Freon (b) CO_2
(c) SO_2 (d) NH_3

(*iii*) Throttle value is used in refrigerator to,

(a) Compress refrigerant (b) Expand the refrigerant
(c) Absorb the heat from the refrigerant (d) Condense the refrigerant

(*iv*) An ideal refrigerant should have,

(a) Low viscosity (b) Low freezing point
(c) Low boiling point (d) All of the above

(*b*) What are the desirable properties of refrigerant? (06 Marks)

(*c*) With a neat sketch, explain the construction and working of vapour absorption refrigeration system. (10 Marks)

Part – B

5. (*a*) Choose the correct answer for the following : (04 Marks)

(*i*) The process of enlarging an already drilled hole is

(a) Spot facing (b) Reaming
(c) Tapping (d) Boring

(*ii*) ________ is the process of generating internal threads.

(a) Tapping (b) Turning
(c) Knurling (d) None of these

(*iii*) The slowest speed in Lathe is adopted for the following operation :

(a) Turning (b) Thread cutting
(c) Tapper turning (d) Knurling

(*iv*) Twist drills are usually made of

(a) HSS (b) Diamond
(c) Carbides (d) MS

(*b*) List the four elements which specify the size of the Lathe. (06 Marks)

(*c*) Explain the difference between facing and turning operations. (04 Marks)

(*d*) Draw the neat sketch of radial drilling machine and label all its parts. (06 Marks)

6. (*a*) Choose the correct answers for the following: (04 Marks)

(*i*) The cutting tool in a milling machine is mounted on _______.

(a) Tool holder (b) Arbor
(c) Column (d) Table

(*ii*) ________ is the one of the abrasive materials used in grinding machine.

(a) Aluminum chloride (b) Calcium chloride
(c) Silicon carbide (d) Tungsten carbide

(*iii*) The thickness of chip is maximum at the beginning of the cut and minimum at the end of the cut, cut in each case of

(a) Up milling (b) Down milling
(c) Straddle milling (d) None of these

(*iv*) One of the milling operations used to produce dovetail groove is

(a) Slot milling (b) Straddle milling
(c) End milling (d) Angular milling

(*b*) Differentiate between Up milling and Down milling. (04 Marks)

(*c*) With a neat diagram, explain the working of a vertical milling machine. (06 Marks)

(*d*) With suitable sketches, explain the operation of centre less grinding machine. (06 Marks)

7. (*a*) Choose the correct answer for the following: (04 Marks)

(*i*) The hard filler material used in brazing

(a) Solder (b) Flux
(c) Shelter (d) Electrode

(*ii*) Support provided for rotating shaft is ______

(a) Bearing (b) Lubricant
(c) Axle (d) Hook

(*iii*) Carburizing flame has ______

(a) One Zone (b) Two Zone
(c) Three Zone (d) No Zone

(*iv*) In arc welding, the electrode which melt along with the work pieces and fill the Joint is called as

(a) Consumable electrode (b) Non consumable electrode
(c) Both (a) and (b) (d) None of these

(*b*) Sketch and explain electric arc welding process. (06 Marks)

(*c*) With a neat sketch, explain the different types of flames used in gas welding and specify their application. (04 Marks)

(*d*) Explain with a neat sketch, the method of splash lubrication. (06 Marks)

8. (*a*) Choose the correct answers for the following: (04 Marks)

(*i*) For converting rotary motion into rectilinear motion type of gear used is

(a) Spur gear (b) Rack and pinion
(c) Spiral gear (d) Bevel gear

(*ii*) The ratio of diameter of driver and driven pulley is called

(a) Module (b) Pitch circle diameter
(c) Ratio of tension (d) Velocity ratio.

(*iii*) The gear used to connect perpendicular axle shaft is

(a) Helical gear (b) Spur gear
(c) Bevel gear (d) Worm gear

(*iv*) The ratio of pitch circle diameter to number of teeth is

(a) Pitch (b) Circular pitch
(c) Module (d) Addendum

(*b*) List five advantages of gear drives over belt drives. (05 Marks)

(*c*) Define slip and creep with respect to belt drives. (05 Marks)

(*d*) Write the different types of gear trains with their applications. (06 Marks)